Managing Forest Ecosystems

Volume 28

Series editors

Klaus von Gadow, Göttingen, Germany
Timo Pukkala, Joensuu, Finland
Margarida Tomé, Lisbon, Portugal

Well-managed forests and woodlands are a renewable resource, producing essential raw material with minimum waste and energy use. Rich in habitat and species diversity, forests may contribute to increased ecosystem stability. They can absorb the effects of unwanted deposition and other disturbances and protect neighbouring ecosystems by maintaining stable nutrient and energy cycles and by preventing soil degradation and erosion. They provide much-needed recreation and their continued existence contributes to stabilizing rural communities.

Forests are managed for timber production and species, habitat and process conservation. A subtle shift from multiple-use management to ecosystems management is being observed and the new ecological perspective of multi-functional forest management is based on the principles of ecosystem diversity, stability and elasticity, and the dynamic equilibrium of primary and secondary production.

Making full use of new technology is one of the challenges facing forest management today. Resource information must be obtained with a limited budget. This requires better timing of resource assessment activities and improved use of multiple data sources. Sound ecosystems management, like any other management activity, relies on effective forecasting and operational control.

The aim of the book series *Managing Forest Ecosystems* is to present state-of-the-art research results relating to the practice of forest management. Contributions are solicited from prominent authors. Each reference book, monograph or proceedings volume will be focused to deal with a specific context. Typical issues of the series are: resource assessment techniques, evaluating sustainability for even-aged and uneven-aged forests, multi-objective management, predicting forest development, optimizing forest management, biodiversity management and monitoring, risk assessment and economic analysis.

More information about this series at http://www.springer.com/series/6247

Richard A. Fournier · Ronald J. Hall
Editors

Hemispherical Photography in Forest Science: Theory, Methods, Applications

Springer

Editors
Richard A. Fournier
Centre d'Applications et de Recherches en Télédétection (CARTEL)
Université de Sherbrooke
Sherbrooke, QC
Canada

Ronald J. Hall
Canadian Forest Services
Natural Resources Canada
Edmonton, AB
Canada

ISSN 1568-1319 ISSN 2352-3956 (electronic)
Managing Forest Ecosystems
ISBN 978-94-024-1487-5 ISBN 978-94-024-1098-3 (eBook)
DOI 10.1007/978-94-024-1098-3

Softcover reprint of the hardcover 1st edition 2017

Printed on acid-free paper

This Springer imprint is published by Springer Nature
The registered company is Springer Science+Business Media B.V.
The registered company address is: Van Godewijckstraat 30, 3311 GX Dordrecht, The Netherlands

© Photograph taken by Sylvain Leblanc and Richard Fournier
Taken with a Nikon D1X and a fisheye lens Nikkor 10.5mm 1:2.8
Trembling aspen stand in Abitibi, Quebec, Canada

To those that can see the magic beyond the technique

Richard A. Fournier

To Judy Hall
Kevin, Angel, Avery
Amanda, Jerry

Ronald J Hall

Preface

Hemispherical photography (HP) is a field-based, remote-sensing approach for acquiring a permanent record of observations from which the light regime and biophysical structure within a forest canopy can be characterized. It is a technique that has seen increasing use, as evidenced from the published literature, in part because of the increasing need to gather information for quantifying and monitoring the status, health, composition, and structure of forested ecosystems as well as the light regime in forest stands. It has been difficult for prospective users to gather information about the theoretical concepts and about how HP can be acquired and processed, because this information has been fragmented and dispersed across a variety of publication media. To date, a single, definitive reference source has not been available. As a result, a book is needed that describes the theory, methods, and applications of HP for acquiring forest biophysical measurements. The purpose of the present volume is to address the information void that exists in this field.

HP is applicable to many fields in forestry, including ecology (e.g., light, moisture and nutrient distribution, plant diversity, seedling regeneration), inventory (e.g., growth and yield, stand dynamics), hydrology (e.g., snowpack, precipitation applications, flooding, stream temperature), and biophysical parameter estimation (e.g., leaf area index, leaf area duration, openness), to name a few. The information that is extracted from hemispherical photographs provides a useful complement to information that is gathered by standard inventory methods, and HP also supplies a series of biophysical measurements that would otherwise be challenging to acquire. For example, variables such as wood volume, biomass, canopy gap distribution, sapwood area, and leaf area can be derived from hemispherical photographs. The other application of high interest is the ability to characterize the solar radiation regime with measures such as gap fraction and fraction of diffuse and direct radiation.

Although most of the earlier work in HP was done with analog film and single-lens reflex cameras, the advent of digital photography has made it easier to acquire photographs, a development that has been accompanied by new digital camera systems and data acquisition opportunities. The availability of software programs that can readily process large numbers of individual photographs has further amplified the adoption of digital technology. These advances have driven increasing demands for knowledge about the theory and equipment of HP and about the methods for acquiring, processing, and analysis. This book has been designed to meet these information needs by compiling information from established researchers into a single source that will be of value to a broad range of users, including academics, senior undergraduate and graduate students, natural resource professionals, and researchers.

This book presents practical information about three domains of HP: (1) theory and concepts; (2) equipment and methods for field data acquisition and image processing; and (3) applications. The first section of the book is an introduction to HP and the fundamental elements of forest structure and light interactions within the forest canopy. The second section provides practical information about the equipment and procedures for procuring and processing the photographs. Example applications using HP in forestry and natural resource assessment are described in the third section. The book concludes with a discussion about modeling tools and future directions for this rapidly growing field.

It has taken several years to bring this book from its inception to completion, a period far longer than initially anticipated. We have greatly appreciated the patience and persistence of the international group of contributors and reviewers. It was particularly challenging to design a structure for the book that would be of most value to our target readers and to organize the topics within that structure. It is our hope that we have produced a volume that will meet the information needs of those who are interested in using HP in forested environments.

Sherbrooke, Canada Richard A. Fournier
Edmonton, Canada Ronald J. Hall

Contents

Chapter 1
Introduction

Ronald J. Hall, Richard A. Fournier and Paul Rich

Abstract This introductory chapter presents the rationale for assembling a book on hemispherical photography (HP) for forest science. The book was designed to provide prospective users with information about theoretical concepts of HP, how to both acquire and analyze hemispherical photo images, and a sampling of real world applications. This chapter provides an historical review of HP and presents HP as a study of two perspectives, the solar regime and canopy structure. A brief summary of each chapter provides the reader with an overview of the contents for each chapter.

Keywords Canopy architecture · Canopy structure · Hemispherical photography history · Leaf area index (LAI) · Solar radiation

1.1 Introduction

Hemispherical or fisheye photography is a technique to characterize plant and tree canopies from analysis of photographs that are taken with an extreme wide-angle lens (Anderson 1964a). Hemispherical photographs (HP) are typically acquired looking upward from beneath a forest or other plant canopy, and geometric analysis

R.J. Hall (✉)
Canadian Forest Service, Northern Forestry Centre, Natural Resources Canada, 5320—122 Street, Edmonton, AB T6H 3S5, Canada
e-mail: Ron.Hall@Canada.ca

R.A. Fournier
Département de géomatique appliquée, Centre d'Applications et de Recherches en Télédétection (CARTEL), Université de Sherbrooke, Sherbrooke, QC J1K 2R1, Canada
e-mail: Richard.Fournier@USherbrooke.ca

P. Rich
Creekside Center for Earth Observation, Menlo Park, CA 94025, USA
e-mail: Paul@professorpaul.com

R.A. Fournier and R.J. Hall (eds.), *Hemispherical Photography in Forest Science: Theory, Methods, Applications*, Managing Forest Ecosystems 28,
DOI 10.1007/978-94-024-1098-3_1

of HP is used to characterize the light environment under the canopy or the attributes of the canopy itself (Rich 1990). The hemispherical lens, also known as a fisheye lens, produces a circular image that is essentially a projection of a hemisphere onto a planar surface (Rich 1990). The term "fisheye" was coined in 1906 by the American physicist and inventor Robert W. Wood to describe the ultra wide hemispherical view that a fish would see from beneath the water, a phenomenon that is known as "Snell's window" (Kingslake 1989). When pointed upward, a fisheye lens captures the full hemisphere of visible sky in a single image. Forest and ecological researchers have been greatly interested in how best to acquire these images and how to analyze and interpret the canopy openings in the images.

Early work using HP focused primarily on the light environment, with researchers recognizing that the technique provided a means of calculating the spatial and temporal pattern of the light environment on the basis of rudimentary first principles of polar geometry (Anderson 1964a; Rich 1990). Subsequent work placed increasing attention on characteristics of the forest canopy. The plant community depends upon incident solar radiation as the source of radiant energy that drives photosynthesis, evapotranspiration and other physiological processes (Gates et al. 1965; Fu and Rich 2002). Solar radiation penetrating through (or intercepted by) a forest canopy results in a geometric distribution of openings that can be precisely quantified to derive attributes of forest structure such as canopy openness, leaf area index (LAI), and leaf angle distribution (Evans and Coombe 1959; Rich 1990). By analyzing HP to obtain information about these kinds of attributes, researchers have been able to study canopy structure and its interaction with the solar radiation regime.

The application of HP for characterizing the solar radiation regime and forest canopy structure has undergone a considerable evolution since Evans and Coombe (1959) and Anderson (1964a) first reported using HP to study forest ecology in the late 1950s and 1960s. Early studies focused on the theoretical foundation (Anderson 1964a, b) from which an improved understanding could be generated about the penetration of solar radiation into the forest canopy and about canopy architecture (Rich 1990). Much of the work in recent decades has focused on the transition from the use of a film-based system for photo acquisition and manual analysis to the use of digital photo acquisition systems and image processing software (Frazer et al. 2001; Hale and Edwards 2002). This technological evolution from film to digital acquisition, together with the array of HP processing software that has become available, has made it challenging for prospective users (1) to obtain knowledge about the theory of HP, (2) to learn how to acquire and analyze images, and (3) to learn how to apply HP to studies in forest science and ecology. This book was compiled to address this challenge: the goal was to produce a single reference source that synthesizes the state of the science in these three domains. It concludes with the lessons that have been learned, future directions, and opportunities for further advancement.

1.2 Brief History

In the 1920s, the British plant biochemist Robert Hill (known as Robin Hill), who was most familiarly known for his work on the mechanism of photosynthesis, developed a camera with a hemispherical or fisheye lens to record cloud patterns in three dimensions using paired stereoscopic whole-sky images (Hill 1924). When the original proof plates of the lens were destroyed, R. and J. Beck Ltd. became the first manufacturer to produce a camera with a fisheye lens, which was built to Hill's specifications. It became known as the Robin Hill camera (Evans and Coombe 1959). The Hill camera resolved many of the difficulties of canopy photography and facilitated the early work of Evans and Coombe (1959), and Coombe and Evans (1960), who pioneered the method to characterize the light conditions within a forest canopy. Taking advantage of the fisheye lens' polar projection, they superimposed sun path diagrams on HP to predict periods of direct sunlight and the fraction of light that was transmitted to the ground. This early work was followed by more specific applications to forestry by Brown (1962), Brown and Wordley (1965), and Madgwick and Brumfield (1969). Anderson (1964a, 1971) built upon this work and developed much of the theoretical foundation for calculating the penetration of sunlight (direct) and skylight (indirect or diffuse) components of solar radiation through canopy openings using a manual analysis process. This foundation served as the basis for a number of additional applications in forestry and natural resources by Becker (1971), Bonhomme (1973, 1976), Lemeur (1973), Bonhomme et al. (1974), Ducrey and Bartoli (1974), Pope and Lloyd (1974), Ducrey (1975a, b), Hutchison (1975), Burger (1976, 1977), Hutchison and Matt (1977), and Nilson (1977).

Manual methods of interpreting HP were extremely tedious, subjective and relatively slow and, thus, the method languished for a number of years. In the mid-1970s, forest ecologists began to focus on forest regeneration in gaps, so-called "gap-phase dynamics," which grew to become a major focus of research for both temperate and tropical forest ecology, building upon insights of the Cambridge botanist A.S. Watt (Watt 1947; Whitmore et al. 1993). Continued interest in HP helped to spur research into improving the efficiency of analysis with digital methods. The value of using HP is greatest when many photographs can efficiently be analyzed to assess the magnitude and variability of the canopy structure and the solar radiation conditions of sites. As a result, from the late 1960s to early 1980s, a growing number of automated image analysis methods and systems were introduced. One of the early systems used a densitometer to measure the brightness at various points on a photograph (Madgwick and Brumfield 1969). Indirect measurements of canopy structure were carried out in apple orchards by Lakso (1976, 1980), in which the percent sky that was derived from false-color densitometry was correlated to total and diffuse photosynthetically active radiation (PAR), the percentage of sunflecks, and specific leaf mass. Bonhomme and Chartier (1972) developed an apparatus that consisted of nine phototransistors to scan HPs that were mounted onto slides, from which gap frequencies were derived from transmitted impulses. Parallel with these developments was the introduction of methods to estimate particular attributes from the HP. Chazdon and Field (1987) described a computerized technique to estimate canopy openings, direct and diffuse site factors, and photosynthetic photon flux density (PPFD) as key descriptors of the light environment. Other studies reported parameters such as mean monthly radiation (Anderson 1964a) and the fraction of available PAR reaching different points in the canopy (Lakso 1980). Prototype computerized fisheye analysis systems were developed by Jones and Campbell (1979), and Miller (1981). Olsson et al. (1982) developed and built a system that employed a microcomputer-controlled diode-array image scanner (OSIRIS) at the Royal Institute of Technology in Stockholm. HP negatives were scanned to compute the sky-obscuration factor, and the digital scanning approach resulted in lower sampling errors and required half the evaluation time. Chan et al. (1986) modified the systems of Jones and Campbell (1979), and Miller (1981) by developing a computerized system in which the negative of the fisheye image was projected with a slide projector onto a flat surface from which the light was then sensed, converted to a digital signal, stored, analyzed and printed. Their system calculated both the percentage of sky that was visible and the average probability of diffuse radiation penetration. No statistical differences were found when the data produced by their computerized system were compared with the results of manual analysis by three different observers. Jupp et al. (1980) developed a FORTRAN IV computer program to process and analyze data that were coded from HP, from which parameters such as LAI, and diffuse and direct PAR were computed. A common objective of these early systems was to achieve some level of automation for more efficient and consistent processing and analysis of HP.

As computers and software generally advanced, so did HP software, which in turn accelerated the development of new applications for HP. One of the first widely available computer programs for processing HP was SOLARCALC (Chazdon and Field 1987), which is still in use today; version 7.0 was recently released (Mailly et al. 2013). Several other HP computer programs followed, including CANOPY (Rich 1989), SYLVA (Becker et al. 1989), SUNSHINE (Smith and Somers 1991), HEMIPHOT/WINPHOT (ter Steege 1994, 1997), GLI/C (Canham 1995), hemiIMAGE (Brunner 1998), CIMES (Walter 1999, 2009; Gonsamo et al. 2011), GLA (Frazer et al. 1999), CAN-EYE (Baret and Weiss 2004), and LIA32 (Yamamoto 2004, http://www.agr.nagoya-u.ac.jp/~shinkan/LIA32/index-e.html). Although these programs were made available free of charge, commercial software has also been produced, including HemiView (Rich et al. 1999) and WinSCANOPY (Regent Instruments 2013).

As software improved for processing HP, attention turned to the quality of the photographs, which led to recommendations for photo acquisition specifications in the field pertaining to types of equipment, use of high-contrast film, exposure control, and to standardize how film was developed and how equipment was calibrated (Olsson et al. 1982; Chan et al. 1986; Rich 1990). It has long been known that HP provides a permanent record of the geometry of canopy openings at the time of observation (Rich 1990). Although film-based systems were prevalent in the past, the emerging availability of digital cameras greatly increased the ease and feasibility of both acquiring and storing HP, because the images were captured in digital format and, thus, the need to scan analog negatives was removed.

The emergence of digital cameras led to several studies that compared digital and film HP. Englund et al. (2000) compared digital and film HP to characterize forest light environments and canopy openness and found that differences in digital camera light sensitivity, film type and speed, and the lens' field of view (FOV) contributed to the differences between results reported for digital and film HP. Frazer et al. (2001) compared digital HP that were taken with a Nikon Coolpix 950 (with FC-E8 fisheye) with film HP that were taken with a Nikon F (with Nikkor 8 mm fisheye) film camera over different stand structures and sky conditions. Differences were reported in the estimates of the size, shape and distribution of canopy gaps, and canopy openness was 1.4 times larger in digital HP than in film HP in 60% of the photos that were analyzed. Hale and Edwards (2002) compared digital and film HP of conifer plantations in southern and central Scotland under a wide range of light environments; similar to Frazer et al. (2001), they used a Nikon Coolpix 950 (with FC-E8 fisheye) to obtain digital photos. Although the results that they obtained for canopy openness, indirect site factor, direct site factor and LAI were highly correlated between digital and film systems, they recommended following a strict protocol for image acquisition. Despite the differences between digital and film HP results that were reported in these three studies, there has been a trend toward increasing use of digital cameras in the decade since these studies were completed. The ease with which images are acquired using digital photography has accelerated the rate of research on applications of HP, and the realization that caution must be exercised when using cameras for acquiring canopy images. In particular, careful attention to image acquisition procedures and camera operation is necessary to generate consistent results under a variety of illumination conditions. Users must undertake controlled experimentation with a given system to gain the experience necessary to ensure consistency and successful results. To help the prospective user along this path, Fournier et al. (2017, Chap. 4) explain the differences between digital and film cameras and provide detailed information on how to generate a protocol for image acquisition in the field.

Of particular interest has been the comparability of software for processing and analyzing HP, but only a few studies have been completed to date. Bréda (2003) compared two commercial programs, HemiView and WinSCANOPY, but did not undertake an analysis of HP that were processed with these programs. Jarčuška et al. (2010) compared several parameters from GLA and WinSCANOPY, and Promis et al. (2011) compared canopy structure estimates from HemiView, GLA, hemiIMAGE and WINPHOT. Hall et al. (2017, Chap. 7) provide a descriptive comparison of several software packages for processing HP and present results from a controlled study. The study consisted of a set of photographs representing a range of forest canopies collected at four sites that were processed by each of the software programs. Across these studies, it can be difficult to draw conclusions about which software may be most suitable for use for a given forest type. The selection of a program remains very much influenced by the preference of the user, the camera system being employed, and cost: "free" access, more than software features, for example, has been a primary motivator for software selection (Hall et al. 2017, Chap. 7).

The future development of computer and digital camera technologies will no doubt continue to increase the availability of low-cost, high-quality products, but the implementation of HP now depends much more upon the availability of proven analysis methods. Digital camera systems and HP processing software clearly offer the opportunity for more efficient acquisition, processing and analysis of HP than conventional systems. In some respects, these technological developments have advanced more quickly than our ability to properly employ these tools and, thus, this book was conceived to help bridge the knowledge and practice gap.

1.3 Application of Hemispherical Photography in Forest Science: A Study of Two Perspectives

Most analyses involving HP involve a study of either the solar regime or the canopy structure. Incoming solar radiation influences the energy and water balance of the soil/landscape surface and is thus fundamental to most physical and biophysical processes (Fu and Rich 2002). The study of solar radiation through a forest canopy is fundamentally a study of the openings or gaps in the forest canopy, as a function of where it is fully or partially blocked by the canopy and where it is transmitted. The study of canopy architecture uses the occlusion of incoming solar radiation to reconstruct the geometry of the forest canopy and, hence, provides information about forest structure. Knowledge of canopy architecture can be used to derive various forest and biophysical attributes (Fournier et al. 2003). Understanding the two perspectives of solar radiation and forest structure provides the foundation for use of HP; these two perspectives are discussed by Schleppi and Paquette (2017, Chap. 2) and Leblanc and Fournier (2017, Chap. 3), respectively.

There is a need to understand the assumptions and limitations that are associated with the methods for HP acquisition and analysis (Roxburgh and Kelly 1995). Among the most significant of these is the assumption that all foliage will completely block the incoming penetration of light to allow a study of light penetration through canopy gaps or openings (Roxburgh and Kelly 1995). A classified digital image is binary, in that canopy elements are black and openings or gaps are white, and a considerable amount of attention has been placed on approaches to determine the threshold at which canopy and sky are defined (Jonckheere et al. 2005; Macfarlane et al. 2007). There is subjectivity in several components of the HP processing chain from image acquisition and analysis to interpretation of results. An important element is being aware of the uncertainties in measurements that can result from variations in photographic exposure, pixel classification in image thresholding and understanding the differences in gamma values that has arisen during the transition from film to digital HP (Chianucci and Cutini 2012). As a result, it is not an exact science, and considerable research has been directed toward increasing the objectivity and repeatability of the results that are obtained from HP.

To help ensure results that were generated from applications of HP are reliable, several authors have undertaken comparative analyses with other ground-based measurements in studies of LAI. Comparative studies have been undertaken or reviewed by Bréda (2003), Fournier et al. (2003), Jonckheere et al. (2004), and Zheng and Moskal (2009) to derive insights into the performance and accuracy of LAI estimates under a range of canopy conditions. The fact that the woody element is incorporated into photographic images results in a source of differences between different optical instruments for estimation of LAI (Chianucci and Cutini 2012). Other factors such as foliage clumping, sampling strategies, field protocols and scaling from field (point) to landscape levels have emerged from these studies as areas requiring further investigation.

The three primary research foci for HP include advancing basic science, improving methods and tools, and developing applications of HP. Basic science governs the rationale for the application of HP to generate new knowledge about solar radiation dynamics and its contribution to biophysical processes, plant productivity and ecosystem properties. The second area of research has involved the mechanics and processes for acquiring and analyzing HP. Photo acquisition systems have evolved from analog to digital cameras, and several software packages continue to be updated. An example of improving methods and tools is the considerable amount of research that has been undertaken to devise more consistent methods for discriminating sky from foliage using binary thresholds (Ishida 2004; Jonckheere et al. 2005; Nobis and Hunziker 2005; Macfarlane 2011). The third area of research has been in the application of HP to generate forest information. Although traditional measurements of the forest canopy have focused upon measures of tree size, volume and biomass, greater attention is being placed on biophysical processes to improve our understanding of forest productivity, health and succession. Information from HP will help to predict responses to future climate change and to other natural and anthropogenic disturbances that create uncertainty about the future sustainability and provision of ecosystem goods and services (Gitay et al. 2001; Fischlin et al. 2007). Research in these three research foci will advance the science and knowledge for application of HP in forest science, and the book is organized around these three domains.

1.4 Organization of the Book

The book consists of ten chapters, which are organized into three sections. The first section describes HP and the theories behind solar radiation and light interactions with the atmosphere and the forest canopy. It also explains the foundation behind the measurement of forest structure by HP. The second section provides practical information about the equipment, methods and software for procuring and processing HP. The third section discusses several applications of HP to forestry and natural resource assessment. The book concludes with a summary and discussion of current and future perspectives and directions of this rapidly growing field.

The opening chapter in the section on theory, by Schleppi and Paquette (2017, Chap. 2), introduces radiation as the primary source of energy that controls biological and physical processes and that provides the source of illumination recorded on HP. Chapter 2 discusses the theories of how solar radiation interacts with the atmosphere and forest canopies. It also reviews radiation sensors for in situ measurement in the field and discusses their comparability with HP. From a sampling perspective, HP is used to describe the radiation regime at the specific location where the camera is positioned and is one tool among several optical sensors for estimating understory light availability within the forest canopy. The chapter ends with a brief comparison of several studies that employed different methods for estimating understory light in different forest types.

Chapter 3 (Leblanc and Fournier 2017) presents the concept of forest structure and its estimation from HP under the premise that openings in the canopy result in measurable gaps on the image that can be related to forest structure. The architecture of a forest canopy, however, is complex and spatially variable, and there is considerable interest from the natural resource communities in estimating and modeling various attributes that describe forest structure. The chapter provides an overview of structural attributes commonly measured in a forest inventory and discusses the theory and problems behind gap fraction and application of Beer's law for estimating parameters such as canopy closure, openness and LAI.

Chapter 4 (Fournier et al. 2017), which opens the section of the book on methods, provides practical information on which equipment to purchase for HP, how to plan a field protocol, acquire HP, and it outlines the workflow for preparing the images for both archiving and analysis. The capabilities and features of computer systems and cameras have evolved rapidly, and digital systems have supplanted the analog film systems. Chapter 4 has been designed to fill the need for a definitive reference to facilitate the acquisition of consistent, high-quality images that addresses a call for a definitive protocol for image acquisition (Chianucci and Cutini 2012).

Chapter 5 (Jonckheere et al. 2017) focuses upon image analysis and includes the calculation of canopy solar radiation and structure. It presents the three primary components of image analysis methods: processing, classification and calculation. The content of Chap. 5 logically follows the content of Chap. 4 (acquisition of HP), as it begins with image selection and quality control, followed by classification, which involves thresholding to discriminate foliage from sky. Foliage clumping, lack of reference measurements, and the need to distinguish woody material from foliage are identified as areas of uncertainty requiring further investigation.

Chapter 6 (Frazer et al. 2017) focuses on methods for estimating LAI, with a particular emphasis on foliage clumping. Although HP is a convenient, non-destructive method for estimating LAI, a primary assumption of the original Beer-Lambert equation is that foliage is randomly distributed. This assumption is violated under most canopies in temperate and boreal forests. The chapter reviews the science behind the methods for deriving clumping indices and compares four methods through the use of simulated fisheye photographs.

Chapter 7 (Hall et al. 2017) acknowledges the problem that although there are now several computer programs available for analyzing HP, past studies have not made it obvious to users about which program they should use. To address this problem, the authors of this chapter compare the features of five software programs for processing HP and compare the estimation of selected light regime and canopy structure variables. A set of ten photographs from four test sites representing open and closed density coniferous and deciduous forests were processed by each of four HP programs. In addition to the results of the empirical comparison, the chapter includes a discussion of software functionality, the ability of software to threshold or classify the image, and the applicability of software for estimating LAI, one of the attributes that is most often estimated from analysis of HP.

Chapter 8 (Mailly 2017) focuses on the application of HP in support of forest inventory and silviculture. Although conventional forest inventories are focused on attributes such as tree diameter, height and volume, a more ecologically based inventory would include attributes such as light transmittance, canopy cover and LAI. Mailly provides considerations for the use of HP to measure attributes from tree- and stand-level perspectives and discusses its potential role in support of silviculture specific to regeneration, thinning and partial cutting. The chapter also reviews other field-based instrumentation for measuring canopy cover and light transmittance through forest canopies.

Chapter 9 (Côté et al. 2017) is the final applications chapter and describes the characterization of canopy structure through canopy architectural models. It acknowledges the complexity of canopy structure given its three-dimensional character and its spatial and temporal heterogeneity over a range of scales. Côté et al. (2017) addresses the problem of characterizing canopy structure by reviewing architectural modeling approaches to understand the extent to which they can account for the spatial variability observed in HP. This review is combined with a comparative analysis of four canopy architecture models that are based on simulations of a HP that represents coniferous stands. The results provide a state-of-the-art perspective on the modeling of canopy structure.

The book ends with Chapter 10, which contains a summary of the key messages on the theory, methods and applications of HP in forest science. There are several key questions that drive continuing research, and multiple lessons can be learned from the vast array of studies that have been published in this domain.

The book, which is comprised of this introductory chapter, the eight following chapters and the summary chapter, provides, for the first time, comprehensive coverage of HP in forest science in a single volume. The application of HP is anticipated to increase as the need to integrate knowledge from plot to landscape scales continues to grow, together with our need to assess and monitor the state and changing dynamics of our forests. Meeting these demands will require information that is obtained from a diversity of tools at a range of scales. It is within this context that we anticipate HP will become an increasingly important data collection tool.

Acknowledgements The assistance in the preparation of the references by M. Filiatrault is greatly appreciated. The reviews of an earlier version of this chapter undertaken by C. Macfarlane and J.-F. Côté are gratefully acknowledged.

References

Anderson MC (1964a) Studies of the woodland light climate. I. The photographic computation of light condition. J Ecol 52:27–41

Anderson MC (1964b) Studies of the woodland light climate. II. Seasonal variation in the light climate. J Ecol 52:643–663

Anderson MC (1971) Radiation and crop structure. In: Sestak Z, Catsky J, Jarvis PG (eds) Plant photosynthetic production manual of methods. Dr. W. Junk N.V., The Hague, pp 77–90

Baret F, Weiss M (2004) CAN-EYE: processing digital photographs for canopy structure characterization. INRA-EMMAH UMR 114, Avignon, France

Becker M (1971) Une technique nouvelle d'utilisation des photographies hémisphériques pour la mesure du climat lumineux en forêt (A new technique using hemispherical photographs for measuring the light climate in a forest). Ann Sci For 28:425–442

Becker P, Erhart DW, Smith AP (1989) Analysis of forest light environments. I. Computerized estimation of solar radiation from hemispherical canopy photographs. Agric For Meteorol 44:3–4

Bonhomme R (1973) Analyse de la surface des taches de soleil, de l'indice foliaire et de l'inclinaison moyenne des feuilles à l'aide de photographies hemispheriques. In: Slatyer RO (ed) Plant response to climatic factors, proceedings of Uppsala symposium. UNESCO, Paris, pp 369–376

Bonhomme R (1976) Détermination des profils d'indice foliaire et de rayonnement dans un couvert végétal à l'aide de photographies hémisphériques faites in situ. Ann Agron 27:33–59

Bonhomme R, Chartier P (1972) The interpretation and automatic measurement of hemispherical photographs to obtain sunlit foliage area and gap frequency. Israel J Agric Res 22:53–61

Bonhomme R, Varlet-Grancher C, Chartier P (1974) The use of hemispherical photographs for determining the leaf area index of young crops. Photosynthetica 8:299–301

Bréda NJJ (2003) Ground-based measurements of leaf area index: a review of methods, instruments and current controversies. J Exp Bot 54:2403–2417

Brown HE (1962) The canopy camera. USDA Forest Service, Rocky Mountain Forest and Range Experiment Station, Ft. Collins, Colorado. Station Paper No. 72, 22 pp

Brown HE, Wordley DP (1965) Some applications of the canopy camera in forestry. J Forest 63:674–680

Brunner A (1998) A light model for spatially explicit forest stand models. For Ecol Manag 107:19–46

Burger J (1976) Nest density of the black-headed gull in relation to vegetation. Bird Study 23:27–32

Burger J (1977) Role of visibility in nesting behavior of *Larus* gulls. J Comp Physiol Psychol 91:1347–1358

Canham CD (1995) GLI/C: software for calculation of light transmission through forest canopies using color fisheye photography (unpublished manuscript). Institute of Ecosystem Studies, Millbrook, New York, 7 pp

Chan SS, McCreight RW, Walstad JD, Spies TA (1986) Evaluating forest vegetative cover with computerized analysis of fisheye photographs. For Sci 32:1085–1091

Chazdon RL, Field CB (1987) Determinants of photosynthetic capacity in six rainforest *Piper* species. Oecologia 73:222–230

Chianucci F, Cutini A (2012) Digital hemispherical photography for estimating forest canopy properties: current controversies and opportunities. iForest 5:290–295

Coombe DE, Evans GC (1960) Hemispherical photography in studies of plants. Med Biol Illus 10:68–75

Côté JF, Fournier RA, Verstraete MM (2017) Canopy architectural models in support of methods using hemispherical photography. In: Fournier RA, Hall RJ (eds) Hemispherical photography in forest science: theory, methods, applications. Springer, Berlin

Ducrey M (1975a) Utilisation des photographies hémisphériques pour le calcul de la perméabilité des couverts forestiers au rayonnement solaire I. Analyse théorique de l'interception. Ann Sci For 32:73–92

Ducrey M (1975b) Utilisation des photographies hémisphériques pour le calcul de la perméabilité des couverts forestiers au rayonnement solaire II. Étude expérimentale. Ann Sci For 32:205–221

Ducrey M, Bartoli M (1974) Utilisation des photographies hémisphériques pour le cubage des arbres sur pied. Ann Sci For 31:229–248

Englund SR, O'Brien JJ, Clark DB (2000) Evaluation of digital and film hemispherical photography and spherical densiometry for measuring forest light environments. Can J For Res 30:1999–2005

Evans GC, Coombe DE (1959) Hemispherical and woodland canopy photography and the light climate. J Ecol 47:103–113

Fischlin A, Midgley GF, Price JT, Leemans R, Gopal B, Turley C, Rounsevell MDA, Dube OP, Tarazona J, Velichko AA (2007) Ecosystems, their properties, goods, and services. In: Parry ML, Canziani OF, Palutikof JP, van der Linden PJ, Hanson CE (eds) Climate change 2007: impacts, adaptation and vulnerability. Contribution of Working Group II to the fourth assessment report of the Intergovernmental Panel on Climate Change. Cambridge University Press, Cambridge, UK, pp 211–227

Fournier RA, Mailly D, Walter J-MN, Soudani K (2003) Indirect measurement of forest structure from in situ optical sensors. Section 1. Data collection and pre-processing. In: Wulder MA, Franklin SE (eds) Methods and applications for remote sensing of forests: concepts and case studies. Kluwer, Dordrecht, pp 77–113

Fournier RA, Mailly D, Walter J-MN, Jonckheere IGC (2017) Acquiring hemispherical photographs in forest environments: from planning to archiving photographs. In: Fournier RA, Hall RJ (eds) Hemispherical photography in forest science: theory, methods, applications. Springer, Berlin

Frazer GW, Canham CD, Lertzman KP (1999) Gap Light Analyzer (GLA), version 2.0: imaging software to extract canopy structure and gap light transmission indices from true-colour fisheye photographs, users manual and program documentation. Simon Fraser University, Burnaby, British Columbia, and the Institute of Ecosystem Studies, Millbrook, New York, 52 pp

Frazer GW, Fournier RA, Trofymow JA, Hall RJ (2001) A comparison of digital and film fisheye photography for analysis of forest canopy structure and gap light transmission Agric For Meteorol 109:249–263

Frazer GW, Fournier RA, Leblanc SG, Walter J-MN (2017) View angle-dependent clumping indices for indirect LAI estimation. In: Fournier RA, Hall RJ (eds) Hemispherical photography in forest science: theory, methods, applications. Springer, Berlin

Fu P, Rich PM (2002) A geometric solar radiation model with applications in agriculture and forestry. Comput Electron Agric 37:25–35

Gates DM, Keegan HJ, Schleter JC, Weidner VR (1965) Spectral properties of plants. Appl Opt 4:11–20

Gitay H, Brown S, Easterling W, Jallow B, Antle J, Apps MJ, Beamish R, Chapin T, Cramer W, Frangi J, Laine J, Erda L, Magnuson, J, Noble I, Price, J, Prowse T, Root T, Schulze E, Sirotenko O, Sohngen B, Soussana J (2001) Ecosystems and their goods and services, pp 235–342 (Chap. 5) in McCarthy JJ (ed) Climate change 2001: impacts, adaptation and vulnerability: contribution of Working Group II to the third assessment report of the Intergovernmental Panel on Climate Change. Cambridge University Press, New York

Gonsamo A, Walter J-MN, Pellikka P (2011) CIMES: a package of programs for determining canopy geometry and solar radiation regimes through hemispherical photographs. Comput Electron Agric 79:207–215

Hale SE, Edwards C (2002) Comparison of film and digital hemispherical photography across a wide range of canopy densities. Agric For Meteorol 112:51–56

Hall RJ, Côté JF, Fournier RA, Mailly D (2017) Comparison of software tools for analysis of hemispherical photographs. In: Fournier RA, Hall RJ (eds) Hemispherical photography in forest science: theory, methods, applications. Springer, Berlin

Hill R (1924) A lens for whole sky photographs. Q J Roy Meteor Soc 50:227–235

Hutchison BA (1975) Photographic assessment of deciduous forest radiation regimes. Dissertation Abstracts International B 36:2046

Hutchison BA, Matt DR (1977) The distribution of solar radiation within a deciduous forest. Ecol Monogr 47:185–207

Ishida M (2004) Automatic thresholding for digital hemispherical photography. Can J For Res 34:2208–2216

Jarčuška B, Kucbel S, Jaloviar P (2010) Comparison of output results from two programmes for hemispherical image analysis: Gap Light Analyser and WinScanopy. J For Sci 56:147–153

Jonckheere I, Fleck S, Nackaerts K, Muys B, Coppin P, Weiss M, Baret F (2004) Review of methods for in situ leaf area index (LAI) determination. Part I. Theories, sensors and hemispherical photography. Agric For Meteorol 121:19–35

Jonckheere I, Nackaerts K, Muys B, Coppin P (2005) Assessment of automatic gap fraction estimation of forests from digital hemispherical photography. Agric For Meteorol 132:96–114

Jonckheere IGC, Macfarlane C, Walter J-MN (2017) Image analysis of hemispherical photographs, algorithms and calculation. In: Fournier RA, Hall RJ (eds) Hemispherical photography in forest science: theory, methods, applications. Springer, Berlin

Jones AJ, Campbell GS (1979) Machine analysis of fisheye photographs for assessment of radiation penetration in plant canopies. Paper presented at the Western Society of Soil Science, Moscow, Idaho, 6 June 1979

Jupp DLB, Anderson MC, Adomeit EM, Witts SJ (1980) PISCES—a computer program for analysing hemispherical canopy photographs. CSIRO, Institute of Earth Resources, Division of Land Use Research. Technical Memorandum 80/23, 22 pp

Kingslake R (1989) A history of the photographic lens. Academic Press, San Diego

Lakso AN (1976) Characterizing apple tree canopies by fisheye photography. HortScience 11:404–405

Lakso AN (1980) Correlations of fisheye photography to canopy structure, light climate, and biological responses to light in apple trees. J Am Soc Hortic Sci 105:43–46

Leblanc SG, Fournier RA (2017) Measurement of forest structure with hemispherical photography. In: Fournier RA, Hall RJ (eds) Hemispherical photography in forest science: theory, methods, applications. Springer, Berlin

Lemeur R (1973) A method for simulating the direct solar radiation regime in sunflower, Jerusalem artichoke, corn and soybean canopies using actual stand structure data. Agric Meteorol 12:229–247

Macfarlane C (2011) Classification method of mixed pixels does not affect canopy metrics from digital images of forest overstory. Agric For Meteorol 151:833–840

Macfarlane C, Grigg A, Evangelista C (2007) Estimating forest leaf area using cover and fullframe fisheye photography: thinking inside the circle. Agric For Meteorol 146:1–12

Madgwick HAI, Brumfield GL (1969) The use of hemispherical photographs to assess light climate in the forest. J Ecol 57:537–542

Mailly D (2017) Hemispherical photography in support of forest inventory and silviculture. In: Fournier RA, Hall RJ (eds) Hemispherical photography in forest science: theory, methods, applications. Springer, Berlin

Mailly D, Turbis S, Chazdon R (2013) SOLARCALC 7.0: an enhanced version of a program for the analysis of hemispherical canopy photographs. Comput Electron Agric 97:15–20

Miller GP (1981) Analysis of radiant environment in forest canopies: technique and application. MS thesis, Washington State University, Pullman, 36 pp

Nilson T (1977) On estimating radiation characteristics of plant stands from hemispherical photographs. In: Phytoactinometrical investigations in mountain forests. Academy of Science, USSR, Pacific Ocean Institute of Geography, Vladivostok

Nobis M, Hunziker U (2005) Automatic thresholding for hemispherical canopy-photographs based on edge detection. Agric For Meteorol 128:243–250

Olsson L, Carlsson D, Grip H, Perttu K (1982) Evaluation of forest-canopy photography with diode-array scanner OSIRIS. Can J For Res 12:822–828

Pope DJ, Lloyd PS (1974) Hemispherical photography, topography and plant distribution. In: Evans GC, Bainbridge R, Rackham O (eds) Light as an ecological factor II. Blackwell, Oxford

Promis A, Gärtner S, Butler-Manning D, Durán-Rangel C, Reif A, Cruz G, Hernández L (2011) Comparison of four different programs for the analysis of hemispherical photographs using parameters of canopy structure and solar radiation transmittance. Kombinierte Methoden aus Modellierung, Messung und Geländearbeit 11:19–33

Regent Instruments (2013) Image analysis for plant science. Regent Instruments Inc., Quebec City. http://www.regent.qc.ca/. Accessed 10 Jan 2013

Rich PM (1989) A manual for analysis of hemispherical canopy photography. Los Alamos National Laboratory Report LA-11733-M, Los Alamos, New Mexico

Rich PM (1990) Characterizing plant canopies with hemispherical photographs. Remote Sens Rev 5:13–29

Rich PM, Wood J, Vieglais DA, Burek K, Webb N (1999) HemiView user manual. Delta-T Devices Ltd., Cambridge

Roxburgh JR, Kelly D (1995) Uses and limitations of hemispherical photography for estimating forest light environments. N Z J Ecol 19:213–217

Schleppi P, Paquette A (2017) Solar radiation in forests: theory for hemispherical photography. In: Fournier RA, Hall RJ (eds) Hemispherical photography in forest science: theory, methods, applications. Springer, Berlin

Smith WR, Somers GL (1991) SUNSHINE: a light environment simulation system based on hemispherical photographs. USDA Forest Service, Southern Forest Experiment Station, New Orleans, Louisiana. Research Paper SO-267, 17 pp

ter Steege H (1994) HEMIPHOT, a programme to analyze vegetation indices, light and light quality from hemispherical photographs. Tropenbos Documents 3. The Tropenbos Foundation, Wageningen, The Netherlands

ter Steege H (1997) WINPHOT 5: a programme to analyze vegetation indices, light, and light quality from hemispherical photographs. Tropenbos-Guyana Programme and Utrecht University, Utrecht, 52 pp

Walter J-MN (1999) CIMES. A package of programs for the assessment of canopy geometry by hemispherical photographs. Université Louis Pasteur, Strasbourg

Walter J-MN (2009) CIMES-FISHEYE. Hemispherical photography of forest canopies: a package of programs for the assessment of canopy geometry and solar radiation regimes through hemispherical photographs [online]. Université de Strasbourg, Strasbourg, France. Available from http://equinoxe.u-strasbg.fr/cimes

Watt AS (1947) Pattern and process in the plant community. J Ecol 35:1–22

Whitmore TC, Brown ND, Swaine MD, Kennedy D, Goodwin-Bailey CI, Gong W-K (1993) Use of hemispherical photographs in forest ecology: measurement of gap size and radiation totals in a Bornean tropical rainforest. J Trop Ecol 9:131–151

Yamamoto K (2004) LIA for Win32, an image analysis software intended to use in the forest science and ecology. http://www.agr.nagoya-u.ac.jp/~shinkan/LIA32/index-e.html. Accessed 9 Sept 2013

Zheng G, Moskal LM (2009) Retrieving leaf area index (LAI) using remote sensing: theories, methods and sensors. Sensors 9:2719–2745

Chapter 2
Solar Radiation in Forests: Theory for Hemispherical Photography

Patrick Schleppi and Alain Paquette

Abstract Solar radiation is not only the main source of energy for life on Earth, but it is also the natural source of light for the optical acquisition of information, such as for vision and photography. Natural sunlight provides the illumination required for taking hemispherical photographs in forests. In turn, such photographs can then be interpreted to provide a description of the plant canopy and its interactions with light, or more generally, its radiation regime. The techniques used for both the acquisition and the analysis of hemispherical photographs are based on the theory of radiation in plant canopies. The goal of this chapter is to present the theoretical foundations describing how solar radiation reaches Earth's surface and interacts with plant canopies. Different sensors for measuring radiation are described, and the principles of hemispherical photography are reviewed. Finally, a comparison of the techniques used to assess the radiation regime of forest canopies is presented.

Keywords Energy budget · Forest canopy · Light sensor · Solar radiation · Overcast sky models · Atmospheric effects

2.1 Introduction

Hemispherical photography (HP) consists of taking pictures in an upward direction through an extreme wide-angle lens (i.e., a lens having a field of view close to 180°) (Rich 1990). By accounting for the geometry of light transmission and interception,

P. Schleppi (✉)
Swiss Federal Institute for Forest, Snow and Landscape Research (WSL), Zürcherstr. 111, 8903 Birmensdorf, Switzerland
e-mail: schleppi@wsl.ch

A. Paquette
Centre for Forest Research, Université du Québec à Montréal (UQAM), P.O. Box 8888 Centre-Ville Station, Montréal, QC H3C 3P8, Canada
e-mail: alain.paquette@gmail.com

R.A. Fournier and R.J. Hall (eds.), *Hemispherical Photography in Forest Science: Theory, Methods, Applications*, Managing Forest Ecosystems 28,
DOI 10.1007/978-94-024-1098-3_2

HP has been successfully used to characterize the structure of forest canopies and the radiation regime of forests. The interpretation of hemispherical photographs is based on theoretical assumptions concerning the interactions between light and the canopy. Conversely, ecosystem modeling requires empirical information on the quantity of foliage, the structure of the canopy, and the properties of incoming solar radiation (e.g., Running and Coughlan 1988). Understanding these patterns and processes is necessary for the correct application and interpretation of HP. Our objective in this chapter is to provide the reader with an introduction to solar radiation in forests with an emphasis on HP.

The Sun is the main source of radiation on Earth; for this reason, we first review the basic physics of solar radiation and its interactions with the atmosphere. Second, we examine the interactions between radiation and forest canopies, which are the main processes that HP aims to describe. Third, we present different types of radiation sensors and discuss their main applications, especially in relation to HP. Last, we briefly discuss the strengths and limitations of HP for assessing forest canopies and their radiation regimes.

2.2 Solar Radiation

Solar radiation, the electromagnetic energy that originates from the Sun, is the essential source of energy that drives climate and life processes on Earth (Gates 1980). In forests, solar radiation is important for diverse biological and physical processes such as photosynthesis, growth, carbon cycling, evapotranspiration, and temperature regimes. In spite of this essential role, solar radiation has received less attention compared to other meteorological parameters such as temperature, precipitation and wind. This disparity can be explained, in part, by the difficulties of measuring radiation. Furthermore, while other meteorological factors can lead to accidents or disasters (e.g., frosts, droughts, floods, windstorms), solar radiation always remains within a range to which ecosystems are adapted, thus making it "good-natured" and easily forgotten. Nevertheless, foresters and ecologists have long recognized that differences exist among species regarding their tolerance to shade (e.g., Engler 1924). While shade-intolerant species are typically pioneer trees, shade-tolerant species appear later in succession because they can grow under denser canopies or at least survive until a gap opens above them. The degree of shade-tolerance of shrubs, herbaceous species, and mosses is also important in determining their occurrence in forests. A favorable light regime is important for tree establishment; therefore, optimizing light availability for seedlings is an important objective for forest management practices, especially in uneven-aged stands and in mixed forests (Lieffers et al. 1999; Drever and Lertzman 2003; Paquette et al. 2006). Here, we provide an introduction to this subject by examining how solar radiation varies in time and space. Numerous textbooks describe the influence of solar radiation on the Earth; for example, Bohren and Clothiaux (2006) and Timofeyev and Vasil'ev (2008) provide reviews covering the physical concepts that are presented in this chapter. The Sun's radiation is maximal around a wavelength of 500 nm (Fig. 2.1). According to Wien's law, this corresponds to a temperature of nearly 5800 K for the light-emitting layer (i.e., the photosphere). The emission of solar radiation covers the whole spectrum that is visible to the human eye (380–750 nm) and extends into the ultraviolet and near-infrared bands. This spectrum is often referred to as shortwave radiation. In contrast, longwave infrared radiation is emitted by surfaces at near-ambient temperatures. By definition, the boundary between shortwave and longwave radiation is set at 4000 nm.

When solar radiation reaches Earth's upper atmosphere, it delivers an average power of 1367 W m^{-2} when measured perpendicularly to the Sun's rays. This "solar constant" varies by a few W m^{-2}, depending upon the amount of solar activity during a solar cycle. Stronger variations occur due to the eccentricity of the Earth's orbit (Fig. 2.2) because radiation flux is inversely proportional to the squared distance between the source and target (i.e., between the Sun and Earth). The effect of orbital eccentricity on the solar constant is given by

$$E_o = (r_o/r)^2 = 1.00011 + 0.03423\cos\Gamma + 0.00092\sin\Gamma + 0.00072\cos 2\Gamma + 0.00006\sin 2\Gamma \quad (2.1)$$

with

$$\Gamma = 2\pi d/365.25, \quad (2.2)$$

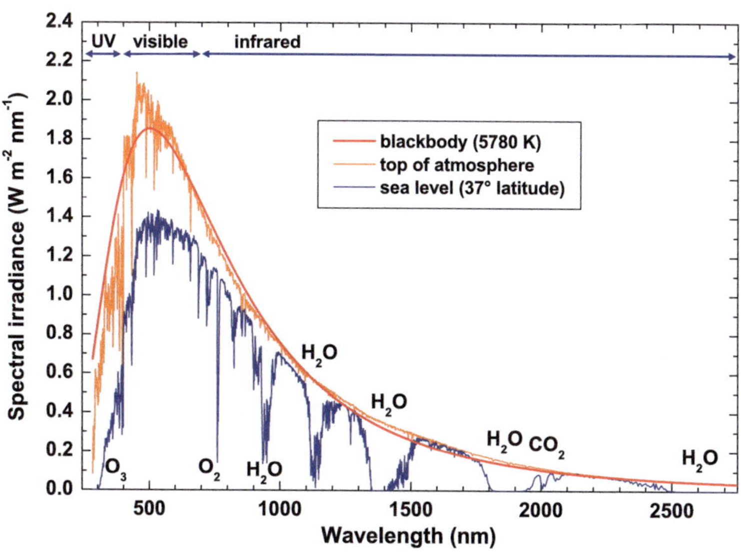

Fig. 2.1 Solar radiation spectrum at the top of Earth's atmosphere and at sea level for direct radiation. Emission of a black body after Planck's law. The main absorption bands of atmospheric gases are indicated by their chemical formulas. *Data source* Emery and Myers (2008)

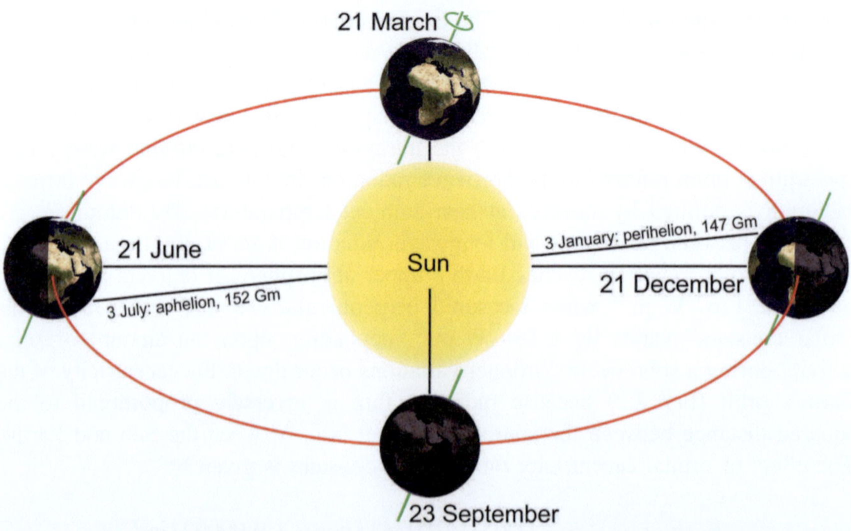

Fig. 2.2 Scheme of Earth's rotation around the Sun, showing the declination of Earth's axis that explains the seasons

where r_o is the average distance and r is the current distance to the Sun. The term Γ is referred to as the day angle and d is the number of days since 12:00 UTC (Coordinated Universal Time), 1 January of the last intercalary or leap year. In addition to precession, long-term changes occur in eccentricity and obliquity, together with long-term changes in solar activity. According to the theory of Milankovitch (1941), these effects (which are amplified by feedbacks involving Earth's albedo) are responsible for the ice age cycles of the last few million years. Such long-term variations will not be developed here.

Because the total area of a sphere is four times its cross-section (i.e., its projected area), the average incoming radiation at the top of the atmosphere is 1367/4 = 342 W m^{-2} when measured horizontally. Its value varies between 0 for Earth's nighttime hemisphere and the solar constant for the subsolar point, which is the location within the tropics where the Sun is directly overhead (i.e., in zenith). Averaged over a year, solar radiation I_0 at the top of the atmosphere is about 417 W m^{-2} at the equator and 173 W m^{-2} at the poles, or

$$I_0 = I_s(0.219 + 0.092\cos 2\Phi - 0.005\cos 4\Phi - 0.002\cos 6\Phi + 0.002\cos 8\Phi) \tag{2.3}$$

where Φ is the latitude and I_s is the solar constant. Note that in spite of Earth's orbital eccentricity, this relationship is practically identical for the southern and northern hemispheres of our planet. This is because the length of the seasons (longer summer in the northern hemisphere) compensates for differences in radiation (shorter Sun-Earth distance when it is summer in the southern hemisphere).

The position of the Sun in the sky can be plotted on a sky map showing the daily sun tracks over the course of a year. Such a map has a geometry that is similar to a hemispherical photograph, where the zenith and azimuth angles are presented as polar coordinates. A map of sun tracks can be combined with a hemispherical photograph, which is used as a mask in order to show the potential duration of direct radiation at the point where the photograph was taken. In a forest, this information is especially useful for characterizing the prevailing under-canopy light regime for the ground vegetation, including tree recruitment. This simple approach, however, does not take into account atmospheric effects, which is the topic of the next section.

2.3 Atmospheric Absorption and Scattering

Solar radiation is partly absorbed by gas molecules, mainly triatomic compounds, in traveling through the atmosphere. Water vapor (H_2O) and carbon dioxide (CO_2) primarily absorb energy in the infrared band, while ozone (O_3) absorbs in the ultraviolet (Fig. 2.1). Solar radiation is further absorbed and scattered by other gases, aerosol particles, and clouds. Very small-diameter particles, which are much smaller than the wavelength of light, or microscopic fluctuations in air density can cause Rayleigh scattering. Blue light is scattered more than other visible colors due to its shorter wavelength. The blue color of a clear sky is due to diffuse sky radiation resulting from Rayleigh scattering of sunlight in the atmosphere. When the Sun is low on the horizon, the missing blue light that has been scattered away leaves the direct sunlight with an orangish or reddish color. The angle between the original and scattered radiation is the scattering angle. In Rayleigh scattering, this angle has a statistical distribution that is maximal at 0 and π rad; in other words, scattered radiation is preferentially directed forward or backward as opposed to the sideward directions. Aerosol particles with diameters of the same order of magnitude as the wavelength of radiation produce Mie scattering. In contrast to Rayleigh scattering, the Mie scattering pattern is relatively independent of wavelength and is preferentially in a forward direction.

Solar radiation also interacts with ice crystals and water droplets in clouds. Multiple reflections and refractions cause scattering. Statistically, all wavelengths diffuse similarly, resulting in the whitish or grayish color of clouds. However, some separation of wavelengths occurs when the number of refractions and reflections is small. This is the case for rainbows, where sunlight is refracted when entering a water droplet, reflected inside, and refracted again upon exiting (or two reflections for a secondary rainbow).

Because much of the solar radiation is scattered back into space by air and clouds, only a fraction of it reaches Earth's surface. The atmosphere affects incoming solar radiation according to the following proportions (Timofeyev and Vasil'ev 2008):

26% is absorbed and heats the stratosphere and thermosphere,
24% is scattered back by the atmosphere and is reflected to space,

22% is scattered and reaches Earth's surface as diffuse radiation,
28% is transmitted and reaches Earth's surface as direct radiation.

Together with corresponding values for near- and longwave infrared radiation, these numbers determine Earth's radiative balance and its average temperature. Direct radiation dominates under clear skies, but it can be completely blocked by clouds that only allow diffuse radiation to pass through. For example, direct radiation dominates over deserts. Over most forests of the world, direct and diffuse radiation roughly contribute equally to global radiation on an annual basis. However, direct radiation is more variable over time and space compared to diffuse radiation because the former is more dependent upon the season, time of day, topography, and weather. In meteorology and climatology, various atmospheric radiative transfer models have been developed to take into account the effects of weather conditions; these are the so-called atmospheric radiative transfer codes.

For a particular time and point on Earth, the transmission (or transmittance, a dimensionless ratio) of radiation through a cloudless atmosphere can be described by the Beer-Lambert-Bouguer law as a function of the atmospheric thickness along the path that the radiation is traveling. Transmission depends on the elevation of the Sun (i.e., the angle between the Sun and horizon, see Eq. 2.13 below for its calculation). The transmittance T is given by

$$T = \tau^{p/p_0 \sin\beta} = \tau^{p/p_0 \cos\theta} = e^{-kp/p_0 \cos\theta}, \tag{2.4}$$

where τ is transmittance at normal pressure (p_0 = 1013 hPa), p is local air pressure, β is the elevation of the Sun above the horizon, and θ is its zenith angle (β + = 90° = π/2 rad). The absorption coefficient is defined as $k = -\ln \tau$, which can be considered as the sum of the absorption coefficients of the different atmospheric constituents (i.e., gases and aerosols). Air pressure is used here as a measure of the air mass above the point of interest, which is referred to as the optical mass of the atmosphere. At higher altitudes, the pressure is lower and the corresponding transmission is higher.

At Sun elevations below 5° (or 0.1 rad), some refraction occurs and the Sun appears to be slightly higher in the sky than it really is. At the same time, the path of sunlight through the atmosphere is shortened because of the latter's curvature. These effects can be approximated by

$$T = \tau^{p/p_0 \cos(0.99\theta)} = e^{-kp/p_0 \cos(0.99\theta)} \tag{2.5}$$

The Beer-Lambert-Bouguer law is essentially valid for any wavelength or spectral band but with variable absorption coefficients. For a clear atmosphere, τ is close to 0.8 for visible light, but this value can be affected by air pollution (Jacobson 2002).

Local measurements that have been made consecutively over years are necessary to obtain reliable estimates of the true transparency of the atmosphere and its variation by taking both direct and diffuse radiation into account. Where sunlight duration is available but no radiation measurements have been made, it is possible to estimate the latter by regional adaptation of an Angström-Prescott model (Angström 1924; Prescott 1940), such that

$$T = a + b(D/D_0) + c(D/D_0)^2. \tag{2.6}$$

Equation 2.6 is the quadratic or second-order model; the linear or first-order model would have $c = 0$. The variable D is the measured duration of sunlight (often calculated per month) and D_0 is the maximal duration without any cloudiness. Estimates of the regression parameters (i.e., a, b and c) are available for many regions of the world and there is a large body of corresponding technical publications from studies on solar energy. In a recent review, Akinoglu (2008) gives the following ranges for the linear model: $0.06 \leq a \leq 0.46$ and $0.19 \leq b \leq 0.87$. The value of D_0 is obtained by calculating the hour angle at sunrise and sunset. The difference is transformed from an angle into a duration, such that

$$\cos \omega_1 = \cos \omega_2 = -\tan \Phi \tan \delta, \tag{2.7}$$

where Φ is the latitude, and δ is the declination of the Sun. The respective hour angles at sunrise and sunset are ω_1 and ω_2, where the first term is the negative arccosine and the second term is the positive arccosine. Because they are approximately symmetrical around $\omega = 0$ (solar noon), the day length is $2\omega_2$; in hours it is

$$D_0 = 24\omega_2/\pi = 24 \arccos(-\tan \Phi \tan \delta)/\pi \tag{2.8}$$

Note that no solutions are found if the Sun remains either above or below the horizon. In these cases, $D_0 = 24$ or $D_0 = 0$, referring to midnight sun and polar night, respectively. The declination δ is the angle between Earth's orbit around the Sun (i.e., orbital plane) and Earth's equator (i.e., equatorial plane). It varies with the season between −23.45° (−0.409 rad) at the December solstice and +23.45° (+0.409 rad) at the June solstice. It can be calculated (in rad) approximately as

$$\delta = -0.409 \cos[2\pi/365(d + 10)], \tag{2.9}$$

where d is the day of the year. A more precise formula takes the eccentricity of Earth's orbit into account (Spencer 1971). We recalculated Spencer's coefficients also to take into account leap years (via Eq. 2.2 for the calculation of Γ), and optimized them for the years 2011–2030. From this development, declination can be calculated as

$$\begin{aligned}\delta = {} & 0.006918 - 0.399912 \cos \Gamma + 0.070257 \sin \Gamma - 0.006758 \cos 2\Gamma \\ & + 0.000907 \sin 2\Gamma - 0.002697 \cos 3\Gamma + 0.00148 \sin 3\Gamma,\end{aligned} \tag{2.10}$$

with a standard deviation (SD) of 0.00025 rad (0.014°), compared to 0.017 rad (1°) for Eq. 2.9. Angström-Prescott models were developed from an intuitive point-of-view using statistical methods. A theoretical basis for these models was recently proposed by Pelkowski (2008). A three-layer atmospheric model is used, where a layer containing clouds is sandwiched between two clear layers. The radiation fluxes in two streams (i.e., direct and diffuse) were modeled through these layers, including reflection from Earth's surface. Pelkowski (2008) concluded that an Angström-Prescott polynomial is a valid approximation and it should generally be at least of a second-order form.

In this section, we have examined how the atmosphere quantitatively affects the radiation traversing it. For most applications, including HP, we also need to consider the angles between the radiation and the objects interacting with it on Earth, which is the topic of the next section.

2.4 Geometry of Solar Radiation on Earth's Surface

The radiative power that is received by a flat surface is maximal when it directly faces the source, i.e., when the angle of incidence is 0. When radiation is parallel to the surface, it delivers no energy. Between these extremes, Lambert's Cosine law applies (Fig. 2.3), such that the radiative flux I onto the surface area is given by

$$I = I_0 \cos \gamma = E_o I_s T \cos \gamma, \tag{2.11}$$

where I_0 is the incident flux (measured perpendicularly), and γ is the angle of incidence. The term I_s is the solar constant, E_o is the correction for the eccentricity of Earth's orbit, and T is the transmittance of the atmosphere. On flat ground, the angle of incidence is equal to the zenith angle, but this is not the case on a slope.

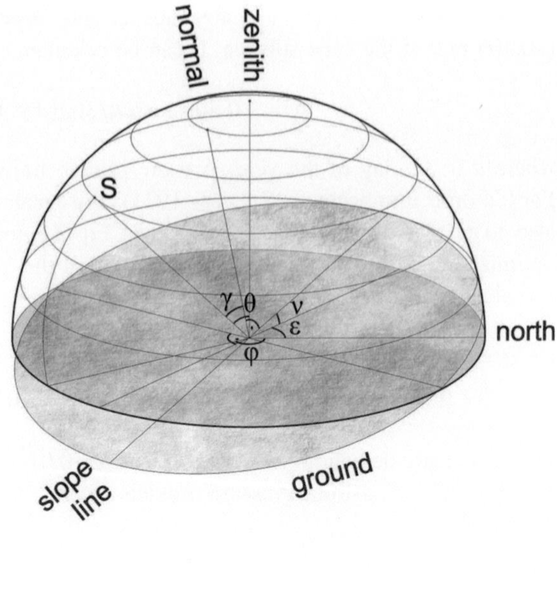

Fig. 2.3 Geometry of solar radiation on an inclined surface. S = position of the Sun in the sky, θ = its zenith angle, φ = its azimuth, ν = slope angle, ε = aspect (orientation of the slope), γ = angle of incidence of the direct radiation on the ground

For sunlight, γ is taken through the center of the Sun. Strictly speaking, it should be integrated over the apparent sun disk, which covers an angle of about 0.009 rad (or slightly more than 0.5°). As long as the Sun is entirely visible, using only the angle through the center results in a negligible source of error. The power delivered by direct solar radiation onto a slope is calculated by

$$\cos \gamma = \sin \beta \cos v + \cos \beta \cos(\varepsilon - \varphi) \sin v, \tag{2.12}$$

where β is the elevation of the Sun, φ is its azimuth, v is the slope angle, and ε is the aspect (i.e., orientation of the slope).

To calculate the position of the Sun, it is necessary to know the local coordinates upon Earth's surface and the time, such that

$$\sin \beta = \cos \theta = \sin \Phi \sin \delta + \cos \Phi \cos \delta \cos \omega \tag{2.13}$$

and

$$\cos \varphi = (\cos \Phi \sin \delta - \sin \Phi \cos \delta \cos \omega) / \cos \beta, \tag{2.14}$$

where Φ is the latitude, and δ is the declination of the Sun. Taking the arccosine from Eq. 2.14 yields two alternate solutions, one valid for the morning ($\omega < 0$ $0 < \varphi < \pi$) and the other for the afternoon ($\omega > 0 \Rightarrow \pi < \varphi < 2\pi$). The hour angle ω is the local solar time expressed as an angle instead of hours, minutes, and seconds; a value of zero indicates solar noon. The hour angle is calculated as

$$\omega = (t - 12)\pi/12 - \lambda - \Delta\omega \tag{2.15}$$

with

$$\Delta\omega = 0.0002 \cos \Gamma - 0.0335 \sin \Gamma - 0.0159 \cos 2\Gamma - 0.0400 \sin 2\Gamma, \tag{2.16}$$

where λ is the longitude, t is the UTC (Coordinated Universal Time) in hours, and $\Delta\omega$ is the equation for time, i.e., the difference between true solar time and average solar time, expressed here as an angle. This difference arises from the obliquity of Earth's axis and from the eccentricity of its orbit. The term $\Delta\omega$ varies roughly by ±0.07 rad, equivalent to ±¼ h, with two minima and two maxima over the year. (See Eq. 2.2 for the calculation of Γ.) Given this formula, ω is obtained in rad. Note that the equation for time is defined with the opposite sign in some countries, such as France.

When Eqs. 2.11–2.14 are combined and the expressions are rearranged, they yield Iqbal's formula (Iqbal 1983), such that

$$\begin{aligned} I = E_o I_s T(&(\sin \Phi \cos v - \cos \Phi \sin v \cos \varepsilon) \sin \delta \\ &+ (\cos \Phi \cos v + \sin \Phi \sin v \cos \varepsilon) \cos \delta \cos \omega \\ &+ \cos \delta \sin v \sin \varepsilon \sin \omega). \end{aligned} \tag{2.17}$$

This formula is deterministic except for its T term: atmospheric transmission is indeed subject to stochastic weather conditions. If certain restrictions or assumptions are applied to T (e.g., considering only clear skies or average overcast conditions), then the equation can be integrated numerically over time to obtain the average direct radiation. Using $\tau = 0.8$ in Eq. 2.5, we calculated yearly values over a large range of latitudes (−72° to 72°) and slopes (0° to 60°), and for all aspects. We subjected the results to multiple regression and, consequently, yearly average radiation under a clear sky I_a can be approximated as

$$\begin{aligned} I_a = I_s(&0.0631 + 0.1964(\cos \Phi)^2 \cos v \\ &- 0.1063 \sin v \cos \varepsilon \sin 2\Phi - 0.0238 \cos \Phi \cos 2\Phi \\ &- 0.0117 \sin v \cos \Phi \cos 2\varepsilon). \end{aligned} \tag{2.18}$$

The coefficient of determination is $R^2 = 0.998$, and the precision of the integration is 0.4% (results not previously published). This formula covers almost all forests of the world, but it only takes a single slope into account rather than considering complex topography that would modify the radiation, such as the opposing slope of a valley.

The geometry of diffuse radiation is important for its interaction with tilted surfaces or with complex structures such as a plant canopy (see review by Torres and Torres 2008). This geometry can be described separately for clear and overcast skies. Diffusion in clear air and in clouds produces different patterns of radiance (i.e., radiation per unit of solid angle) and, consequently, different colors. In a clear sky, photons are not scattered in all directions but preferentially travel forward. Radiance is thus anisotropic (Brunger and Hooper 1993): it is higher near the Sun and decreases with the scattering angle. The pattern of this decrease depends upon the relative importance of Rayleigh and Mie scattering; in other words, it varies with the turbidity of the atmosphere (e.g., Vida et al. 1999). Some models also account for the effect of zenith angle since it is related to the optical mass of the atmosphere (Hooper and Brunger 1980; Harrison and Coombes 1988). Because scattering is a function of wavelength, different parameters have to be used in the model, depending upon which band of the spectrum is being considered. For example, scattering of near-infrared radiation is not the same as scattering of photosynthetically active radiation (PAR).

Under a uniformly overcast sky where the Sun is not visible, a first approximation considers the clouds as a diffusive horizontal surface with radiance subjected to Lambert's Cosine law. This results in isotropic radiance (i.e., radiance with the same value over the whole visible sky). This model is referred to as "uniform overcast sky" (UOC). Consideration of the thickness of clouds and reflections between clouds and Earth's surface leads to a model referred to as a "standard overcast sky" (SOC) in which radiance is a function of the zenith angle θ (Steven and Unsworth 1980), such that

$$I(\theta) = I(0)(1 + b\cos\theta)/(1 + b), \tag{2.19}$$

where b is an empirically determined factor, typically between 0.5 and 1.5. On a slope v, total radiation is reduced compared to a flat surface because part of the sky is obstructed (Steven and Unsworth 1980), such that

$$I = I_0(\cos v/2)^2 + 2b\left[\sin v - v\cos v - \pi(\sin v/2)^2\right]/[\pi(3 + 2b)]. \tag{2.20}$$

Since the UOC model corresponds to $b = 0$, the effect of slope simplifies in that case to the factor $(\cos v/2)^2$.

There are also models that describe partly overcast skies where the Sun is visible through uniform clouds (Hooper and Brunger 1980; Grant et al. 1996). Such models combine a distribution function that is based on theory with parameters that are fitted empirically. Like the Angström-Prescott polynomials that were discussed in the previous section, these models have regional validity and need to be adapted to different climatic conditions (Siala and Hooper 1990). An additional model describes luminance, which is the brightness of light that is emitted or reflected from a surface as perceived by the human eye, which is more sensitive to green than to red and blue wavelengths. In this model (Kittler 1994; CIE 2003), luminance is characterized as combining a function of the zenith angle ("gradation") and a function of the scattering angle ("indicatrix"). For each of these functions, standard parameter sets are provided to match most typical situations ranging between clear and overcast skies. This model was developed primarily for industrial and architectural lighting applications, but it can also be adapted to environmental radiance measurements, viz., the analysis of hemispherical images (Lang et al. 2010). The method consists of using a radiance model to interpolate between sky patches that are visible through canopy openings so as to obtain a virtual reference image. Image analysis is then performed by comparing the actual hemispherical photograph with the interpolated reference.

On a slope, the same geometrical rules apply as for direct radiation, with the difference being that they have to be integrated over all sky directions, which means a double integration over the zenith and azimuth angles. Liu and Jordan (1961) have provided an integrated formula based on isotropic diffusion and isotropic reflection by the surrounding ground. Hay's model (Hay and McKay 1985) is similar but is based on anisotropic diffusion. In complex terrain, the integration of diffuse sky radiation is obviously limited to the visible portion of the sky (e.g., Wang et al. 2005), but reflection from surrounding slopes may also have to be considered. Like the case of sun tracks being combined with a hemispherical photograph to derive potential direct radiation (see Sect. 2.2), a model of sky radiation can be used to calculate diffuse radiation that is transmitted by the pictured canopy and to integrate it over time. There is, however, more information about the structure of the canopy to be gained from a hemispherical photograph than just creating a sky mask. In the next section, the theoretical background for the extraction of this information will be examined.

2.5 Interactions of Solar Radiation with Forest Canopies

In the preceding sections, we have considered how much solar radiation reaches a given surface on Earth. Compared to a solar panel, a forest canopy has a complex geometry with receptors (the leaves) that do not all face the same direction and are not all exposed to direct radiation. Their power yields (photosynthesis) and maintenance costs (transpiration and respiration) are thus highly variable in time and space, and much more difficult to predict than the output of a solar panel. Except in cases when the gas exchange of a single leaf is measured under known light conditions, it is necessary to consider that radiation is heterogeneous within the canopy and that photosynthesis shows a non-linear response to irradiance (e.g., Sinclair et al. 1976). Photosynthesis is further limited by the radiation spectrum that it can use. Since they both evolved under the same radiation spectrum, photosynthesis in terrestrial plants and human vision cover about the same spectrum. By definition, photosynthetically active radiation (PAR) is the band of wavelengths that ranges from 400 to 700 nm. Blue and red light are preferentially absorbed by photosynthetic pigments, while green light tends to be reflected, giving leaves their characteristic color (Fig. 2.4).

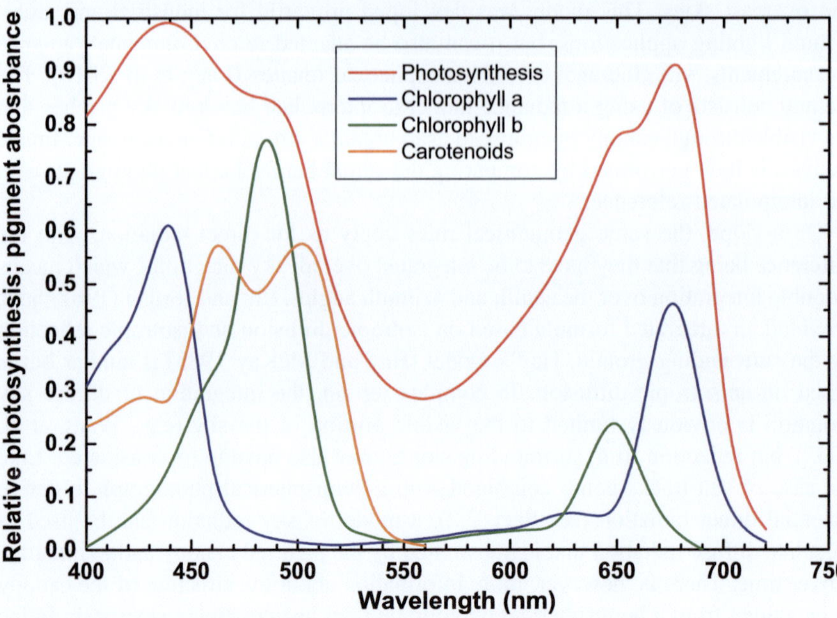

Fig. 2.4 Absorption spectra of photosynthetic pigments and photosynthesis rate as a function of wavelength. Redrawn from Singhal et al. (1999) and Lodish et al. (2000)

2.5.1 Canopy Radiation Models

Absorption, scattering, and transmission modify radiation as it interacts with a leaf or with a plant canopy (Fig. 2.5). The intensity, the spectral distribution, and the ratio of direct-to-diffuse radiation are all affected by these interactions. At the level of single leaves, several dimensionless coefficients can be defined as follows: absorbance, $\alpha = A/I$; reflectance, $\rho = R/I$; and transmittance, $\tau = (T + S)/I$ (see Fig. 2.5 for abbreviations). Note that, depending upon the application, transmission can refer either to T or to $T + S$. To distinguish between these two terms, $T + S$ is sometimes called non-interceptance. Radiation scattering is summarized as a scattering coefficient $\omega = (R + S)/I$ (here, the symbol ω is different from the solar hour angle ω in Eqs. 2.7–2.17). For a leaf, transmission without scattering is very small and, thus, $\omega \approx \tau + \rho = 1 - \alpha$. A typical spectrum for leaves is shown in Fig. 2.6. It can be seen that absorption dominates in the visible range, while scattering (reflection and transmission) dominates in the near-. The amount of radiation that is absorbed or scattered by a leaf also depends upon the angle of incidence, which determines the projected leaf area in the direction of the radiation. Individual leaves are not always flat and their inclination angle within a canopy is not constant but displays a statistical probability distribution. Several models have been proposed to describe this probability distribution; the most common ones include the ellipsoidal distribution, beta distribution, de Wit's function, and the Ross-Goudriaan function (see Wang et al. 2007 for a comparison). If foliage grows or moves in relation to the position of the Sun (i.e., through phototropism and heliotropism), then its projection coefficient (projected area/leaf area) is a function of both the azimuth and zenith angles. If foliage is distributed symmetrically along all azimuths, however, then its

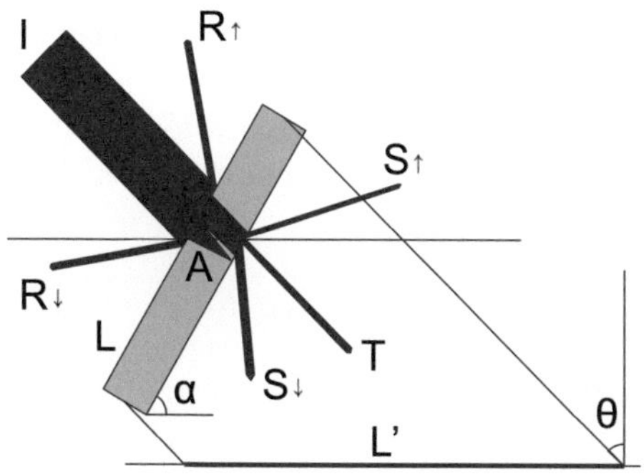

Fig. 2.5 Schematic interaction of radiation with a leaf blade: I = incoming radiation, A = absorption, R = reflection, S = scattering, T = transmission, L = leaf area, L' = projected leaf area on a horizontal surface, θ = zenith angle of the incoming radiation, α = leaf inclination angle. The *subscript arrows* denote the downward or upward components of the diffused radiation (each corresponding to one of the hemispheres delimited by a horizontal surface)

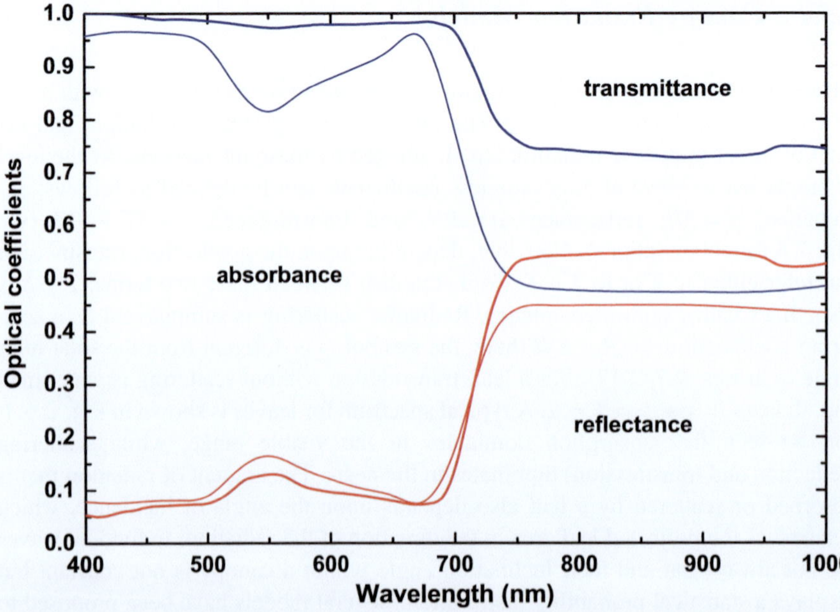

Fig. 2.6 Typical spectra of leaves (*thin lines*) and needles (*thick lines*). Redrawn (simplified) from Berbigier and Bonnefond (1995) and Dai and Sun (2006)

projection coefficient is a function $G(\theta)$ of the zenith angle. The horizontal area L' that is shaded by a leaf of area L is then (Fig. 2.5)

$$L' = LG(\theta)/\cos\theta. \tag{2.21}$$

The leaf area index (LAI) is defined as the leaf area per unit ground area and, therefore, is a dimensionless quantity. Considering that leaves, and especially needles, are not perfectly flat, the leaf area itself is defined either as the maximum projected area or as half of the total three-dimensional surface. Direct radiation is reduced by the leaf area in a manner that can be expressed as a differential equation, such that

$$dI/dL = -(G(\theta)/\cos\theta)I = -kI, \tag{2.22}$$

where k is the extinction coefficient of the canopy for direct radiation, which corresponds to an optical depth per unit leaf area. When integrated, this differential equation yields the Beer-Lambert-Bouguer law, such that

$$I(L) = I_0 e^{-kL}, \tag{2.23}$$

where I_0 is the radiation at the canopy top.

If L is taken as the LAI above any level of the canopy, then this relation describes the attenuation of direct radiation as it enters the canopy. The fate of diffuse radiation is more complicated because it is not only absorbed by the leaves, but also is produced by scattering on them. Moreover, scattering occurs essentially in all directions; therefore, it must be described by an angular distribution that is a function of the incident and scattering angles, which is called a scattering phase function (Ross 1981). To simplify, scattering can be partitioned into an upward and a downward component, where both represent an angular integration over a hemisphere that is either below or above a horizontal surface (Fig. 2.5). If diffuse radiation is assumed to be isotropic, then a two-stream model can be used to model radiative transfer in a canopy (Dickinson 1983). The partitioning of scattering is first calculated via an "upscatter coefficient" β, which is the proportion of downward diffuse radiation that is scattered upward. Similarly, β_o is defined as the upscatter of direct incident radiation. Consequently, the two coefficients are respectively calculated as

$$\beta = \left(\rho + \tau + (\rho - \tau)(\cos\alpha)^2\right)/2\omega \tag{2.24}$$

and

$$\beta_o = \frac{1 + \overline{\mu}k}{\omega\overline{\mu}k}\omega_s(\mu), \tag{2.25}$$

where ρ is the leaf reflectance, τ is its transmittance, $\omega = \rho + \tau$ is its scattering coefficient, and α is the mean leaf inclination angle. The average inverse diffuse optical depth per unit leaf area $\overline{\mu}$ is

$$\overline{\mu} = \frac{1}{k'} = \int_0^1 \frac{\mu'}{G(\mu')}d\mu', \tag{2.26}$$

where μ' is the cosine of the zenith angle of the scattered radiation, and k' is the diffusive extinction coefficient (all quantities in Eq. 2.26 are dimensionless). The single scattering albedo of the canopy ω_s is the proportion of scattering over the sum of scattering and absorption. It is a function of the scattering phase function and the angular distributions of both the radiation and leaves (Dickinson 1983; Sellers 1985). The downscatter coefficients are values that are complementary to the aforementioned upscattering coefficients (viz., $1 - \beta$ and $1 - \beta_o$, respectively). Without going into the details of the calculations, it is important to note that the reflectance (i.e., albedo) of a canopy is always smaller than that of a flat leaf; the former usually is not much more than half of the latter. This is due to radiation being trapped in the canopy, where each leaf is surrounded by other leaves. This offers new chances for the scattered radiation to be absorbed, especially in tall plant canopies like forests (Stanhill 1970). Two separate equations that can be used then to describe upward and downward (isotropic) diffuse radiation are

$$\mu di_{\downarrow}/dL = \beta\omega i_{\uparrow} - (1 - (1 - \beta)\omega)i_{\downarrow} + \omega\mu k(1 - \beta_o)e^{-kL} \quad (2.27)$$

and

$$\mu di_{\uparrow}/dL = \beta\omega i_{\downarrow} - (1 - (1 - \beta)\omega)i_{\uparrow} + \omega\mu k\beta_o e^{-kL}, \quad (2.28)$$

where $i = I/I_0$. In other words, both components (up and down) of diffuse radiation are the result of their own and respective scattering. together with the scattering of direct radiation.

When these differential equations (Dickinson 1983) are integrated (Sellers 1985) by choosing appropriate angular distributions for the foliage and by setting the radiation that is reflected by the soil beneath the canopy, they yield

$$I_{\uparrow} = \rho_{soil}\left(I_{\downarrow} + e^{-kL}\right). \quad (2.29)$$

It is also possible to use a discrete ordination method that consists of defining a finite number of cells in the canopy space and a finite number of directions in space (Myneni 1991). Obviously, this method is computationally intensive. In Eqs. 2.27 and 2.28, the same value of β is used for the upscatter of downward diffuse radiation as for the downscatter of upward radiation. However, the optical properties of both leaf sides can differ. Dai and Sun (2006) introduced this difference in a generalized model, together with anisotropic incoming radiation. Their model is too complicated to be described here in detail. Its main result is a difference in overall reflectance of the canopy compared to the two-stream model. Depending on the leaf angle, on the proportion of direct or diffuse incoming radiation, and on the wavelength, this difference (positive or negative) can be significant (Dai and Sun 2007).

Compared to a canopy in which all leaves would be black, scattering increases transmittance and reduces the absorbance of a canopy. As suggested by Fig. 2.6, reflectance and transmittance of annual leaves are often similar, while persistent leaves and conifer needles exhibit low transmittance. For leaves with $\rho = \tau$, it can be shown that scattering decreases the effective extinction coefficient of a canopy by a factor $\sqrt{\alpha}$. This rule applies to flat leaves and approximately to non-flat ones. The extinction of diffuse radiation through the canopy can thus be described with a single differential equation of the same form as Eq. 2.23, but with an adapted value for k (Goudriaan 1977). A further method consists of using probabilities in describing the interactions of each photon with any leaf and calculating the resulting probabilities through the canopy with a Markov chain. This approach was used, for example, to assess the effect of within-shoot scattering, which accounts for the fact that leaves are not distributed randomly in the whole volume of a canopy but are grouped around shoots (Smolander and Stenberg 2005).

Models of radiation transfer in canopies are often used in remote sensing. In the visible range, the soil and vegetation are both dark. In the near-infrared band, however, vegetation has a lower absorbance than soil. Measuring the reflectance in both bands thus allows the LAI and the fraction of absorbed PAR (FAPAR) of vegetation canopies to be estimated (Tucker 1979; Sellers 1985). In fact, a forest canopy, as seen from above, includes not only its trees but also its shrub, herb, and moss layers. This is especially important when comparing ground-based measurements with aerial or satellite data.

In the analysis of hemispherical photographs, vegetation is often separated from the visible sky by segmentation of the original colors or gray values into black and white. This corresponds to setting the absorbance of canopy elements to $\alpha = 1$ and allows the simpler Eq. 2.23 to be used. It should be noted that I in this equation is an average, but direct radiation actually produces sunflecks and shadows rather than uniform average radiation (Way and Pearcy 2012). The same remark applies when slopes that are more or less exposed to the Sun are compared. In this case, different radiation fluxes are calculated using Iqbal's equation (Eq. 2.17) and, within the canopy, Eq. 2.22 must be replaced with that of Schleppi et al. (2007), i.e.,

$$I(\theta,\gamma) = I_0(\theta,\gamma)\mathrm{e}^{-G(\theta)L\cos\nu/\cos\gamma}, \tag{2.30}$$

where ν is the slope angle and γ is the angle of incidence (see Eq. 2.12 for its calculation). Note that we use here L per projected area while it is considered per inclined area in Schleppi et al. (2007). Varying slopes and aspects will result in different average irradiances within the canopy, but this corresponds to different proportions of sunlit versus shaded leaves and not to a proportional modification of direct radiation (Wang et al. 2002). As depth into the canopy increases, sunflecks and shades are no longer sharp because the sunrays are not exactly parallel. With an apparent sun diameter of 0.009 rad, there is a fringe of penumbra measuring 9 mm at a distance of 1 m from the object casting shade. Because they increase proportionally with distance, these penumbral effects are especially important for herb layers beneath large trees (Smith et al. 1989).

2.5.2 *Photosynthesis Models*

Different models of radiation penetration form the basis of photosynthesis models. Theoretical developments in these respective fields are thus linked. In the simplest case, the entire canopy is lumped together and considered as one entity. This is known as the "big-leaf" model, where incident PAR is equally partitioned over the whole leaf area and photosynthesis is driven according to a saturating response curve. This model does not directly take into account that individual leaves receive different irradiances and that they are adapted to different irradiance regimes over time. Under certain assumptions, both of these sources of variation may indeed cancel one another out and result in a correct model (Sellers et al. 1992). In most cases, however, these effects have to be taken into account indirectly through empirical parameters and response curves. Further, it has been shown (de Pury and

Farquhar 1997) that these corrections are themselves affected by the leaf area index and by the nitrogen content of leaves.

A more mechanistic approach considers sunlit and shaded foliage separately (e.g., Sinclair et al. 1976). This “two-big-leaf” model is quite effective because photosynthesis in sunflecks is light-saturated most of the time, while in the shade it varies with PAR in an approximately linear fashion. Both processes are well described by simple equations. Further, overestimation that is due to the averaging of radiation along the photosynthesis response curve (Smolander 1984) is avoided. Such models are considered to be a good compromise between simplicity and accuracy (de Pury and Farquhar 1997). Yet another way to avoid overestimation of photosynthesis is to use light data with a high temporal resolution in conjunction with a light-response curve (Paquette et al. 2010).

In more sophisticated models, the canopy is divided into several layers (e.g., Norman 1982; Aber and Federer 1992; Berbigier and Bonnefond 1995). In each layer, radiation is partitioned into absorption and scattering (up and down). Direct and diffuse radiation can be treated separately, as has been previously explained. An important advantage of these models is that changes in leaf morphology and physiology can be integrated with canopy depth. Indeed, leaves that are adapted to shade have lower photosynthetic capacity, lower specific mass and lower nitrogen concentrations compared to leaves at the top of the canopy (Posada et al. 2009). These trends are very generally observed even if species and stress factors also affect photosynthesis. Sellers et al. (1992) explained this robustness as an optimization of resource allocation both within the plant and within the canopy. Nitrogen plays a prominent role for plants because it represents, along with carbon, a heavy investment in the photosynthetic machinery (particularly, the Rubisco enzyme). In these models, radiation, water, carbon and nitrogen intimately interact throughout the canopy.

2.5.3 Canopy Structures

In the preceding sections, we implicitly treated canopies as leaves that were distributed randomly in horizontal directions and, therefore, as horizontally homogenous (at scales larger than the size of leaves). Scattered radiation is not very sensitive to this randomness assumption (Ross 1981), but direct radiation is. In reality, plants feature different aboveground organs (stems, branches, leaves, flowers, fruits) and the overall architecture of the stand depends on their morphology (Ross 1981). In forests, two levels of canopy structure can be distinguished: first, the distribution of single trees in the stand, and second, the architecture of single trees (Côté et al. 2015). The distribution of trees over an area is the main feature of so-called “gap-models,” which are individual-based, spatially explicit forest models that mainly describe population dynamics (e.g., SORTIE-ND; Canham and Murphy 2005). These models describe reproduction, growth (including rough morphological parameters), and mortality of trees. All three processes can be directly affected by light, which can thus regulate competition between individuals and between species. Species can be affected by light as early as during germination. While some species need light to germinate, others require darkness, whereas many are light-insensitive. Borthwick et al. (1952) discovered that this response is due to red light (650–680 nm), but germination can be repressed by subsequent exposure to far-red light (710–740 nm; i.e., wavelengths at the border between visible and near-infrared). The pigment phytochrome is responsible for these responses in plant cells (Butler et al. 1959). Phytochrome controls germination as well as different morphological effects of light (i.e., photomorphogenesis), including the influence of day-length on flowering (i.e., photoperiodism). It must be noted that these mechanisms are independent of photosynthesis and take place even at low radiation levels. The spectral band controlling the phytochrome system is narrow, but it corresponds to a sharp change in leaf optical properties (Fig. 2.6); therefore, the red/far-red ratio (R/FR) is strongly reduced in a plant canopy (e.g., Brasseur and de Sloover 1976). The red/far-red ratio is used by plants, along with sensitivity to blue light, to sense the presence of possible competing plants even before they can have an effect on photosynthesis through mutual shading (Ballaré 1999).

The shape of single trees is affected by light: this is easily observed by comparing trees that have been growing together in a forest versus solitary individuals, or by comparing stands on slopes with different aspects (Ameztegui et al. 2012). Three mechanisms are responsible for these differences: photomorphogenesis, growth as affected by photosynthesis, and mortality of branches due to a lack of light. The effect of light on tree seedlings, for example, was studied in situ by Roussel (1972), who showed etiolation at low light levels (i.e., growth that is characterized by longer but weaker shoots, and smaller roots), and by Paquette et al. (2007a), who showed that species adaptations to shade were expressed through phenotypic plasticity. Etiolation is a photomorphogenetic effect that can be understood as a strategy for increasing the chances that a seedling reaches a height with more available light (e.g., above litter or mosses). From their second or third year onwards, most coniferous tree seedlings become insensitive to this morphogenetic effect. Their growth form is then largely determined by apical dominance. As a consequence, the effects of radiation become limited to growth rates (via photosynthesis) and to the shedding of lower branches receiving insufficient light. However, some differences exist within conifers: *Pinus* branches, for example, retain the ability to grow asymmetrically in response to light, while species such as *Pinus mugo* or *Taxus canadensis* adopt a shrub-like growth form with multiple stems. As already noted by Engler (1924), broadleaved trees remain affected by photomorphogenesis throughout their lives; therefore, as large trees, they are better able to take advantage of small canopy gaps by growing into them. As a result, at similar stem densities, broadleaved or mixed canopies generally exhibit fewer gaps and capture more incoming light than do coniferous ones (e.g., Gower and Norman 1991; Frazer et al. 2017).

Some models have been developed to simulate the morphology of branches of whole trees based on relatively simple growth rules. Tree morphology can then be used to populate small-volume cells (viz., voxels) with leaves. Myneni (1991), for example, tested his discrete ordination method (see Sect. 2.5.1) on *Populus* trees by simulating their crown structure with fractal geometry. Similarly, Langensiepen et al. (2006) modeled the radiation regimes, photosynthesis and transpiration of *Eucalyptus* trees. Working with very open canopies, they were able to derive the three-dimensional morphology of trees from photographs that were taken from below and from the side. It is also possible to derive allometric equations from detailed destructive measurements (e.g., Porté et al. 2000), but the method is time-consuming, and the acquisition or verification of three-dimensional data remains a challenge. Ground-based LIDAR is also in the process of being developed as a tool for assessing the three-dimensional structure of trees (Côté et al. 2009).

The problems of explicit tree crown models are avoided by the gap-size approach of Chen and Cihlar (1995). Gaps in the canopy can be measured as sunflecks on the ground. Gaps within tree crowns are small; their size is of the same order of magnitude as leaves. Gaps between trees, in contrast, are much larger. From the probability distribution of their sizes, between-crown gaps can thus be separated from within-crown gaps. The whole canopy is then regarded as a juxtaposition of leaf-populated areas (i.e., the crowns) and empty spaces. In the original method using sunflecks to estimate leaf sizes and canopy clumping, measurements are taken only in the direction of the Sun. Using HP, however, it is possible to extend this approach to all zenith angles; this can be done by using several software packages (Frazer et al. 2017; Hall et al. 2017). The structure of the forest stand (i.e., tree pattern type and size) can even be inferred from the variance of gaps detected in hemispherical photographs (Montes et al. 2008).

2.5.4 Energy Budgets

In addition to photosynthesis, growth, and morphogenetic effects, solar radiation has a crucial effect on forests by determining their energy budgets and, consequently, their temperatures. As shown above, leaves absorb mainly the visible portion of incoming solar radiation, and they largely scatter its near-infrared component. The thermal radiation that is emitted by plants and soils is in the longwave (thermal) infrared range. According to the Stefan-Boltzmann law, emitted power is proportional to the fourth power of the absolute temperature T. The emitted radiative flux j is defined as

$$j = \varepsilon \sigma T^4, \tag{2.31}$$

where ε is the emissivity of the surface relative to an ideal black body, and σ is the Stefan-Boltzmann constant. Foliage has a high emissivity (typically between 0.94 and 0.99); it also has a high absorbance at these wavelengths (typically near 0.95) and practically no transmittance. The absorbed energy that heats the leaf tissues can be reemitted in all directions as though it were being scattered. As a result, a plant canopy acts as greenhouse for thermal radiation because it partly reemits the thermal radiation that is coming from the soil.

The radiation budget at Earth's surface is the sum of shortwave and longwave budgets (Geiger et al. 2009). The net radiation budget Q^* is

$$Q^* = K + L = K_\downarrow - K_\uparrow + L_\downarrow - L_\uparrow, \tag{2.32}$$

where $K_\downarrow$ is the solar shortwave radiation, $K_\uparrow$ is the reflected shortwave radiation, $L_\downarrow$ is the longwave radiation from the atmosphere, and $L_\uparrow$ is the emitted longwave radiation. The total energy budget at the canopy top also includes the exchange of heat. The energy accumulating in the ecosystem Q_S is represented as

$$Q_S = Q^* + Q_H + Q_E + Q_R, \tag{2.33}$$

where Q^* is the net radiation budget, Q_H is the heat transported by air convection and advection, Q_E is the latent heat of water vapor, and Q_R is the heat brought by precipitation. Most of these energy fluxes are bidirectional, and their sign has to be defined by convention. In Eq. 2.33, fluxes into the system are considered positive and fluxes flowing out of the system are negative. For example, evapotranspiration represents a negative Q_E. Within the ecosystem, there are further exchanges between above- and belowground components, which can be summarized as

$$Q_S = Q_G + Q_B + Q_A + Q_V + Q_P, \tag{2.34}$$

where Q_G, Q_B, and Q_A are the net heat storage into the soil, the above-ground biomass, and the air, respectively. The term Q_V is the latent heat of water vapor within the canopy and Q_P is the energy used by net photosynthesis (i.e., gross photosynthesis minus ecosystem respiration). The net energy that is absorbed by photosynthesis and accumulated as chemical biomass energy is usually very small compared to radiative and thermal fluxes. It is not generally possible or necessary to measure all components of the energy balance. Rather, for most applications, it is sufficient to directly measure or model major components and to use coarser approximations or assumptions for minor fluxes. HP can provide important parameters (e.g., gap fraction, LAI, mean leaf angle) for such applications (Leblanc and Fournier 2014, 2017).

The energy budget of a forest is important in that it determines the temperature regime of the trees, other plants and the soil. The energy budget affects all physiological processes including photosynthesis, transpiration, translocation, and growth. The temperature regime is important for all microbiological soil processes, for root growth, and for the uptake of water and nutrients. Over an entire day, Q_S is approximately zero (Eq. 2.33). During the day, Q^* is positive while Q_S, Q_H, and Q_E are negative: radiation is a source of energy, which goes mainly into the ecosystem (viz., the soil) or is released as latent heat into the atmosphere. At night, Q^*

becomes negative, as do Q_S and Q_G: heat that is stored in the soil is lost by longwave radiation. During the day, the air temperature is generally lower under a forest canopy than in open sites. This effect is especially marked in the morning and during hot, sunny days (Renaud and Rebetez 2009). Because of its height, a forest canopy absorbs incoming radiation in a large volume where convection and evapotranspiration can effectively remove part of this energy (i.e., negative $Q_H + Q_E$). In the canopy of low-statured plants, in contrast, radiation energy is kept within a small volume, which is aerodynamically less tightly coupled to the free atmosphere. The consequence is a higher temperature in both the vegetation and soil. This difference between trees and herbaceous plants was shown to be the main cause of altitudinal tree lines (Körner 2007). At night, when there is no incident shortwave radiation, the temperature is usually higher under a canopy Canopydue to the reflection of emitted thermal radiation (i.e., lower $L_{\uparrow}$). For example, the presence of trees can effectively prevent grass from freezing in a silvopastoral system (Feldhake 2002). The combination of day and night effects results in lower soil temperatures under trees compared to open land (e.g., Hawke and Wedderburn 1994).

An important application of energy budgets is for understanding and predicting snowmelt. Snow interception, sublimation, and melt are all affected by the presence of trees. For example, LaMalfa and Ryle (2008) observed that more snow accumulates in meadows than in coniferous forests, with deciduous forests taking an intermediate position. Burch et al. (1996) found comparable differences and could explain the reduced snow accumulation in forests only by a combination of interception and sublimation from tree crowns over the winter. The necessary energy is provided by solar radiation, which is much better absorbed by a forest canopy (even without leaves and partly under snow) than by continuous snow cover where the albedo is very high. Within a forest stand, variability in the canopy cover and topography affect the patterns of both snow accumulation and melt (Essery 1998; Stähli et al. 2009).

2.6 Radiation Sensors and Their Use in Forests

Radiation above, within, and under forest canopies is variable in time, space, direction, and wavelength spectrum. There is no measuring device that takes all of these seven dimensions into account at once: one dimension for time, three for space, two for direction, and one for wavelength. Different instruments are available for different applications. In this section, we give a very short overview of their possibilities and limitations. While a more detailed review can be found in Pearcy (1989), here we include a description of new sensor technologies that have recently become available. Table 2.1 summarizes the main strengths and limitations of different instruments.

Table 2.1 Strengths and limitations of different instruments and techniques that are used to assess radiation, especially in forests

Instrument	Precision and accuracy	Time dimension	Space dimensions	Directions (azimuth and elevation)	Defined wave-length band(s)
Actinometers	++	+	±	–	+
Spectroradiometers	+	+	±	–	++
PAR sensors	+	+	±	–	+
Ceptometers[a]	+	±	+	–	+
Mobile systems[b]	+	±	++	–	±
Plant canopy analyzer[c]	+	–	±	±	±
Multi-angular systems[d]	±	±	±	+	±
Photography	–	–	±	++	–

We compare how well the different dimensions of a radiation measurement are covered, with respect to time and location of the measurement, direction of the radiation, and its wavelength bands. See Sect. 2.8 for more details

[a]E.g., AccuPAR (Decagon, Pullman, WA, USA)

[b]Based on actinometers or PAR sensors, e.g., TRAC (Decagon), DEMON (CSIRO, Canberra, ACT, Australia)

[c]E.g., LAI-2000 (LI-COR Biosciences, Lincoln, NE, USA)

[d]E.g., BF5 (Delta-T Devices, Burwell, Cambridge, UK)

2.6.1 Actinometers

Actinometers are broadband radiation energy sensors. Shortwave and longwave radiation can be precisely measured by heating a black thermopile (i.e., a sensor that converts heat into electrical energy). The radiation emitted by the sensor itself can be corrected using the Stefan-Boltzmann equation (Eq. 2.31). An actinometer following the track of the Sun that measures only direct radiation is called a pyrheliometer; a sensor that is used to measure global radiation from the sky is called a pyranometer; a pyrgeometer is an actinometer that is designed to measure longwave radiation from the Earth's surface. Net radiometers are composed of either two or four sensors, one up and one down (or two of each), for shortwave and longwave radiation, respectively (Fig. 2.7). Shortwave sensors have a glass dome that is transparent to visible and near-infrared radiation, while longwave sensors are equipped with a silicon surface that is transparent only to longwave radiation. The star pyranometer has a special design, where the difference between its black and white sectors is measured; as a consequence, the instrument is less sensitive to temperature changes (Table 2.2).

Thermopile actinometers are relatively expensive but precise instruments. Their calibration is subject to international standardization (ISO 9060), with the reference being held by the World Radiation Centre in Davos, Switzerland. Actinometers are

Fig. 2.7 Net radiometer measuring shortwave and longwave radiation, downward and upward. Note the glass dome of the pyranometers and the flat silicon window of the pyrgeometers. *Photograph* Patrick Schleppi

typically used in standard weather stations (i.e., open-field stations). For shortwave radiation, photodiodes can be used instead of thermopiles. Photodiodes are electronic components that directly convert light into an electrical current. They are less expensive than thermopiles, but are also less precise.

2.6.2 Spectroradiometers

Studying a radiation spectrum requires an instrument that is able to scan the spectrum in typically one or several steps per nm, as shown in Fig. 2.1. Spectroradiometers that are used in ecological studies measure wavelengths ranging from 300–350 nm to 1000–1100 nm (i.e., from the near-ultraviolet into the near-infrared). In some models, the receptor is linked to the measurement unit by fiber optics. Spectra that are produced by such instruments are useful in describing radiation as it affects plants (e.g., photosynthesis, photomorphogenesis) and animals (e.g., vision). In forests, spectra can differ appreciably according to the form of the incoming radiation (direct or diffuse) and to the presence and size of gaps (Endler 1993).

Table 2.2 Comparison of light availability indices derived from HP analysis (e.g., GLI) and other methods for estimating understory light availability in different forest types

Source	Forest type	Benchmark	Method	r[a]	HP performance
Paquette et al. (2007b)	Temperate deciduous (mature to pioneer)	Overcast T^{b}_{PAR}	HP	0.93	Among best—less accurate in deep shade
			BF2[c]	0.94	
			Densiometer[d]	0.91	
Gendron et al. (1998)	Young deciduous coastal forest	Growing season T^{e}_{PAR}	HP	0.97	Among best
			DIFN[f]	0.95	
			Overcast T_{PAR}	~0.95	
			Instantaneous sunny day T_{PAR}	0.82	
			Average sunny day T_{PAR}	0.92	
Comeau et al. (1998)	Mixedwood (with a deciduous overstory)	Growing season T_{PAR}	HP	0.98	Among best
			Average sunny day T_{PAR}	~0.96	
			Overcast T_{PAR}	0.99	
			DIFN	0.99	
			Densiometer	0.96	
Rich et al. (1993)	Tropical rainforest	Growing season T_{PAR}	HP	~0.87	Not available
Easter and Spies (1994)	Mature and old-growth conifer coastal forests	Growing season T_{PAR}	HP	~0.96	Not available
Battaglia et al. (2003)	Longleaf pine forest	Growing season T_{PAR}	HP	0.88	Best
			Average sunny day T_{PAR}	~0.39	
			Overcast T_{PAR}	~0.76	
Machado and Reich (1999)	Conifer-dominated temperate forest	Mean daily T_{PAR}	HP	0.74	Good overall—less accurate in deep shade
			Overcast T_{PAR}	0.96	
			DIFN	0.96	

When available, the authors' evaluations of the relative performance of HP are also given
[a]When r^2 coefficients were published, they were converted to correlations (r) for uniformity's sake. ~ denotes an estimated average from several similar results
[b]PAR transmittance estimated instantaneously using quantum sensors under an overcast sky (Messier and Puttonen 1995; Parent and Messier 1996)
[c]Developed as an alternative to the overcast method using a hemispherical PAR sensor array (BF2, Delta-T Devices, Burwell, Cambridge, UK). Now replaced by model BF5
[d]Results from a spherical densiometer (Forest Densiometers, Arlington, VA, USA) were published only online: https://www.researchgate.net/publication/248428385_A_less_restrictive_technique_for_the_estimation_of_understory_light_under_variable_weather_conditions
[e]Growing season PAR transmittance standard (see text)
[f]Diffuse non-interceptance of blue light as measured with a plant canopy analyzer (LAI-2000, LI-COR Biosciences, Lincoln, NE, USA)

2.6.3 PAR Sensors

Photosynthetically active radiation (PAR, defined as the radiation between 400 and 700 nm; see Sect. 2.4) is commonly measured with sensors that are based on photodiodes and equipped with filters to restrict their response range. Instruments from different manufacturers may differ in their actual response curves to these wavelengths (for an example, see Fig. 2.8). Over its spectrum, photosynthesis is more closely related to the number than to the total energy of photons in the radiation flux (McCree 1972). For this reason, PAR is commonly measured as a photosynthetic photon flux density (PPFD), which is expressed in moles of photons (or Einsteins); the corresponding instruments are called quantum sensors.

2.6.4 Distributed Systems

Because of their relative affordability, multiple photodiodes (i.e., PAR sensors or others) can be used in an array within or under a plant canopy, thereby capturing the spatial and temporal heterogeneity of radiation. To measure sunflecks in small canopies, it is possible to group sensors along a bar to create what is referred to as a

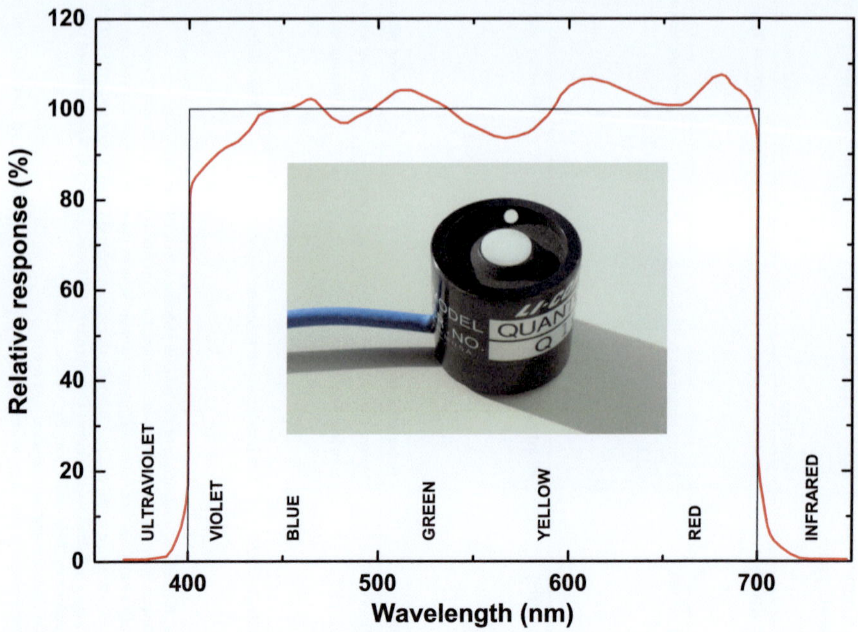

Fig. 2.8 A quantum sensor and its response curve with respect to the ideal response (equal for all photons in the 400–700 nm waveband—*thin black line*). Graph redrawn from LI-COR Biosciences (Lincoln, NE, USA); *photograph* Patrick Schleppi

"ceptometer" (e.g., AccuPAR, Decagon, Pullman, WA, USA). In forest canopies, it is often necessary to perform measurements at distances that are greater than the length of a ceptometer. Thus, mobile devices have been developed to simultaneously measure both the radiation and position of the sensor. While some systems are commercially available (e.g., TRAC, Decagon, Pullman, WA, USA; DEMON, CSIRO, Canberra, ACT, Australia), it is also possible to fit one or more sensors onto a small wagon that travels along a rail (e.g., Stähli et al. 2009). Each measurement point is then visited at defined time intervals and a single sensor is able to capture variability in both time and space.

2.6.5 Angular Systems Including Photographic Sensors

Different systems exist that simultaneously measure radiation coming from different directions. These so-called plant canopy analyzers (e.g., LAI-2200, LI-COR, Lincoln, NE, USA) measure blue light within a series of concentric rings and are specially designed to measure LAI of the canopy.

Photodiodes mounted on platforms that rotate horizontally and vertically can scan the entire sky vault and are called sky scanners. Typically, they take measurements at more than 100 discrete angular positions and are mainly used to measure sky luminance (see Sect. 2.3). A static counterpart of the sky scanner is the so-called Solar-Igel, which has 135 sensors covering the sky hemisphere (Appelbaum 1987).

One of the major challenges in measuring the light that is available below forest canopies is accounting for direct light (i.e., sunflecks through gaps) in one or several instantaneous measurements. Two handheld devices enable the observer to view an image of the canopy above them: the spherical densiometer (Lemmon 1956) and the horizontoscope (Schütz and Brang 1995). Their main advantage lies in in situ assessment of canopy coverage, which makes them useful as decision-making tools for practical purposes such as tree planting or felling. Fisheye photographs provide essentially the same information but record it; this allows more precise analysis as will be shown in the remainder of this chapter and in Leblanc and Fournier (2014). Several other methods have also been developed using light sensors (see Gendron et al. 1998). One of the most recent attempts relies on a new sensor array that uses seven photodiodes located under a hemispheric dome with a shading pattern arranged in such a way that at least one sensor is always measuring both direct and diffuse light. The BF5 and SPN1 sunshine sensors (Delta-T Devices, Burwell, Cambridge, UK) require no adjustment of a shade-ring to track the Sun; thus, they can be used at any latitude and under any sky condition (e.g., overcast to sunny). To evaluate available light, two simultaneous measurements are taken (a) under the canopy to be evaluated, and (b) in a nearby open field. Using only the diffuse signal from both sensors, available light is simply the *a*/*b* ratio, thereby integrating direct light from all gaps throughout the daily and seasonal course of the Sun in one single measurement (Paquette et al. 2007b).

All photographic instruments can also be considered as angular measuring devices since they record radiation in an array of angles and then project these measurements onto an even surface. The quantitative analysis of photographs is done by computer software that uses digital or scanned images as input. Sensors in digital cameras are arrays of photodetectors. Two main types exist: charge-coupled devices (CCD) and active-pixel sensors (APS). While the first requires separate amplification of the signals that are obtained, the second integrates acquisition and amplification within a single integrated circuit. Most color cameras use an interlaced grid of sensors for red, green, and blue colors (RGB) in a ratio of 1:2:1. Special designs use three layered grids or three separate arrays. To produce conventional file formats, the raw readings are converted to bring all three color values into single pixels. While consumer-grade cameras are equipped with RGB sensors, more specific detectors are used in astronomy and remote sensing; for example, they can measure different bands of radiation.

The first purpose of consumer-grade photography is to record subjective color patterns to form images and not to produce absolute radiance readings. As a consequence, the RGB values of the files that are produced should not be considered as an accurate measure of radiation. There is especially loss of brightness (or saturation) from the center towards the edges of the photograph. This effect, which is called "vignetting," is due to the geometry of the lens and sensor. It increases with the aperture of the lens. There is also a limitation in the dynamic range because RGB values are limited in common formats to values between 0 and 255 (8 bits per color). Photographs are encoded on a scale that is adapted to the sensitivity of human vision through a power function, where the exponent is denoted γ (hence "gamma encoding"), such that

$$Y = X^{\gamma}, \tag{2.35}$$

where Y is the encoded brightness and X is the physical radiance (in arbitrary units). Typical values that are used for γ are 1.8 or 2.2.

When light saturation occurs on the sensor, light patches can appear larger than they really are. This "blooming" effect (Leblanc et al. 2005; Thimonier et al. 2010) is due to the electronic excitation of sensor pixels that spills over onto neighboring pixels.

In addition to conventional photography in the visible range, near-infrared (NIR) photography can also be used in forests. Combining both techniques into a single instrument permits what is referred to as "multiband vegetation imaging" (MVI, Kucharik et al. 1997). The main advantage of this technique is that it can differentiate between branches, shaded leaves, and sunlit leaves under different sky conditions.

2.7 Hemispherical Photography

2.7.1 Principles and Development

HP uses a lens, which is called a "fisheye," with a very short focal distance. Thus, the field of view (FOV) is very broad, essentially 180° (π rad). The hemisphere that is seen by the lens is projected as a disk onto the film or sensor of the camera. When the camera is directed upward, the circle on the image corresponds to the horizon. Each point within this circle is then defined by its polar coordinates. The angle on the image corresponds to the azimuth in reality, while the relative radius corresponds to the zenith angle via

$$\theta = f(r/R)\pi/2, \tag{2.36}$$

where r is the distance to the center and R is the radius of the image (i.e., the distance between the projected zenith and horizon). The function f is a monotonically increasing function describing the geometry of the image that is produced by the fisheye lens, where $f(0) = 0$ and $f(1) = 1$. Each lens type has its own geometry, which is usually determined by fitting a polynomial regression to measurements taken in a known setup of the camera and targets.

Interestingly, the SORTIE individual-based forest simulator (Canham and Murphy 2005) models the canopy as a hemispherical photograph. Indeed, for every simulation run, fisheye pictures are generated over every tree for which the model needs to compute light availability, and it analyzes them in the same manner as the actual fisheye photographs that are used to compute light availability indices (e.g., the Gap Light Index, see Sect. 2.7.2 below and in Jonckeere et al. (2017). The SORTIE model has been or is being parameterized for many forest ecosystems around the world using real hemispherical photographs to determine species-specific crown openness and to validate the model by comparing simulated and observed light (e.g., Pacala et al. 1996; Canham et al. 1999; Beaudet et al. 2002).

Historically, upward HP was first used by Hill (1924) to take pictures of the sky with clouds. Evans and Coombe (1959) drew sun tracks on pictures taken in forests to study the direct radiation below the canopy. Anderson (1964) provided the theoretical basis for using hemispherical photographs to estimate diffuse radiation. Bonhomme and Chartier (1972) introduced the automatic digitization of photographs, which enabled the first wave of research in forests. The theory was refined in the late 1980s (Pearcy 1989; Rich 1990). A second wave of research and applications arose when digital photography became broadly available around the year 2000.

Compared to other techniques, HP has the advantage of providing a permanent record of canopy geometry. This record is essentially valid as long as the geometry does not change (e.g., foliage growth or shedding, tree growth or removal). The directional geometry of transmission through canopy openings is identified as visible sky directions, while interception by canopy elements is identified as obscured sky directions. The objective of image analysis is therefore not to measure the prevailing light conditions directly but to infer them from the obtained photograph (Rich 1990). HP can easily be repeated over time to cover seasonal or interannual variation. It can also be repeated at different locations, thereby taking spatial variability into account. Two main fields of application can be distinguished (Fig. 2.9): local analysis (e.g., Zlatanov et al. 2013) versus canopy analysis (e.g., Thimonier et al. 2010).

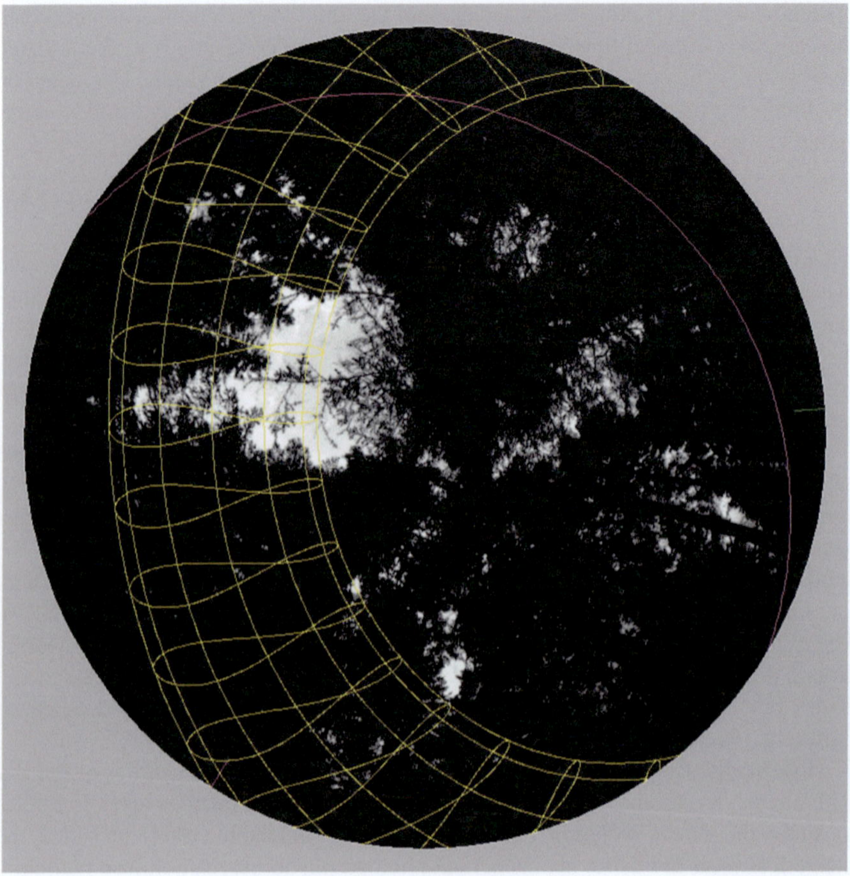

Fig. 2.9 Hemispherical photograph of a *Picea abies* forest in Alptal, Switzerland. The *parallel lines* indicate the path of the Sun every month and the *eight-shaped lines* show its position at full hours. The *top right curve* indicates the position of the ground (slope of 11°). Over the year, assuming that ¼ of the solar radiation directly reaches the canopy, it delivers 10 W m^{-2} at the position of the photograph. Solving the model of Campbell (1986) using a weighted least-squares approach (Schleppi et al. 2007) gives an effective LAI of 3.30 and a clumping-corrected LAI (Chen and Cihlar 1995) of 3.67, with an average foliage angle of 61°. These values correspond to a direct solar radiation value of 5 W m^{-2} on average under this canopy. The fact that local direct radiation is twice as high as the average is due to the large gap to the south. For diffuse radiation, depending upon the selected model, the local value and the canopy average are both around 5 W m^{-2}. *Photograph* Patrick Schleppi

2.7.2 Local Analysis

Local analysis, or point analysis, of hemispherical photographs aims to describe the radiation regime at the point at which each image is taken. The visible part (viz., the black-and-white "viewshed") of the canopy is first extracted from the image to determine for each pixel whether radiation is transmitted or blocked. This viewshed is then combined with a model of above-canopy incoming radiation to calculate the transmitted radiation for a given period (e.g., a growing season). This calculation is done over time based on the position of the Sun, its direct radiation, and patterns of diffuse radiation (see Sects. 2.2 and 2.3). A relatively simple index is often used for the evaluation of the understory light climate and is referred to as the gap light index (GLI, Eq. 2.37), which is also called the global site factor (GSF) (Canham 1988; Rich et al. 1993). The gap light index T is given by

$$T = p_{dir}T_{dir} + p_{dif}T_{dif} \tag{2.37}$$

where p_{dir} and p_{dif} represent the proportions of incident light (above canopy) that are respectively direct and diffuse, and T_{dir} and T_{dif} are the proportions of each fraction reaching the understory (i.e., transmitted through gaps). The term T_{dif} is often called diffusive non-interceptance (abbreviated DIFN or DNI). The p factors can be either measured or modeled, and the corresponding transmittances T are derived from photographs (see Hall et al. 2017 for details). GLI has been used to evaluate and model the effect of forest management on canopy structure (Beaudet and Messier 2002; Drever and Lertzman 2003), and to predict seedling growth and morphology (Paquette et al. 2007b). It is used extensively in forest ecology research (e.g., Canham et al. 1990; Rich et al. 1993; Easter and Spies 1994). The direct and diffuse components can also be assessed separately, for example, to test their respective effects on seedling growth (Paquette et al. 2007b) or to characterize the duration and frequency distribution of sunflecks (Canham et al. 1990).

A local analysis can be computed for a flat surface and, where required, for an inclined one (Ameztegui et al. 2012). In a mountainous landscape, it can also be combined with a digital elevation model to take topographic shading and reflection into account. Scattering within the canopy cannot be directly estimated from the images as it varies too greatly over time. It is therefore usually neglected in these kinds of analyses, which can be a source of error, especially when the Sun is low over the horizon (e.g., morning, evening, winter, high latitudes) and canopy scattering is large compared to transmittance. Accordingly, HP has been found to be less precise in estimating the radiation regime under dense canopies (below about 6% light transmission) than under more open canopies (Machado and Reich 1999; Paquette et al. 2007b; also see Sect. 2.8). Local analyses are essentially used to study small-scale structures and processes. Some examples are forest regeneration in canopy gaps, reaction of photosynthesis to variable light, patterns of ground vegetation, habitat studies for rare or invasive plant species, and intensive plantation forestry (Paquette et al. 2008; Martin et al. 2010; Zlatanov et al. 2013).

2.7.3 Canopy Analysis

Canopy analysis does not look at photographic images in and of themselves but only as samples of the tree stand (or, in some cases, a single tree). Interpretation is thus not performed on a point-by-point basis; rather, it is integrated over the entire canopy. The properties that are derived from this analysis typically include canopy openness, LAI, mean leaf angle, and a clumping factor (Bréda 2003). The information that is contained in one, or usually several photographs, is used to characterize the canopy. The radiation regime can then be evaluated indirectly, i.e., by combining the incident radiation with a canopy model (and no longer with the pictures themselves). Such analyses are often conducted with the objective of obtaining important input parameters for ecosystem (process) models (Boivin et al. 2011; Ameztegui et al. 2012). They apply to processes that act at the stand-level, such as forest productivity, forest hydrology and climate, and suitability as habitat for animal species (Dobbertin et al. 2010; Manetti et al. 2010).

2.8 Comparison of HP with Other Methods Used to Estimate Understory Light Availability in Forests

As stated in this chapter's Introduction (Sect. 2.1), understory light plays an important role in forest dynamics, especially for the survival and growth of tree seedlings. Over the years, researchers have developed techniques for evaluating understory light availability as accurately as possible. A standard and direct method is to assess the transmittance of PAR, sometimes called %PAR or %PPFD, over the whole growing season (hereafter, T_{PAR}). This involves continuous measurements at given locations in the understory (e.g., using quantum sensors). An additional sensor that is positioned above the canopy to record incident PAR is used as a reference for computing T_{PAR} ratios (Comeau et al. 1998; Gendron et al. 1998). This method is simple in design and exhaustive in rendering the time variability. Nevertheless, it has limited use because it requires several sensors and data loggers over a long period of time (Rich et al. 1993; Machado and Reich 1999). A faster method is to measure the instantaneous transmittance under an overcast sky (Messier and Puttonen 1995; Parent and Messier 1996). It is fast, reliable, and has become a popular alternative in recent years; however, the overcast sky conditions that are required can be restrictive (Paquette et al. 2007b). HP is also fast but, for local analysis, it allows one to reconstruct how light transmission varies through time following the position of the Sun. We reviewed comparisons of direct PAR measurements with the gap light index (GLI, see Sect. 2.7.2) that was derived from hemispherical photographs (Table 2.2). For the different forest types that have been studied, GLI correlates well with most of the other available methods. It compares particularly well with the standard continuous measurement of T_{PAR} over the growing season.

2.9 Conclusion

In this chapter, we have presented theoretical aspects that are important for the application of HP in forests. We examined how solar radiation interacts with the atmosphere, vegetation and soil, as well as how reflected and reemitted radiation behaves in this system. The concepts and physical laws that have been introduced will enable the reader to better understand the following chapters in this book. They clarify the relationships that exist within a hemispherical image, the structures of the canopy that it depicts, and the function of the canopy. In many cases, HP is a useful technique. Like all other measurements that are used to assess radiation in forests, it has both strengths and limitations. Its main strength is undoubtedly its ability to cover all directions of the hemisphere in one measurement. Spatial and temporal dimensions can be taken into account by taking additional photographs across space and time; however, the dimension of wavelength is not measured in a useful manner. The precision, accuracy, and repeatability of HP are limited compared to other instruments that are specifically designed to measure radiation. Depending upon the application, HP may therefore be used in conjunction with models or with other instruments.

Acknowledgements We thank William F.J. Parsons and Catherine A. Brown for editing the language of our manuscript.

References

Aber JD, Federer CA (1992) A generalized, lumped-parameter model of photosynthesis, evapotranspiration and net primary production in temperate and boreal forest ecosystems. Oecologia 92:463–474

Akinoglu BG (2008) Recent advances in the relations between bright sunshine hours and solar irradiation. In: Badescu V (ed) Modeling solar radiation at the Earth surface, recent advances. Springer, Berlin, pp 115–143

Ameztegui A, Coll L, Benavides R, Valladares F, Paquette A (2012) Understory light predictions in mixed conifer mountain forests: role of aspect-induced variation in crown geometry and openness. For Ecol Manag 276:52–61

Anderson MC (1964) Studies of the woodland light climate I. The photographic computation of light condition. J Ecol 52:27–41

Angström A (1924) Solar and terrestrial radiation. Q J Roy Meteor Soc 50:121–126

Appelbaum J (1987) A solar radiation distribution sensor. Sol Energy 39:1–10

Ballaré CL (1999) Keeping up with the neighbours: phytochrome sensing and other signalling mechanisms. Trends Plant Sci 4:97–102

Battaglia MA, Mitchell RJ, Mou PP, Pecot SD (2003) Light transmittance estimates in a longleaf pine woodland. For Sci 49:752–762

Beaudet M, Messier C (2002) Variation in canopy openness and light transmission following selection cutting in northern hardwood stands: an assessment based on hemispherical photographs. Agric For Meteorol 110:217–228

Beaudet M, Messier C, Canham CD (2002) Predictions of understorey light conditions in northern hardwood forests following parameterization, sensitivity analysis, and tests of the SORTIE light model. For Ecol Manag 165:235–248

Berbigier P, Bonnefond JM (1995) Measurement and modelling of radiation transmission within a stand of maritime pine (*Pinus pinaster* Ait.). Ann Sci For 52:23–42

Bohren CF, Clothiaux EE (2006) Fundamentals of atmospheric radiation. Wiley-VCH, Weinheim

Boivin F, Paquette A, Racine P, Messier C (2011) A fast and reliable method for the delineation of tree crown outlines for the computation of crown openness values and other tree crown parameters. Can J For Res 41:1827–1835

Bonhomme R, Chartier P (1972) The interpretation and automatic measurement of hemispherical photographs to obtain sunlit foliage area and gap frequency. Israel J Agric Res 22:53–61

Borthwick HA, Hendricks SB, Parker MW, Toole EH, Toole VK (1952) A reversible photoreaction controlling seed germination. Proc Natl Acad Sci U S A 38:662–666

Brasseur F, de Sloover JR (1976) L'extinction du rayonnement dans les gammes spectrales bleu, rouge et rouge lointain. Comparaison de deux peuplements forestiers de Haute-Ardenne. Bull Soc R Bot Belg 109:319–334

Bréda NJJ (2003) Ground-based measurements of leaf area index: a review of methods, instruments and current controversies. J Exp Bot 54:2403–2417

Brunger AP, Hooper FC (1993) Anisotropic sky radiance model based on narrow field of view measurements of shortwave radiance. Sol Energy 51:53–64

Burch H, Forster F, Schleppi P (1996) Zum Einfluss des Waldes auf die Hydrologie der Flysch-Einzugsgebiete des Alptals. Schweiz Z Forstwes 147:925–938

Butler WL, Norris KH, Siegelman HW, Hendricks SB (1959) Detection, assay, and preliminary purification of the pigment controlling photoresponsive development of plants. Proc Natl Acad Sci U S A 45:1703–1708

Campbell GS (1986) Extinction coefficients for radiation in plant canopies calculated using an ellipsoidal inclination angle distribution. Agric For Meteorol 36:317–321

Canham CD (1988) An index for understory light levels in and around canopy gaps. Ecology 69:1634–1638

Canham CD, Denslow JS, Platt WJ, Runkle JR, Spies TA, White PS (1990) Light regimes beneath closed canopies and tree-fall gaps in temperate and tropical forests. Can J For Res 20:620–631

Canham CD, Coates KD, Bartemucci P, Quaglia S (1999) Measurement and modeling of spatially explicit variation in light transmission through interior cedar-hemlock forests of British Columbia. Can J For Res 29:1775–1783

Canham CD, Murphy LE (2005) SORTIE-ND. Cary Institute of Ecosystem Studies, Millbrook, New York. http://www.sortie-nd.org/

Chen JM, Cihlar J (1995) Plant canopy gap-size analysis theory for improving optical measurements of leaf-area index. Appl Opt 34:6211–6222

CIE (2003) Spatial distribution of daylight—CIE standard general sky (ISO 15469:2004 (E)/CIE S 011/E:2003). Commission Internationale de l'Éclairage, Vienna, Austria

Comeau PG, Gendron F, Letchford T (1998) A comparison of several methods for estimating light under a paper birch mixedwood stand. Can J For Res 28:1843–1850

Côté J-F, Widlowski J-L, Fournier RA, Verstraete MM (2009) The structural and radiative consistency of three-dimensional tree reconstructions from terrestrial lidar. Remote Sens Environ 113:1067–1081

Côté J-F, Fournier RA, Verstraete MM (2017) Canopy architectural models in support of methods using hemispherical photography. In: Fournier RA, Hall RJ (eds) Hemispherical photography in forest science: theory, methods, applications. Springer, Berlin

Dai Q, Sun S (2006) A generalized layered radiative transfer model in the vegetation canopy. Adv Atmos Sci 23:243–257

Dai Q, Sun S (2007) A comparison of two canopy radiative models in land surface processes. Adv Atmos Sci 24:421–434

de Pury DGG, Farquhar GD (1997) Simple scaling of photosynthesis from leaves to canopies without the errors of big-leaf models. Plant Cell Environ 20:537–557

Dickinson RE (1983) Land surface processes and climate—surface albedos and energy-balance. Adv Geophys 25:305–353
Dobbertin M, Eilmann B, Bleuler P, Giuggiola A, Pannatier EG, Landholt W, Schleppi P, Rigling A (2010) Effect of irrigation on needle morphology, shoot and stem growth in a drought-exposed *Pinus sylvestris* forest. Tree Physiol 30:346–360
Drever CR, Lertzman KP (2003) Effects of a wide gradient of retained tree structure on understory light in coastal Douglas-fir forests. Can J For Res 33:137–146
Easter MJ, Spies TA (1994) Using hemispherical photography for estimating photosynthetic photon flux density under canopies and in gaps in Douglas-fir forests of the Pacific Northwest. Can J For Res 24:2050–2058
Emery K, Myers D (2008) Reference solar spectral irradiance: air mass 1.5. http://rredc.nrel.gov/solar/spectra/am1.5. Accessed 5 Mar 2013
Engler A (1924) Heliotropismus und Geotropismus der Bäume und deren waldbauliche Bedeutung. Mitteilungen der Schweizerischen Zentralanstalt für das forstliche Versuchswesen 13:225–283
Endler JA (1993) The color of light in forests and its implications. Ecol Monogr 63:1–27
Essery R (1998) Boreal forests and snow in climate models. Hydrol Process 12:1561–1567
Evans GD, Coombe DE (1959) Hemispherical and woodland canopy photography and the light climate. J Ecol 47:103–113
Feldhake CM (2002) Forage frost protection potential of conifer silvopastures. Agric For Meteorol 112:123–130
Frazer GW, Fournier RA, Leblanc SG, Walter J-MN (2017) View angle-dependent clumping indices for indirect LAI estimation. In: Fournier RA, Hall RJ (eds) Hemispherical photography in forest science: theory, methods, applications. Springer, Berlin
Gates DM (1980) Biophysical ecology. Springer, Berlin
Geiger R, Aron RH, Todhunter P (2009) The climate near the ground, 7th edn. Rowman & Littlefield, Lanham
Gendron F, Messier C, Comeau PG (1998) Comparison of various methods for estimating the mean growing season percent photosynthetic photon flux density in forests. Agric For Meteorol 92:55–70
Goudriaan J (1977) Crop micrometeorology: a simulation study. Dissertation, Center for Agricultural Publishing and Documentation, University of Wageningen, Wageningen
Gower ST, Norman JM (1991) Rapid estimation of leaf area index in conifer and broad-leaf plantations. Ecology 72:1896–1900
Grant RH, Heisler GM, Gao W (1996) Photosynthetically-active radiation: sky radiance distributions under clear and overcast conditions. Agric For Meteorol 82:267–292
Hall RJ, Fournier RA, Côté J-F, Mailly D (2017) Comparison of software tools for analysis of hemispherical photography. In: Fournier RA, Hall RJ (eds) Hemispherical photography in forest science: theory, methods, applications. Springer, Berlin
Harrison AW, Coombes CA (1988) Angular distribution of clear sky short wavelength radiance. Sol Energy 40:57–63
Hawke MF, Wedderburn ME (1994) Microclimate changes under *Pinus radiata* agroforestry regimes in New Zealand. Agric For Meteorol 71:133–145
Hay JE, McKay DC (1985) Estimating solar irradiance on inclined surfaces: a review and assessment of methodologies. Int J Sol Energy 3:203–240
Hill R (1924) A lens for whole sky photographs. Q J R Meteorol Soc 50:227–235
Hooper FC, Brunger AP (1980) A model for the angular distribution of sky radiance. J Sol Energy-T ASME 102:196–202
Iqbal M (1983) An introduction to solar radiation. Academic Press, Toronto
Jacobson MZ (2002) Atmospheric pollution: history, science, and regulation. Cambridge University Press, Cambridge, p 399 pp
Jonckheere IGC, Macfarlane C, Walter J-MW (2017) Image Analysis of hemispherical photographs, algorithms and calculation. In: Fournier RA, Hall RJ (eds) Hemispherical photography in forest science: Theory, methods, applications. Springer, New York

Kittler R (1994) Some qualities of scattering functions defining sky radiance distributions. Sol Energy 53:511–516

Körner C (2007) Climatic treelines: conventions, global patterns, causes. Erdkunde 61:316–324

Kucharik CJ, Norman JM, Murdock LM, Gower ST (1997) Characterizing canopy nonrandomness with a multiband vegetation imager (MVI). J Geophys Res 102(D24):29455–29473

LaMalfa EM, Ryle R (2008) Differential snowpack accumulation and water dynamics in aspen and conifer communities: implications for water yield and ecosystem function. Ecosystems 11:569–581

Lang M, Kuusk A, Mõttus M, Rautiainen M, Nilson T (2010) Canopy gap fraction estimation from digital hemispherical images using sky radiance models and a linear conversion method. Agric For Meteorol 150:20–29

Langensiepen M, Burgess S, Lambers H, Mitchell P, Veneklass E (2006) A model for simulating transpiration of *Eucalyptus salmonophloia* trees. Physiol Plant 127:465–477

Leblanc SG, Fournier RA (2014) Hemispherical photography simulations with an architectural model to assess retrieval of leaf area index. Agric For Meteorol 194:64–76

Leblanc SG, Fournier RA (2017) Measurement of forest structure with hemispherical photography. In: Fournier RA, Hall RJ (eds) Hemispherical photography in forest science: theory, methods, applications. Springer, Berlin

Leblanc SG, Chen JM, Fernandes R, Deering DW, Conley A (2005) Methodology comparison for canopy structure parameters extraction from digital hemispherical photography in boreal forests. Agric For Meteorol 129:187–207

Lemmon PE (1956) A spherical densiometer for estimating forest overstory density. For Sci 2:314–320

Lieffers VJ, Messier C, Stadt KJ, Gendron F, Comeau PG (1999) Predicting and managing light in the understory of boreal forests. Can J For Res 29:796–811

Liu BYH, Jordan RC (1961) Daily insolation on surfaces tilted towards the equator. Trans ASHRAE 67:526–541

Lodish H, Berk A, Zipursky SL, Matsudaira P, Baltimore D, Darnell J (2000) Molecular cell biology, 4th edn. Freeman, New York

Machado J-L, Reich PB (1999) Evaluation of several measures of canopy openness as predictors of photosynthetic photon flux density in deeply shaded conifer-dominated forest understory. Can J For Res 29:1438–1444

Manetti MC, Amorini E, Becagli C, Pelleri F, Pividori M, Schleppi P, Zingg A, Conedara M (2010) Quality wood production from chestnut (Castanea sativa Mill.) coppice forests—comparison between different silvicultural approaches. Acta Hortic 866:683–692

Martin PH, Canham CD, Kobe RK (2010) Divergence from the growth-survival trade-off and extreme high growth rates drive patterns of exotic tree invasions in closed-canopy forests. J Ecol 98:778–789

McCree KJ (1972) Test of current definitions of photosynthetically active radiation against leaf photosynthesis data. Agric For Meteorol 10:443–453

Messier C, Puttonen P (1995) Spatial and temporal variation in the light environment of developing Scots pine stands: the basis for a quick and efficient method of characterizing light. Can J For Res 25:343–354

Milankovitch M (1941) Kanon der Erdbestrahlungen und seine Anwendung auf das Eiszeitenproblem. Königlich Serbische Akademie, Belgrad

Montes F, Rubio A, Barbeito I, Cañellas I (2008) Characterization of the spatial structure of the canopy in *Pinus silvestris* L. stands in Central Spain from hemispherical photographs. For Ecol Manag 255:580–590

Myneni RB (1991) Modeling radiative transfer and photosynthesis in three-dimensional vegetation canopies. Agric For Meteorol 55:323–344

Norman JM (1982) Simulation of microclimates. In: Hatfield JL, Thomason IJ (eds) Biometeorology in integrated pest management. Academic Press, New York, pp 65–99

Pacala SW, Canham CD, Saponara J, Silander JA Jr, Kobe RK, Ribbens E (1996) Forest models defined by field measurements: estimation, error analysis and dynamics. Ecol Monogr 66:1–43

Paquette A, Bouchard A, Cogliastro A (2006) Survival and growth of under-planted trees: a meta-analysis across four biomes. Ecol Appl 16:1575–1589

Paquette A, Bouchard A, Cogliastro A (2007a) Morphological plasticity in seedlings of three deciduous species under shelterwood under-planting management does not correspond to shade tolerance ranks. For Ecol Manag 241:278–287

Paquette A, Bouchard A, Cogliastro A (2007b) A less restrictive technique for the estimation of understory light under variable weather conditions. For Ecol Manag 242:800–804

Paquette A, Messier C, Périnet P, Cogliastro A (2008) Simulating light availability under different hybrid poplar clones in a mixed intensive plantation system. For Sci 54:481–489

Paquette A, Fontaine B, Messier C, Brisson J (2010) Homogeneous light regime in shade-house experiment overestimates carbon gains in Norway maple seedlings. J Hortic For 2:117–121

Parent S, Messier C (1996) A simple and efficient method to estimate microsite light availability under a forest canopy. Can J For Res 26:151–154

Pearcy RW (1989) Radiation and light measurements. In: Pearcy RW, Ehleringer J, Mooney HA, Rundel PW (eds) Plant physiological ecology: field methods and instrumentation. Chapman & Hall, New York, pp 95–116

Pelkowski J (2008) Insolation at the Earth's surface. Rev Acad Colomb Ci Exact 32:25–35

Porté A, Bosc A, Champion I, Loustau D (2000) Estimating the foliage area of maritime pine (*Pinus pinaster* Ait.) branches and crowns with application to modelling the foliage area distribution in the crown. Ann For Sci 57:73–86

Posada JM, Lechowicz MJ, Kitajima K (2009) Optimal photosynthetic use of light by tropical tree crowns achieved by adjustment of individual leaf angles and nitrogen content. Ann Bot 103:795–805

Prescott JA (1940) Evaporation from a water surface in relation to solar radiation. Trans R Soc South Aust 64:114–116

Renaud V, Rebetez M (2009) Comparison between open-site and below-canopy climatic conditions in Switzerland during the exceptionally hot summer of 2003. Agric For Meteorol 149:873–880

Rich PM (1990) Characterizing plant canopies with hemispherical photographs. In: Goel NS, Norman JM (eds) Instrumentation for studying vegetation canopies for remote sensing in optical and thermal infrared regions. Remote Sens Rev 5:13–29

Rich PM, Clark DB, Clark DA, Oberbauer SF (1993) Long-term study of solar radiation regimes in a tropical wet forest using quantum sensors and hemispherical photography. Agric For Meteorol 65:107–127

Ross J (1981) The radiation regime and architecture of plant stands. Dr W. Junk publishers, The Hague

Roussel L (1972) Photologie forestière. Masson, Paris, p 144 pp

Running SW, Coughlan JC (1988) A general model of forest ecosystem processes for regional applications. I. Hydrologic balance, canopy gas-exchange and primary production processes. Ecol Model 42:125–154

Schleppi P, Conedera M, Sedivy I, Thimonier A (2007) Correcting non-linearity and slope effects in the estimation of the leaf area index of forests from hemispherical photographs. Agric For Meteorol 144:236–242

Schütz J, Brang P (1995) L'horizontoscope: un étonnant outil pratique de sylviculture, notamment en haute montagne. Office National des Forêts, Bulletin Technique 28

Sellers PJ (1985) Canopy reflectance, photosynthesis and transpiration. Int J Remote Sens 6:1335–1372

Sellers PJ, Berry JA, Collatz GJ, Field CB, Hall FG (1992) Canopy reflectance, photosynthesis, and transpiration. III. A reanalysis using improved leaf models and a new canopy integration scheme. Remote Sens Environ 42:187–216

Siala FMF, Hooper FC (1990) A model for the directional distribution of the diffusive sky radiance with an application to a CPC collector. Sol Energy 44:291–296

Sinclair TR, Murphy CE Jr, Knoerr KR (1976) Development and evaluation of simplified models for simulating canopy photosynthesis and transpiration. J Appl Ecol 13:813–829

Singhal GS, Renger G, Sopory SK, Irrgang KD, Govindjee (eds) (1999) Concepts in photobiology: photosynthesis and photomorphogenesis. Narosa Publishing House, New Delhi, p 1019 pp

Smith WK, Knapp AK, Reiners WA (1989) Penumbral effects on sunlight penetration in plant communities. Ecology 70:1603–1609

Smolander S (1984) Measurement of fluctuating irradiance in field studies of photosynthesis. Acta For Fenn 187:1–56

Smolander S, Stenberg P (2005) Simple parameterizations of the radiation budget of uniform broadleaved and coniferous canopies. Remote Sens Environ 94:355–363

Spencer JW (1971) Fourier series representation of the position of the Sun. Search 2:172

Stähli M, Jonas T, Gustafsson D (2009) The role of snow interception in winter-time radiation processes of a coniferous sub-alpine forest. Hydrol Process 23:2498–2512

Stanhill G (1970) Some results of helicopter measurements of the albedo of different land surfaces. Sol Energy 13:59–66

Steven MD, Unsworth MH (1980) The angular distribution and interception of diffuse solar radiation below overcast skies. Q J R Meteorol Soc 106:57–61

Thimonier A, Sedivy I, Schleppi P (2010) Estimating leaf area index in different types of mature forest stands in Switzerland: a comparison of methods. Eur J For Res 129:543–562

Timofeyev YM, Vasil'ev AV (2008) Theoretical fundamentals of atmospheric optics. CABI, Cambridge

Torres JL, Torres LM (2008) Angular distribution of sky diffuse radiance and luminance. In: Badescu V (ed) Modeling solar radiation at the Earth's surface: recent advances. Springer, Berlin, pp 427–448

Tucker CJ (1979) Red and photographic infrared linear combinations for monitoring vegetation. Remote Sens Environ 8:127–150

Vida J, Foyo-Moreno I, Alados-Arboledas L (1999) Performance validation of MURAC, a cloudless sky radiance model proposal. Energy 24:705–721

Wang S, Chen W, Cihlar J (2002) New calculation methods of diurnal distribution of solar radiation and its interception by canopy over complex terrain. Ecol Model 155:191–204

Wang Q, Tenhunen J, Schmidt M, Otieno D, Kolcun O, Droesler M (2005) Diffuse PAR irradiance under clear skies in complex alpine terrain. Agric For Meteorol 128:1–15

Wang W-M, Li Z-L, Su H-B (2007) Comparison of leaf angle distribution functions: effects on extinction coefficient and fraction of sunlit foliage. Agric For Meteorol 143:106–122

Way DA, Pearcy RW (2012) Sunflecks in trees and forests: From photosynthetic physiology to global change biology. Tree Physiol 32:1066–1081

Zlatanov T, Schleppi P, Velichkov I, Hinkov G, Georgieva M, Eggertsson O, Zlatanova M, Vacik H (2013) Structural diversity of abandoned chestnut (*Castanea sativa* Mill.) dominated forests: Implications for forest management. For Ecol Manag 291:326–335

Chapter 3
Measurement of Forest Structure with Hemispherical Photography

Sylvain G. Leblanc and Richard A. Fournier

Abstract This chapter presents the theoretical concepts necessary to link optical sensor data to forest structural attributes. Forest stand architecture is relatively complex and spatially variable, and many approaches can be used for its measurement: from traditional forest measurements in ground plots, to the statistical representation of forest attributes with the use of allometric relationships, and to the use of optical sensors. The forest parameters routinely measured are presented followed by the specific measurements that optical field instruments provide. Emphasis is placed on the optical sensors measuring light transmission through the forest canopy using the hemispherical view approach and, more specifically, on the use of hemispherical photographs. Strengths and weaknesses of forest canopy structural attributes measured based on Beer's law, in particular from canopy gaps as seen from a hemispherical sensor or photograph, are presented. Finally, scaling issues are introduced from single trees to a stand and from stands to area mapping with satellite sensing.

Keywords Allometric relationships · Beer's Law · Canopy structure · Foliage clumping · Gap fraction · Leaf area index (LAI)

List of Symbols

α	Woody to total area ratio (W/L_t)
α_f	Elevation angle of a leaf
$\bar{\alpha}_f$	Mean leaf elevation angle
CC	Chen and Cihlar (1995) clumping method
CLX	Leblanc et al. (2005a) clumping method

S.G. Leblanc (✉)
Canada Centre for Mapping and Earth Observation, Natural Resources Canada, 560 Rochester Street, Ottawa, ON K1A 0E4, Canada
e-mail: Sylvain.Leblanc@canada.ca

S.G. Leblanc · R.A. Fournier
Département de géomatique appliquée, Université de Sherbrooke, 2500, boul. de l'Université, Sherbrooke, QC J1K 2R1, Canada

R.A. Fournier and R.J. Hall (eds.), *Hemispherical Photography in Forest Science: Theory, Methods, Applications*, Managing Forest Ecosystems 28,
DOI 10.1007/978-94-024-1098-3_3

CO	Canopy openness (%)
G(θ)	Projection coefficient at view (zenith) angle θ
γ_E	Needle-to-shoot-area ratio
L	Leaf Area Index defined as one-half the total leaf area per unit of horizontal ground (m^2/m^2)
$L_e(\theta)$	'Effective' leaf area index at view (zenith) angle θ (m^2/m^2)
$L_{et}(\theta)$	'Effective' plant area index at view (zenith) angle θ (m^2/m^2)
L_t	Plant area index defined as one-half the total plant area per unit of horizontal ground (m^2/m^2)
$L_t(\theta)$	Plant area index at view (zenith) angle θ (m^2/m^2)
W	Woody (twig, branch, and bole) area index (m^2/m^2)
PCA LAI-2000	Plant Canopy Analyzer LAI-2000 from LI-COR ©
LX	Lang and Xiang (1986) clumping method
P(θ)	Gap fraction at view (zenith) angle θ
θ	Zenith angle (degree)
TRAC	Tracing Radiation and Architecture of Canopies
Ω(θ)	Element clumping index at spatial scales greater than the conifer shoot or broadleaf at view (zenith) angle θ

3.1 Introduction

Forest structure refers to the architectural and functional elements that constitute the forest canopy (Fournier et al. 2003). Knowledge of this structure is required to derive a multitude of forest and biophysical attributes that are relevant to forest inventories, ecology, and management (Franklin 2001). There is also a growing demand for the measurement and monitoring of the state of forested landscapes for the purposes of conservation, biodiversity, and sustainable development (Layton et al. 2004). As a consequence, efficient and accurate methods are required for the measurement of forest structure, regardless of whether these are made in the field or by using remote sensing approaches. This chapter reviews theoretical concepts that are necessary for understanding the relationships between optical sensor data and forest structural attributes. This background knowledge is relevant to the measurement of forest structure, and is a useful reference for the other chapters in this volume.

Forest stand architecture is complex and spatially variable. Important elements in the vertical dimension include the heights of stands and those of trees within the canopy layer. Species composition, canopy closure, and the spatial distribution of trees and stands are important elements in the horizontal dimension (Fournier et al. 2003). Measurements of forest architecture have been made using various approaches, including traditional forest measurements within ground plots (i.e., sample plots in a forest inventory), statistical representations of forest attributes using allometric relationships, and optical sensors. A growing emphasis on conservation, biodiversity, and sustainable development has widened the focus of traditional forest inventories to include more diverse measurements (Layton et al. 2004). In addition, the tools that are used to measure forest attributes are improving with technological advances. For instance, laser instruments that have been designed for use outside the laboratory have been primarily utilized by land surveyors; however, technological advances have led to the development of affordable instruments for taking precise measurements in other fields such as forestry (Hopkinson et al. 2004). In this chapter, a short description of the forest parameters that are routinely measured is followed by a review of the most commonly used allometric approaches. The specific measurements that optical field instruments provide are subsequently discussed, with an emphasis on measurements of light transmission through the forest canopy from a hemispherical view and, more specifically, on hemispherical photography (HP).

Biophysical attributes are related to physiological and functional characteristics of the forest canopy. For example, they include elements such as canopy openness. The leaf area index (LAI) is also a biophysical attribute, but it is more closely related to ecological processes such as photosynthesis, evapotranspiration, and net primary productivity (NPP) (Coops and Waring 2001). These attributes can be measured using a variety of direct and indirect methods. Optical instruments that are frequently used for these purposes include the LAI-2000 or 2200 Plant Canopy Analyzer (LI-COR Biosciences, Lincoln, NE, USA; LI-COR 1991), the LA-80 LAI Ceptometer (Decagon Devices, Pullman, WA, USA), and TRAC (Tracing Radiation and Architecture of Canopies; Gower and Norman 1991; Welles and Norman 1991; Chen et al. 1997).

The acquisition and use of hemispherical photographs to derive estimates of forest structure has also been demonstrated in the literature (Anderson 1964; Rich 1990; Chen et al. 1991; Frazer et al. 2001; Walter et al. 2003; Jonckheere et al. 2004). Understanding how incident radiation penetrates the forest canopy (Schleppi and Paquette 2017) is fundamental to the use of hemispherical photographs for estimating forest structure.

Beer's Law is a theoretical representation used to calculate the transmission of light in a turbid medium. As the path length increases through the canopy, the reduction in light transmission depends greatly upon the number and spatial distribution of forest elements. Forest canopy structural attributes, such as canopy gaps in hemispherical photographs, can be extracted using Beer's Law. However, Beer's Law assumes a turbid medium that is a random distribution of obstructing elements of negligible size compared to the volume that is being studied, which is rarely the case in forest canopies. The original equation for Beer's Law has been modified for its application to vegetation canopies. The main changes are the inclusion of complex structural features of the forest environment such as the following: the apparent projection of conifer shoots; relative contribution of wood and foliage structures to light obstruction; foliage transmittance and reflectance; and clumping of foliage components at the shoot, branch, tree, and canopy levels.

The theory that is used is common to most optical sensors but is particularly well-suited to the use of hemispherical photographs. Many of these concepts have been presented in scientific papers on measurement methods using optical sensors. Here, we wish to summarize these concepts and describe them in more detail. Thus, this chapter is a handy reference for those seeking a review of the theoretical foundations of gap fraction and the adaptation of Beer's Law to measure forest structure. In addition, we discuss the optical instruments that are typically used for these purposes. Finally, we review several issues concerning the application of this theory, such as scaling from single trees to stands and from stands to landscape mapping with satellite remote sensing.

3.2 Forest Structural Attributes

Forest canopies are very complex and their structural representation depends upon the level of detail that is required by specific applications or end-users. A forest canopy can be considered as a hierarchical suite of interrelated attributes, which can be divided, for convenience, into the following levels: stand, tree, whorl/branch, shoot, and leaf/needle. In most cases where forest canopy measurements are necessary, the notion of the forest stand prevails. This notion assumes that given values for cover type, canopy height and density, we can describe an area relative to its composition and structure, as often depicted in forest inventory maps of Canada (Gillis and Leckie 1993). Those who are responsible for the inventory have defined the minimum spatial extent and the criteria for maximum acceptable variability for an area to be considered homogeneous. Spatial extent is described by starting with a detailed description of the forest canopy at the tree level and going up to the stand level, which usually reaches several hectares. Spatial extent is generally defined by categorical classes and one set usually dominates (Avery and Burkhart 2002). Constraints that are inherent to most inventories dictate that only a few tree samples be taken within a stand. Therefore, the sampling strategy that is used for selecting trees is important to obtain meaningful statistics and avoid biases. Foliage organization at the whorl or branch level has an effect on incoming light and canopy gaps, as viewed from a sensor on the ground looking upward. Moreover, the organization of leaves into tufts or into clumps also has an effect on the light regime of the canopy and, incidentally, upon canopy structure. Even if greater emphasis is placed on tree- and stand-level measurements when describing a forest canopy, finer levels also exert a great deal of influence on the patterns and processes that are observed by vegetation. A tool such as HP can depict the effects of canopy attributes over a wide range of hierarchical levels.

3.2.1 Traditional Inventory Measurements

Most inventories that are made within the context of forest management focus upon canopy attributes that determine timber supply and forest productivity. Inventories can also be expanded for other purposes, where a wide array of attributes can provide information for various management issues, such as the identification of suitable habitats, exceptional ecosystems, or ensuring sufficient biodiversity. Two inventory products that are typically used in forestry are ground plots and stand maps. Stand maps are useful for defining areas with a homogeneous set of stand attributes that can be described using categorical classes of cover type, stand height, and tree crown density. They provide the spatial extent and stand boundaries, as identified by an experienced interpreter of aerial photographs following a set of guidelines. Both inventory products are useful for describing the vegetation canopy, but the use of hemispherical photographs is more strongly related to ground plots.

Ground plots usually include measurements of all stand-level attributes that are available for map products; in addition, tree-level measurements are taken for a selection of trees. Stand attributes need not be categorical, in contrast to those that are employed for a map, because they can be measured with more accuracy within the limits of a plot. The principal measurements that are obtained for a stand plot usually include tree species composition, canopy height, and crown closure. The individual tree attributes that are measured include tree height, diameter at breast height (DBH, 1.3 m), crown width, crown length, and tree age. Other attributes that are more descriptive in nature can also be used to characterize forest canopies, such as disturbance, tree health, and contextual estimates of soil and site drainage. A subplot may be established within a ground plot to provide supplementary information about shrubs, smaller plants, seedlings (regeneration), and soil type. Stand-level attributes such as stem density, basal area, wood volume, and stand growth can be calculated from the ground measurements by adding supplementary information that is related to species and site characteristics. As a complement to traditional inventories, ecological studies have been more focused on the quantity of photosynthetically active material and leaf area, which regulate water and carbon cycles (Running et al. 1989; Norman 1993; Liu et al. 1997; Chen et al. 2003; Pypker et al. 2005). Although most forest inventories include basic tree and stand attributes, there are often substantial differences in measurement types or guidelines when moving from one jurisdiction or project to another. For instance, increasing emphasis on carbon accounting, conservation, and habitat modelling has introduced new requirements for forest inventories that extend beyond estimating merchantable volume. If inventory data are available for test areas, great care should be given to the definitions and guidelines that are used. Otherwise, if no inventory data are available, an inventory program can be tailored to the specific requirements of the study within the limits of available resources.

3.2.2 *Leaf Area*

Leaf area is often expressed as the scale-independent leaf area index (LAI), which is defined as one-half of the total leaf surface area per unit of horizontal ground surface (Chen and Black 1992). LAI is a convenient stand attribute that is used to estimate light interception by all types of vegetation canopies, including those with non-flat leaves (e.g., needles). LAI is used by many climate, weather forecasting, ecosystem, biogeochemical cycles, and atmosphere-ecosystem models (FAO 2017).

Stand LAI can be estimated using direct measurement methods that include litter trap (Hall et al. 2003), point-quadrat (Warren Wilson 1960), and destructive sampling (Gower et al. 1999) methods. The direct methods derive LAI by counting leaves and measuring leaf area, weight and shape. Litter traps involve capturing leaves that fall at the end of the growing season, whereas the point-quadrat method provides an estimate of the position and size of leaves in a selected area. Point-quadrat methods are very difficult to use in a forest environment because they require long probes to acquire contact points. Therefore, it is considered to be the least practical of all direct methods. Destructive sampling involves cutting trees to weigh their components. It usually relies upon a limited number of sampled trees and the use of allometric relationships to estimate leaf area from one or several measured attributes, such as tree DBH, height, and crown dimensions. Allometric relationships at the tree level are not universal: local destructive sampling is required if no suitable relationships exist. Scaling relationships can be applied to LAI values for individual trees or for a given area to obtain stand-level estimates (Chen et al. 1997; Gower et al. 1999). Although uncertainties exist in the scaling methods that are used, litter trap and allometric methods are considered more accurate estimates of LAI compared to indirect methods. However, direct methods are time-consuming to the point where it frequently is not possible to use them. Hence, indirect methods have been developed using optical instruments because they can measure stand LAI much more rapidly. Several passive instruments can estimate LAI either from available light while looking directly at the solar corolla or from the spatial distribution of canopy gaps as seen from below the canopy when looking up with a fisheye lens providing a hemispherical perspective.

3.2.3 Towards Complex Descriptions of Forest Structure

Growing interest in the monitoring and prediction of the status of ecosystems has increased the demand for measurements of forest attributes. Fortunately, new technologies have also increased the number of approaches that are available for measuring vegetation canopies. For instance, the explicit description of the 3D distribution of canopy elements has provided new options for mapping its current status and predicting its state according to silvicultural practices. In contrast, current inventory measures cannot provide such detailed descriptions. Active sensors, such as airborne or terrestrial LIDAR (Light Detection and Ranging) instruments, have the potential to measure canopy architecture (Lowell et al. 2003; Hopkinson et al. 2004; St-Onge et al. 2008). The development of approaches for the inclusion of new measurement techniques in forest inventories is progressing rapidly because they can answer more specific questions with greater accuracy. For example, identifying treed wetlands is difficult using only aerial or satellite optical images. However, it is possible with the use of radar images that are taken at the time of high water levels

(Fournier et al. 2007). Model requirements are also a strong driver for the diversification of inputs that are related to forest structure. For example, there is a growing demand for the inclusion of forest structural attributes for many applications, including the following: stand volume models (Sun et al. 2007); measurements of tree and stand biomass (Parresol 1999); canopy reflectance models (Li and Strahler 1985; Nilson and Peterson 1991; Disney et al. 2000; Leblanc and Chen 2000); and architectural models (Godin 2000). Hemispherical photographs have the potential to provide some of these diverse measurements of forest structure.

3.3 Theory on Gap Fraction and Forest Structure

Optical instruments with a hemispherical field of view can record canopy gaps. This allows the position and amount of canopy elements to be identified, in contrast to the sky background. A theoretical foundation has been developed to describe how the amount of forest structure can be evaluated from the gap fraction as seen from a hemispherical field of view. It is based on the premise that canopy elements are darker than the contrasting background; normally, the sky background on the image is bright but not saturated. Several reviews of these concepts exist (Norman and Campbell 1989; Campbell and Norman 1989; Fournier et al. 2003; Jonckheere et al. 2004), and a summary of the relevant theory is given here.

3.3.1 Application of Beer's Law to Measure Forest Structure

The amount of light that is transmitted through a forest canopy depends upon the density of its structure (Monsi and Saeki 1953; Kira et al. 1969). As a first approximation, a forest canopy can be described as a turbid medium with randomly distributed elements. Light penetration through such a medium is described with Beer's Law, discovered in 1852 by August Beer; although it was previously described independently by Johann Heinrich Lambert in 1760 and Pierre Bouguer in 1729. It is expressed as

$$P = \exp\left[-\int_0^s k ds\right], \tag{3.1}$$

where P is the probability of light passing through the medium, s is the path through the medium, and k is the extinction factor. This relationship for light transmission is used extensively in studies on light propagation through the atmosphere (e.g., Liou 1980). Particles in the atmosphere may absorb or diffuse incident light, which is reduced accordingly. For a forest canopy represented as a turbid medium and considering light penetration at any zenith angle, Eq. 3.1 can be re-written as

$$P(\theta) = \exp\left[-\int_0^H G(\theta)L(h)\mathrm{d}h/\cos\theta\right], \tag{3.2}$$

where θ is the zenith angle (see Fig. 3.1). In this way, cosθ is used to represent the increasing path due to penetration at an angle θ as compared with the vertical maximum H (i.e., the canopy height). The term $L(h)\mathrm{d}h$ is the LAI contained in an infinitesimal dh slab at a given height h. The canopy extinction coefficient $G(\theta)$, or G-function, is the mean projection coefficient of a unit leaf element on a plane normal to θ (Warren Wilson 1960; Ross 1981; Campbell and Norman 1989). The G-function not only depends upon the view direction θ, but also upon the distribution functions that are associated with the normal leaf angle α_f. The G-function is conceptually understood as the extinction by canopy elements of the light beam; it is integrated over the whole path in the canopy in the direction θ. For a canopy composed of flat leaves with their normal directed upward, the G-function is the ratio of the projected area of all leaf surfaces onto a horizontal plane to the surface area of that plane; it is expressed by $G(\theta) = \cos\theta$. A general form of the G-function that is valid for a broad range of conditions can be applied if some assumptions are

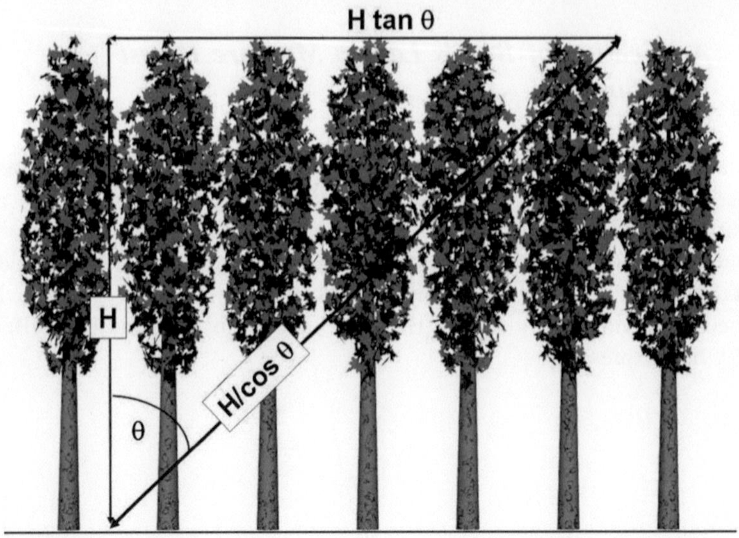

Fig. 3.1 Basic geometry used throughout the book. The term θ is the view zenith angle and H is the canopy height. The path through the canopy at any θ is given by H/cosθ. A gap at the *top* of the canopy is found at H tan θ

made for the leaf angle distribution function and density of canopy elements along the path. This provides

$$G(\theta) = \int_0^{\pi/2} g(\theta, \alpha_f)\mu(\alpha_f)\mathrm{d}\alpha_f, \tag{3.3}$$

where α_f refers to the inclination of the foliage elements to the horizontal. We simplified the notation by keeping only the zenithal component (with the integral from 0 to $\pi/2$) of the angles and assuming complete randomness for the angular distribution in the azimuthal plane. Most cases assume azimuthal symmetry with the exception of foliage with heliotropic behaviour. The function $g(\theta, \alpha_f)$ is the equivalent of the G-function for the case where all leaves are at an elevation of α_f. A series of functions have been developed for idealized cases including the uniform, conical, spherical, and ellipsoidal distributions of leaves' normal (Campbell and Norman 1989), amongst others. Warren Wilson (1960) developed a generic equation which relates the leaf angle and $g(\theta,\alpha_f)$ such that

$$g(\theta, \alpha_f) = \begin{matrix} \sin\alpha_f \cos\theta & \theta \leq \alpha_f \\ \sin\alpha_f \cos\theta\left[1 + \frac{2(\tan x - x)}{\pi}\right] & \theta > \alpha_f \end{matrix}, \tag{3.4}$$

where $x = \cos^{-1}(\tan \alpha_f/\tan \theta)$. Lastly, $\mu(\alpha_f)$ is the probability distribution of leaf elements at angle α_f. Equations 3.3 and 3.4 do not explicitly account for vertical changes in leaf density and orientation. For forest canopies dominated by a single canopy layer, the probability distribution of canopy elements usually follows the shape of a Weibull or triangular function. There is a quick increase from 0 at the top of the canopy towards a maximum, which is followed by a decrease of foliage density in the shaded portion of the crowns towards the ground. For practical reasons, instead of the leaf angle distribution function, in most cases it is more useful to use the extinction coefficient, that is, $G(\theta)$ or the G-function.

The formalism related to the G-function is described in more detail by Ross (1981), Campbell and Norman (1989), and Wang et al. (2006). In most practical situations, foliage is assumed to follow a behaviour described by a simple function. In the majority of cases, the spherical function is assumed; to a much lesser extent, the elliptical function is sometimes adopted (Campbell 1990). Although $G(\theta)$ can be resolved exactly from the theoretical properties of the leaf angle distribution, the inverted problem is mathematically challenging (Wagner and Hagemeier 2006). The inversion of gap fraction measurements to estimate the leaf angular distribution (LAD) can be done based on assumptions, such as in Campbell (1990). In this approach

$$G(\theta) = \frac{(\chi^2 + \tan\theta^2)^{0.5}\cos\theta}{\chi + 1.774(\chi + 1.182)^{-0.73}}, \tag{3.5}$$

where χ can be obtained analytically from a series of measurements of $G(\theta)$ based on gap fraction at various angles from hemispherical photographs. In turn, according to Campbell and Norman (1989), resolving χ allows the possibility to calculate the mean leaf inclination by

$$\overline{\alpha_f} = 90\left(0.1 + 0.9\mathrm{e}^{-0.5\chi}\right). \tag{3.6}$$

The availability of generic cases, such as the spherical or elliptical distribution functions, greatly simplifies the application of the general Eq. 3.3. For the random or spherical distribution of foliage elements, $G(\theta) = 0.5$. In all cases, an assumption must be made for $G(\theta)$ in order to resolve the general equation (Eq. 3.2) for the probability of light transmission. When considering non-flat surfaces such as shoots and woody material, the G-function can be modelled as Eqs. 3.3–3.6. The G-function has been redefined by Lang (1991), Chen and Black (1992), Leblanc et al. (1999), and Stenberg (2006). For the purpose of resolving Eq. 3.2, it is useful to integrate over the length of the canopy height (h from 0 to H) in order to determine the probability of light transmission such that

$$P(\theta) = \exp[-G(\theta)L/\cos\theta], \tag{3.7}$$

where L is the LAI.

Equation 3.7 is most frequently used for calculating the probability of having direct light penetration at a given zenith angle when the foliage is assumed opaque, non-scattering, randomly distributed, and with woody material considered negligible. Figure 3.2 shows how the gap fraction changes with the zenith angle for different LAI when using Eq. 3.7. It can be used to predict the average direct solar radiation penetrating through a canopy at any solar zenith angle. Equation 3.7 is based on a Poisson distribution of foliage (or gaps) requiring a significantly large sample size. This has an important impact on the analysis of hemispherical photographs. For example, it implies that Eq. 3.7 is only valid for an area with a sufficient number of samples (e.g., the complete image of a stand or a sector of an image comprising a sufficient number of pixels and integrating a large stand area). For instance, Eq. 3.7 may not be valid for an image of a single point or a small area of a canopy since the premise of Poisson sampling would be transgressed.

3.3.2 Modification of Beer's Law for Foliage Clumping and Wood Structure

Basic working equations to calculate the amount of canopy elements from thegap fraction, such as Eq. 3.7, assume that canopy elements are randomly distributed. While this assumption may be a useful approximation, it cannot adequately represent a complex system like a forest canopy. An additional assumption used during the development of theoretical equations is that the canopy elements are not

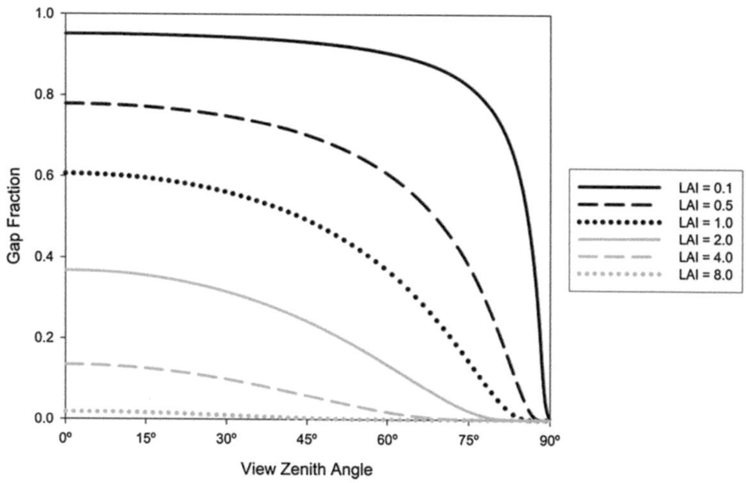

Fig. 3.2 Gap fraction as a function of view zenith angle for different LAI values simulated using Beer's Law (Eq. 3.7). The forest stand is composed of opaque elements that are randomly distributed with a spherical orientation of their normal and without woody material

discriminated, which implies that they can all be considered as foliage. Evidently, the elements of forest canopies are not randomly distributed; furthermore, elements other than foliage, such as trunks and branches, can have an impact on the resulting gap fraction. Consequently, the basic equation presented as Eq. 3.7 that links light penetration to forest structure must be modified accordingly.

For a canopy with foliage elements that are not distributed randomly in space, Nilson (1971) introduced a clumping index; a zenith dependence was later added by Kucharik et al. (1997). Foliage clumping is due to many factors, such as foliage clumped in crowns and branches; needleleaf species have needles clumped into shoots. As a result, light penetration through the canopy is obstructed by both foliage and wood components. To consider the clumping and wood effects, Eq. 3.7 can be rewritten as

$$P(\theta) = \exp[-G(\theta)\Omega(\theta)L_t/\cos\theta], \tag{3.8}$$

where L_t is the plant area index, which is composed of half the total surface area of foliar and woody materials, and $\Omega(\theta)$ is the clumping index (Nilson 1971; Chen et al. 1997). Canopy elements are clumped when $\Omega(\theta)$ is less than unity; they are distributed orderly when $\Omega(\theta)$ is greater than unity; canopy elements are randomly distributed when $\Omega(\theta)$ is unity. Equation 3.8 assumes that all sky obstructions are associated with foliage and woody elements. Assuming that foliage and woody material are statistically independent, the woody material can be removed from the plant area index by

$$L = L_t - W = L_t(1 - \alpha), \tag{3.9}$$

where W is the woody area index, and $\alpha = W/L_t$ is the woody to total area ratio. The term W is specific to a stand, and the ratio α is more constant for a given species than W. In conifer stands, shoots are generally dense opaque objects; thus a separation of the clumping effect of needle into shoot from scale larger than the shoot for conifers as Chen (1996) is required:

$$\Omega(\theta) = \Omega_E(\theta)/\gamma_E, \tag{3.10}$$

where $\Omega_E(\theta)$ is the clumping of discernable elements, which are the shoots within the canopy, and γ_E is the needle-to-shoot ratio (Oker-Blom and Smolander 1988; Chen and Cihlar 1995). The subscript E refers to the smallest discernable element in the canopy. For broadleaf species, γ_E is unity. Although some species-specific values for γ_E have been tabulated (Chen et al. 1997, 2006; Gower et al. 1999), it is preferable to measure this factor for each plot because it can vary significantly within a species (Stenberg 1996).

The LAI calculated from optical sensors without considering the clumping of foliage or wood structures should be called the effective LAI or $L_e(\theta)$. It is usually found by inverting Beer's Law without including the clumping effect $\Omega(\theta)$ at any scale. Mathematically, $L_e(\theta)$ is given by the multiplication of L and $\Omega(\theta)$ (Chen 1996) such that

$$L_e(\theta) = \Omega(\theta)L. \tag{3.11}$$

Unfortunately, in many cases it is not possible to extract the contribution of woody elements from the foliage. It is therefore simpler to suggest an effective plant area index or PAI (L_{et}) including foliage, wood, and all elements composing the canopy. In this way,

$$L_{et}(\theta) = \Omega(\theta)L_t. \tag{3.12}$$

The introduction of an angular dependence in the clumping index makes the effective LAI (L_e) and PAI (L_{et}) with a zenithal dependence. The variables L_e and L_{et} can be evaluated, but they depend upon the inversion of the gap fraction under assumptions that are often known to be erroneous; these assumptions could include a random distribution of elements, a known clumping factor $\Omega(\theta)$, or the extinction coefficient $G(\theta)$. Gap visibility in a hemispherical photograph or hemispherical medium is the only real physical property of the canopy; therefore, it is the only independent variable from which canopy attributes can be calculated.

3.3.3 Calculation of LAI from Gap Fraction

The simplest method to invert Beer's Law in order to retrieve LAI involves isolating the variable L_t in Eq. 3.7 such that

$$L_t = -\frac{\ln[P(\theta)]\cos\theta}{\Omega(\theta)G(\theta)}. \tag{3.13}$$

Solving Eq. 3.13 requires that three values are measured: the gap fraction $P(\theta)$ at given zenith angles, together with their related clumping indices $\Omega(\theta)$ and extinction coefficients $G(\theta)$. For the purposes of this chapter, HPs were treated as if only sky (illuminated) or obstructed (shaded) pixels exist. Therefore, the value of $P(\theta)$ is generally expressed as the ratio of illuminated pixels to the total number of pixels in zenithal sectors of HPs. Photographs must be divided into ranges of zenithal sectors for which θ corresponds to the centre zenith angle. For example, a photograph could be divided into nine zenithal ranges of ten degrees each, which would lead to nine values of $P(\theta)$. The hemispherical view is sometimes subdivided into azimuthal sectors, depending upon the analysis. Values of $P(\theta)$ and, subsequently L_t, are available for a series of θ, which leads to the resolution of L_t by solving a series of linear equations that are associated with each $L_t(\theta)$.

The retrieval of the clumping index $\Omega(\theta)$ is an active research topic (e.g., Stenberg 1996; Walter et al. 2003; Leblanc et al. 2005a; Pisek et al. 2011; Leblanc and Fournier 2014). Some clumping index values are available from the literature for various test sites. Although their use in a different context may not be judicious, these published values could provide rough estimates for sites without suitable measurements. Methods to estimate $\Omega_E(\theta)$ are complex and beyond the scope of this chapter. This topic is addressed in detail by Frazer et al. (2017).

Since the calculation of $L_t(\theta)$ includes both leaf and woody material, $G(\theta)$ represents the angular distribution of all significant plant elements; the measurement of the extinction coefficient for leaves is rarely possible. A $G(\theta)$ of 0.5 from the spherical distribution function is often assumed for the sake of simplicity (e.g., Lang 1991; Chen and Black 1992; Stenberg 2006, 2007). Other alternatives include the assumption of an ellipsoidal distribution (Campbell 1990) or a rotated ellipsoidal distribution (Thomas and Winner 2000), for which a mean leaf angle can be retrieved for each zenithal range. For many ecological models, the information that is contained in $G(\theta)$ is often sufficient (Leblanc et al. 2005a) because some models do not explicitly require leaf angle in their design.

Solving the inversion of Eq. 3.8 is not trivial. Miller (1967) suggested a mathematical approach using the gap fraction at several view-zenith angles to estimate L_t, but with no explicit clumping being considered. When the concept of clumping was subsequently introduced, independently of θ, the derivation was unchanged; however, the results were renamed effective LAI or PAI (Chen 1996). Miller's Theorem is based on the fact that integrating $G(\theta)$, which is weighted by $\sin\theta$, remains constant when integrated from 0° to 90° such that

$$\int_0^{\pi/2} G(\theta)\sin\theta\,\mathrm{d}\theta = 0.5. \tag{3.14}$$

This transformation is assumed valid for any type of foliage distribution. Isolating $G(\theta)$ times L_t from Eq. 3.8 and using Miller's Theorem permits the removal of $G(\theta)$ (Leblanc et al. 2005a) and provides

$$L_t = -2\int_0^{\pi/2} \frac{\ln[P(\theta)]\cos\theta}{\Omega(\theta)} d\theta. \tag{3.15}$$

To use Miller's Theorem properly, both the gap fraction and clumping index are needed for all angles that are being used. One approach that facilitates calculations assumes a constant clumping index with zenithal angles. However, this may not be suitable in all cases. Although estimating these two variables is possible with hemispherical photographs, another inversion method exists. From the work of Warren Wilson and Reeves (1959) and Warren Wilson (1960) (Eq. 3.4), it can be shown that $G(\theta)$ is always very close to 0.5 at a view zenith angle of one radian (57.3°). Equation 3.15 can then be inverted at 57.3° (Bonhomme and Chartier 1972; Neumann et al. 1989) with only one value of the gap fraction and clumping index (Walter et al. 2003; Leblanc et al. 2005a; Leblanc and Fournier 2014). Moreover, this can also be shown using an approximation that was demonstrated by Lang (1987). As a first approximation, and assuming that the variable within the integral of Eq. 3.15 is a linear function, we obtain

$$\frac{-\ln P(\theta)\cos(\theta)}{\Omega(\theta)} = A\theta + B. \tag{3.16}$$

Then Eq. 3.7 becomes

$$L_t = 2\int_0^{\pi/2} (A\theta + B)\sin\theta\,\mathrm{d}\theta = 2(A + B). \tag{3.17}$$

Assuming that L_t can be estimated from a single θ (equating Eqs. 3.8–3.13 with $G(\theta) = 0.5$) yields

$$2(A\theta + B) = 2(A + B), \tag{3.18}$$

which is only true when θ is equal to unit radian (or 57.3°) for values of A that are different from zero. By either using the approach of Warren Wilson and Reeves (1959) for $G(\theta)$ or by using Lang (1987), it can be shown mathematically that using measurements near 57.3° is sufficient to estimate LAI, in addition to using Miller's

Theorem, over a large range of angles. This has also been empirically demonstrated (e.g., Leblanc et al. 2005a). An alternative to solving Eq. 3.8 is to use CAN-EYE software (Jonckheere et al. 2004; Weiss et al. 2004), which uses a look-up table to fit LAI curves to the observed gap fraction measurements.

3.3.4 *Calculation of Canopy Closure and Openness from Gap Fraction*

Gap fraction can be used to estimate forest canopy architectural variables other than LAI. For example, canopy closure C is defined as the percentage of canopy cover. It can be estimated from the gap fraction near the zenith, such that

$$C = 1 - P(\theta = 0). \tag{3.19}$$

Only a small number of pixels are exactly at $\theta = 0°$ or close enough to be approximated as being at the zenith; thus, canopy closure is often measured over a range of angles near the zenith. At the zenith angle, an estimate of C can be derived using

$$C(\theta) = 1 - e^{\frac{G(\theta=0)\Omega(\theta)}{G(\theta)\Omega(\theta=0)} \ln[P(\theta)] \cdot \cos\theta}. \tag{3.20}$$

When θ is near zero, Eq. 3.20 can be approximated by

$$C(\theta) = 1 - e^{\ln[P(\theta)] \cdot \cos\theta}. \tag{3.21}$$

Averaging $C(\theta)$ over a range of θ can then be used to estimate canopy closure. It is expected that the estimates of $C(\theta)$ will be less accurate as θ increases. Another canopy structural variable is openness O. This variable is also based on the gap fraction and is indicative of the general number of openings. It is not a measure that describes the angular distribution of gaps; instead, O is an integrated value. Therefore, two canopies with very different architecture could have similar openness: this would only be indicative of a similar amount of gaps. Note that there is a gradual importance (weight) for the gaps from the horizon (90°) to nadir (0°). The general equation to calculate openness is

$$O = \int_0^{\pi/2} P(\theta) \sin\theta \, d\theta. \tag{3.22}$$

The integration is generally done numerically and can be limited in the zenith and azimuth view. For more details and definitions of different variations of canopy openness see Gonsamo et al. (2013).

3.4 Optical Instruments with a Hemispherical View to Measure Forest Structure

An efficient way to measure the widest field of view of a canopy with one single point measurement is to use a fisheye lens, or hemispherical lens, attached to a sensor. The sensor can be a camera or a specific sensor such as the one used in the LICOR Plant Canopy Analyzer, hereafter called LAI-2000.

The acquisition principle of the LAI-2000 and an HP system is the same in that both operate with a hemispherical field of view of the canopy (see Fig. 3.3). The LAI-2000 assumes that canopy elements are opaque and randomly distributed. Its field of view is limited to view angles from 0° to 75° from the zenith, which are segmented into five concentric rings. Gap fraction is calculated from the LAI-2000 by the ratio of below canopy measurements to those that are above (or outside) the canopy. If light conditions are changing during the below-canopy measurements, simultaneous measurements are necessary for the below- and above-canopy measurements. When possible, these two measurements are taken with two instruments simultaneously, where one instrument is taking continuous measurements in an open area near the stand and the other one is used by an operator within the stand looking up. If the below- and above-canopy measurements are taken consecutively with one instrument at a time near dawn or dusk, a curve fit between the two

Fig. 3.3 Digital hemispherical photograph taken with the Nikon CoolPix 5000 in a jack pine stand near Thompson, Manitoba

above-canopy measurements might be necessary because light conditions can change rapidly during these times of day. Under variable conditions, such as under a cloudy sky or a long delay between below- and above-canopy measurements, two units are needed. It is possible to reduce the azimuthal viewshed of the instrument with opaque plastic masks over the hemispherical lens. The final values available from the LAI-2000 are gap fraction at every zenith angle range corresponding to the five rings.

HPs do not require an above-canopy reference measurement; however, Chen and Black (1992) and Zhang et al. (2005) recommended measuring the intensity of light outside of the canopy to assess the camera settings (speed and aperture). Furthermore, Cescatti (2007) used the ratio of outside- and below-canopy hemispherical photographs to estimate gap fraction for all pixels. Other methods have been used to classify hemispherical photographs into gap and canopy elements based on colours, but they are usually more common when applied to down-loooking hemispherical photographs (e.g. Chen et al. 2002; Jonckheere et al. 2004; Weiss et al. 2004). The division of the photographs into zenithal or azimuthal sectors is decided by the operator to suit the requirements of the analysis and has an impact on the retrieved LAI (Leblanc et al. 2005a; Ryu et al. 2010). The retrieval of LAI usually assumes that the canopy elements are perfectly black (i.e., they are opaque and non-reflective) or that the light source is perpendicular to the normal to the sensor (Leblanc et al. 2005a). The LAI-2000 collects its signal in the blue ($\sim$400 nm) portion of the electromagnetic spectrum: this is a range for which leaves are known to be nearly opaque. Hemispherical photographs are taken in either panchromatic mode (black and white covering the visible range), three colour composite (red, green, and blue), or near-infrared with either a modified digital camera or infrared film (Chapman 2007). Light transmission through the foliage is significant for some of these wavelengths. As a result, contrast between the foliage and sky is harder to achieve with hemispherical photographs compared with a sensor like the LAI-2000. Some studies have focused on using only the blue channel of digital colour hemispherical photographs in order to produce a perspective that would be similar to that observed by the LAI-2000 (e.g., Leblanc et al. 2005a). Contrast between the sky and canopy is the main reason why image thresholding is a critical step for the use of HP to produce estimates of gap fraction. Its description is beyond the scope of this chapter, but the reader is referred to other studies on this topic (Wagner 1998, 2001; Privette et al. 2004; Weiss et al. 2004; Leblanc et al. 2005a; Nobis and Hunziker 2005) or to a review by Jonckheere et al. (2004). Hemispherical photographs in digital format thus provide a gap fraction value per angular sector.

The selection of appropriate light conditions, instrument settings, and a suitable zenithal range for the reliable calculation of LAI are still subjects of debate. Measurements with hemispherical field of view sensors are generally taken either at sunset or dusk under diffused light conditions. However, practical considerations often override taking measurements in optimal conditions. The overexposure of hemispherical photographs can result in erroneous calculations (Beckschäfer et al. 2013). We prefer to take photographs at nearly the saturation value for the CCD

(255 for an 8-bit sensor), where only a few pixels have the maximum intensity level permitted by the digital count; thus, we avoid the overexposure problem while maximizing the dynamic range of the pixel values in the image. The suitable zenithal range has been a source of debate primarily because of the difficulty in extracting significant information at zenith angles larger than 70°. Hemispherical views within a forest canopy tend to be very dark, if not completely obscured, at large zenith angles. The very low illumination at these large zenith angles is, consequently, more affected by light scattering than smaller zenith angles. Chen et al. (1997) suggested that a 15% increase in estimated LAI may occur because of light scattering from LAI-2000 measurements. One approach to correct the obstruction and scattering problems is to restrict observations solely to rings 1–4 (e.g., Frazer et al. 1999; Jonckheere et al. 2005). In the context of their application, Küßner and Mosandl (2000) employed a further restricted view from rings 1 to 3 because of the adverse conditions of clearings. Even though removal of ring 5 is sometimes used, ring restriction does not follow the premises of mathematical inversion from Miller (1967), which requires applying calculations over the entire zenithal range. Leblanc and Chen (2001) suggest correcting this problem by using only ring number 4, centred at 53°, to invert for LAI while Leblanc and Fournier (2014) suggest using a range of 55°–60°. These suggestions are based on their empirical results and the premise that the view zenith angle of 57.3° is contained in the fourth ring. Overall, the scientific debate over the most suitable zenithal range for the calculation of LAI is on-going. However, restricting the angular range to rings 1–4 (representing zenith angles from 0° to 60°) remains a common practice (e.g., Jonckheere et al. 2005) when inverting because it leads to reasonable results although others are only using rings 1–3 (e.g., Chen et al. 2006). Similarly, the calculation of LAI using a small zenith angular range centred on 57.3° was also suggested as an alternative (e.g., Leblanc 2006; Macfarlane et al. 2007a). A sound theoretical basis for these options still needs to be developed to confirm their validity.

3.5 Problems Associated with the Application of Gap Fraction Theory

Problems directly related to image acquisition and analyses, such as thresholding issues, are treated in Jonckheere et al. (2004). Only problems resulting from the application of theoretical equations are discussed in this chapter. For example, the inversion of Eq. 3.8 in the case of a canopy with randomly distributed elements can be performed at different averaging scales. In theory, results of the inversion should be more accurate as the length (e.g., number of images) increases over which the gap fraction is averaged. However, in the presence of a clumped canopy, the scale at which the gap fraction is averaged becomes very important and has yet to be universally understood. The method proposed by Lang and Xiang (1986) to account for clumping of heterogeneous canopies was based on the fact that the logarithm of a mean value is not the same as the mean of the logarithm. The averaging length in the literature varies from one to a few pixels (Cescatti 2007), to segments (Weiss et al. 2004; Leblanc et al. 2005a), to one image (Chen et al. 2006), and, finally, to a series of images (Leblanc et al. 2005a). The appropriate length to which to apply the inversion method is therefore a relevant scientific question.

Other problems due to gap fraction theory introduce limitations when it is applied to HP. There are five specific problems affecting the practical implementation of the inversion of the gap fraction to calculate LAI: (1) the absence of gaps at high zenith angles ($\theta > 80°$), (2) the low number of pixels for small zenith angles ($\theta < 10°$), (3) the oversimplification of leaf angle distribution (LAD) through a model, (4) the importance of canopy element clumping on the resulting LAI, and (5) the evaluation of canopy elements other than foliage. All studies using hemispherical photographs must address each of these problems according to the available data or canopy configuration. Here, the current understanding of these problems is presented using examples and potential solutions.

3.5.1 The Absence of Gaps at High Zenith Angles

In a dense forest canopy, very few gaps can be seen at large zenith angles because LAI has a logarithmic relationship with the gap fraction (Eq. 3.5 or Eq. 3.6). Figure 3.4 shows this effect using a 12 megapixel camera setting for a simple case when a simulated stand with an infinite extent has foliage that is distributed and oriented randomly (Eq. 3.5). Beer's Law implies that stands will have almost no gap at $\theta > 80°$ for a canopy with LAI >3, while hundreds of pixels could still be found with gaps at θ of 40°. The same situation prevails in real stands, but for larger LAI, because of foliage clumping which favours gaps. This situation prevents the proper calculation of LAI through the inversion formula (Eq. 3.11) because it requires the definition of the logarithm of zero, $P(\theta)$, which is not defined. At a given LAI of 2.8, only 0.6% of the pixels can be identified as gaps at $\theta = 80°$, but 30% were gaps at 40°. Following the realistic configuration used for the simulation, it is impossible to use Eq. 3.11 to calculate LAI above 3.4 with a single photograph as no gaps occurred at $\theta \geq 80°$. A few methods have been used in the literature to avoid the calculation problem related to no gap pixels at high θ. The LAI-2000 uses by default the gap fraction in the fifth ring (62.3°–74.1°) to complete the zenith angle up to 90°. Other methods generally rely upon either reducing the range of the view zenith angle (Jonckheere et al. 2005) to avoid the angular area with no gaps, or simply using more photographs or a camera permitting the acquisition of images at very high resolution to increase the chance of finding a gap at high θ. This limitation has an impact on the length used to estimate the gap fraction: the averaging length should be larger for denser canopies compared with more open canopies.

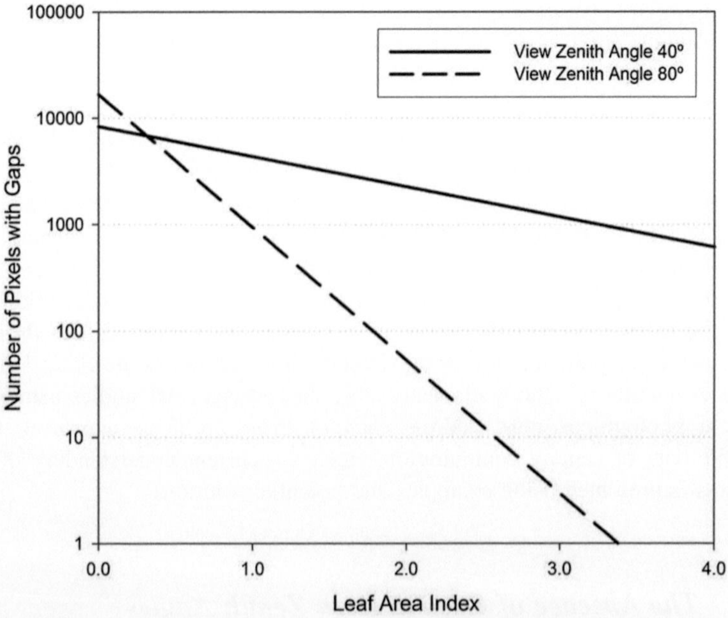

Fig. 3.4 Number of pixels (logarithmic scale) identified as gaps as a function of LAI of a simulated stand at $\theta = 40°$ and 80° on a hemispherical photograph corresponding to circular rings of 1 pixel wide and 8376 and 16,755 pixels long, respectively. This configuration is equivalent to a camera acquiring images with a 12 megapixel resolution

3.5.2 The Low Number of Pixels for Small Zenith Angles

View sectors at very low θ are also a problem when using the inversion method (Eq. 3.11). Although large gaps are generally found near the zenith, a camera placed near or underneath a crown may result in a photograph with no gaps for $\theta < 10°$ (or larger, depending upon crown closure). Mathematically speaking, this example is similar to the one that was described for large θ. An alternative situation is also likely to occur when a camera that is placed near an opening in the canopy may not have pixels obstructed for $\theta < 10°$. Both situations may not be representative of the overall canopy and, thus, can influence the results. The viewshed looking up is influenced primarily by the elements immediately above the camera, which is spatially a more confined view compared with the other zenith angle sectors at larger θ. The low number of pixels for small zenith angles ($\theta < 10°$) places a disproportionate weight on a fewer number of pixels. As a result, the calculations made from hemispherical photographs may easily overestimate or underestimate the gap fraction near the zenith depending upon where the camera is placed, especially in the presence of clumped foliage or in the presence of important understory. A first solution, which is similar to the problem of the absence of gaps, would be to use a camera with a large number of pixels to resolve the problem that is associated with the lower pixel count. However, the limited viewshed can only be resolved by taking a larger number of photographs within the stand, and it relies greatly upon a suitable design for the collection of field data. Miller's theorem (Eq. 3.15) is not affected by this problem since it has a sin θ weight that puts more emphasis on the large θ when estimating LAI.

3.5.3 The Oversimplification of the Leaf Angle Distribution

Problems can arise when considering leaf angular properties, in particular LAD. They are either linked to the inversion of Beer's Law over limited θ or associated with a model that is not representative of a real canopy. The worst range of θ to invert Beer's Law without a realistic knowledge of LAD is near the zenith as it is where $G(\theta)$ can exhibit the largest potential range of values, from near zero to unity (Warren Wilson 1960). Use of the spherical model with $G(0) = 0.5$ results in different errors, depending upon the real LAD of the canopy. If the canopy is perfectly planophile (i.e., all leaves are horizontal) the LAI will be underestimated by a factor of two at most since the maximum value $G(0)$ can have is unity. However, if the canopy is dominantly erectophile, the error can be larger as $G(0)$ can be as small as the thickness of a leaf. Assuming the case where the mean leaf angle is 10° from the vertical with $G(0) = 0.17$ based on the Warren Wilson (1960) equations (Eq. 3.4), the LAI retrieved could be underestimated by a factor of three. Two options are generally used to solve for LAI for canopies with non-random foliage angular distributions; one can either invert the gap fraction near 57.3° where $G(57.3°)$ is near 0.5 or one can use Miller's Theorem. The use of the gap fraction at $\theta = 57.3°$ is a good place to evaluate the most suited LAD model because it is not influenced by it. Experience from both a large number of forest canopies and theoretical $G(\theta)$ functions show that $G(57.3°)$ is frequently near 0.5 and can be used to assess LAI integrated over a limited range of θ. Another approach to find a suitable LAD model is to numerically integrate Miller's Theorem for the calculation of LAI over a smaller range of θ than the full 0°–90°. As an example, we calculated the case of a canopy with leaves all inclined at 20° from the vertical and LAI of 5 (Fig. 3.5). Using Miller's Theorem from 0°–89°, the retrieved LAI is 4.99 using the simplest integration method. In reality, the difficulty in measuring gaps at large θ implies that the integral is done over a smaller range such as 0°–60°. In that case, this canopy would yield an LAI of 4.02. Using angles near 57.3° results in a LAI of 5.31 (55°–60°) compared to 5.00 from the LAI-2000 fourth ring equivalent. So, the best solution when the integral cannot be calculated over the near full 0°–90° range in the presence of a non-spherical leaf distribution is to invert in a range near 57.3°.

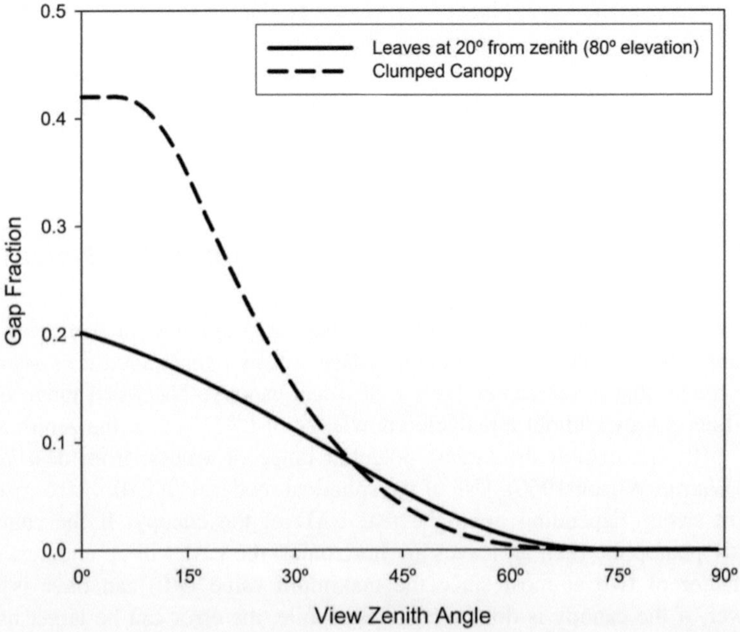

Fig. 3.5 Gap fraction from Beer's law with LAI of 5. For cases where the clumping index is present but with a spherical LAD, and no clumping present but with leaves oriented at 20° from the zenith (80° elevation)

3.5.4 The Importance of Canopy Element Clumping on the Resulting LAI

Forest canopies are rarely a perfect turbid medium and foliage clumping found in most canopies explains, in large part, the difference between the effective (L_e) and true LAI (L): in other words, it explains the relationship between gap fraction and LAI. In clumped canopies, the Miller Theorem can be incorrect because it was defined for foliage that is randomly distributed in space. Assuming a linear variation of the clumping index from $\theta = 0°$ with 0.64 to unity at 90° (0.004 clumping index unit change per degree) and LAI = 5 (see Fig. 3.5 for the resulting gap fraction curve), a retrieval using Miller Theorem gives an L_e of 4.34 if the integral is done from 0°–89° and 3.99 for 0°–60°. Once again, a solution to this problem is to carry out the inversion at $\theta \approx 57.3°$ because it is easier to separate the effect of clumping and LAD as $G(57.3°)$ is nearly constant at 0.5. The averaged L_e from 55° to 60° is close to the full integral at 4.35. This demonstrates that the inversion at a θ range near 57.3° is more accurate than a partial use of Miller's Theorem in the presence of foliage clumping. In cases where there is not much clumping and where leaves are randomly oriented, it should be possible to use any θ for the LAI inversion as long as there are enough measurable gaps.

3.5.5 *Evaluation of Canopy Elements Other Than Foliage*

Canopy elements other than foliage also have an impact on the calculated LAI; unfortunately, estimating their contribution is not simple. Several studies have attempted to evaluate the contribution of the wood structure on LAI from modelling or field measurements (Fournier et al. 1996; Barclay et al. 2000) by estimating the contribution of the woody material through hemispherical photograph simulations based on stem maps. Image processing (Weiss et al. 2004) and adapted image acquisition (Kucharik et al. 1998; Chapman 2007) have also been considered for evaluating woody material. Results of these techniques vary greatly and very little validation has been done. In short, this is an area of optical LAI retrieval that has yet to be studied comprehensively. The best solution to date relies upon destructive sampling of a few trees that are representative of the stand to estimate the woody to total area ratio. In any case, studies should report the results as PAI when woody material is not considered and in the case when it has not been estimated locally to allow better measurement intercomparisons.

Although five intrinsic problems have been mentioned, the use of hemispherical data to estimate canopy architecture has shown great potential and often good correlation with allometric LAI measurements (e.g., Chen et al. 1997). Many alternatives exist to deal with potential problems in applying the inversion formula, which is the basis for estimating LAI (Eq. 3.11). Consequently, the selected methodology used to handle them should always be stated explicitly to facilitate interpretation of the results calculated from hemispherical photographs.

3.6 Discussion

3.6.1 *Sampling and Representativeness*

This chapter has focussed on theory and provides the background knowledge for establishing a quantitative link between gap fraction from HP and canopy structure. Although important issues like sampling strategy, experimental setup, optimal equipment, and image collection are all important for a successful analysis, they are beyond the scope of this chapter. However, we wish to emphasise how important the number of photographs combined with a suitable experimental setup are for a representative dataset of a stand. We raised the importance of a sufficient number of hemispherical photographs as the only alternative to reduce the negative impact associated with small viewing angles. Also, the availability of alternative measurements can either improve the calculation of LAI or provide a convincing validation. As an example, the availability of detailed architectural descriptions (Nilson 1999) or alternate measurements like those from non-hemispherical photographs (Macfarlane et al. 2007b) or the TRAC (Chen 1996; Leblanc 2002) provide strong support for selecting a suitable leaf angle distribution (LAD) model or calculating the clumping index for a given stand. Even though the theory on gap fraction is a good base for the quantitative analysis of canopy structure, all the other aspects of data collection and analysis also need to be inspected carefully.

3.6.2 Scaling Issues

Field measurements of canopy structure also need to be linked with satellite images. Hemispherical photographs, like several other field measurements, are convenient for validating and deriving empirical relationships for regional or large scale mapping using remote sensing products. A good georeference between the locations of datasets is the first critical step in linking both measurements. However, scaling issues are also important to address. Remote sensing data, such as aerial photographs or satellite imagery from IKONOS and QuickBird, all provide very high spatial resolution images of the order of 1 m or better. These images may not be suitable for direct comparison with field measurements of LAI because individual pixels provide much finer measurements than the footprint of a hemispherical photograph. In this case, methods of pixel averaging or image segmentation may provide a means to generalize the information of the images at a scale compatible with the field data footprint. Medium resolution sensors, such as the Landsat TM and ETM series with a pixel spatial resolution of about 30 m, are usually more compatible for combined use with field measurements. Forest field plots can hardly be less than 30 m as the footprint at 57.3° in one direction is 2.6 times the height of trees. This can make the needed plots somewhat larger than typical forestry plots. Low resolution sensors are used in continental and global map products, such as the NOAA-AVHRR series, the MODIS sensor, and SPOT-VEGETATION, which all have a spatial resolution of the order of a kilometre. Although a few attempts have been made (e.g., Privette et al. 2002), these sensors are not appropriate for direct comparison with field measurements because of the gross mismatch between the two spatial footprints. Mapped parameters at 30 m can then be used to validate products at lower resolution (Cihlar et al. 2000; Chen et al. 2003). A link between the field measurements and satellite images is often possible but only if the compatibility of their footprints can be established.

Thus far in the scientific literature, properties such as LAI have been mapped mainly from optical remote sensing data (Myneni et al. 1997; Chen et al. 2003). Active microwave remote sensing, such as synthetic aperture radar (SAR), has shown potential to estimate biomass, especially in combination with optical remote sensing data (e.g., Treuhaft et al. 2002, 2003). In addition, multi-polarization from ENVISAT ASAR has been shown to be sensitive to LAI (Manninen et al. 2005). LIDAR has also been proven to be very efficient to map canopy height (St-Onge et al. 2008) or biomass (Skowronski et al. 2007). Some studies have shown the potential of multi-angular optical data to estimate the clumping index (Lacaze et al. 2002; Chen et al. 2003, 2005; Leblanc et al. 2005b). Unfortunately, only a few satellite instruments can provide enough angular sampling to estimate the clumping index. Overall, many remote sensing sensors have potential for either measuring LAI or its associated parameters. However, only LAI is operationally mapped from spaceborne sensors.

3.6.3 *Ecological Models*

From measurements of forest structure, models can be used to estimate forest properties related to carbon and water cycles. Studies related to climate change have used land cover and vegetation properties such as LAI to estimate CO_2 absorption by vegetation. Mapped LAI can be used as input for models with other structure attributes, such as the clumping index, leaf angular distribution, and landcover (Liu et al. 1997; Chen et al. 2003). Moreover, the water cycle needs vegetation information to estimate the effect of precipitation and runoff. The Farquhar et al. (1980) model is used to estimate leaf-level photosynthesis; the challenge is the scaling of this model to the canopy level (Chen et al. 2003). It was first done based on LAI and the incoming solar radiation reaching the foliage as big leaf models (Monteith 1973). For large-scale studies, it is now very common to separate the radiation regime into direct sunlit and shaded LAI (L_{Sun} and L_{Shaded}) portions, which receives solar radiation indirectly based on Norman (1993) such that

$$L_{Sun} = [1 - \exp(-G(\theta)\Omega(\theta)L/\cos\theta)]\frac{\cos\theta}{G(\theta)}, \tag{3.23}$$

and

$$L_{Shaded} = L - L_{Sun} \tag{3.24}$$

Forest attributes such as LAI, foliage orientation, and foliage clumping are important when using Eqs. 3.23 and 3.24 in ecological models. Some studies indicate that LAI is not the forest characteristic that most influences the optical remote sensing signal: the effective LAI (or gap fraction) may in fact be more related to remote sensing data (Leblanc et al. 2002; Chen et al. 2003) than LAI. Therefore, mapping LAI through satellite data or using it to run models are both very active research areas.

3.7 Conclusions

The theory on the use of HP rests on the principle that gaps present in the image are linked to the amount of forest structure. The theoretical development presented in this chapter is based on light transmission through the forest canopy using Beer's Law. First, how the zenith angular distribution of gap fractions could become a measurement of the surface of canopy elements through the PAI was examined. Effects associated with the leaf angular distribution, foliage clumping, and differentiation between foliage and other canopy elements were further introduced in the mathematical framework. This resulted in a mathematical formula to assess LAI and other canopy architecture variables from gap fractions. Estimation of these variables is a complement to measurements taken in most forest inventories. HP can be useful to estimate variables that would otherwise be time–consuming, or even impossible, to estimate. Overall, HP is simple and can be done with affordable instruments. With proper measurement techniques, it provides a complement to the traditional tree and stand measurements often taken in forestry.

Understanding the limitations of the theory associated with the analysis of hemispherical photographs facilitates data acquisition and ensures realistic interpretation of results. For instance, the uncertainty related to LAD models, foliage clumping, and the wood ratio forces all users of HP to address these points before and during data acquisition. It is generally accepted in the scientific community that none of these points can be resolved entirely. Nevertheless, a reasonable compromise must be reached. There are many precedents in the literature that can help and provide guidelines. Moreover, these limitations leave much room for specialized studies designed to address the current limitations. Overall, meaningful measurements can be calculated from HP if the basis for the theoretical development and the limitations of the technique are understood and addressed. Canopy measurements with HP provide information on the spatial coverage of an extended ground plot. Linking these measurements with satellite remote sensing images can only be done if the georeference and the scale of both datasets are compatible. Describing the theoretical basis for the use of the gap fraction with hemispherical photographs has highlighted several research topics that are associated with current limitations, such as the following: a practical method to evaluate clumping of canopy elements, image processing to improve the estimation of gap fraction, suitable models of LAD, a legitimate theoretical base to apply calculations on a limited zenithal range, and the link between hemispherical photographs and multi-scale datasets for modelling. Regardless, HP is a practical tool to estimate structural canopy variables, which can provide the ability to spatially assess the distribution of canopy elements.

References

Anderson MC (1964) Studies of the woodland light climate I. The photographic computation of light condition. J Ecol 52:27–41

Avery TE, Burkhart HE (2002) Forest measurements, 5th edn. McGraw-Hill, New York, 456 p

Barclay HJ, Trofymov JA, Leach RI (2000) Assessing bias from boles in calculating leaf area index in immature Douglas-fir with the LI-COR canopy analyzer. Agric For Meteorol 100:255–260

Beckschäfer P, Seidel D, Kleinn C, Xu J (2013) On the exposure of hemispherical photographs in forests. iForest 6:228–237 [online 2013-06-13] URL: http://www.sisef.it/iforest/contents/?id=ifor0957-006

Bonhomme R, Chartier P (1972) The interpretation and automatic measurement of hemispherical photographs to obtain sunlit foliage area and gap frequency. Israel J Agr Res 22:53–61
Campbell GS (1990) Derivation of an angle density function for canopies with ellipsoidal leaf angle distributions. Agric For Meteorol 49:173–176
Campbell GS, Norman JM (1989) The description and measurement of plant canopy structure. In: Russell G, Marshall B, Jarvis PG (eds) Plant canopies: their growth, form and function. Cambridge University Press, Cambridge, pp 1–19
Cescatti A (2007) Indirect estimates of canopy gap fraction based on the linear conversion of hemispherical photographs: methodology and comparison with standard thresholding techniques. Agric For Meteorol 143:1–12
Chapman L (2007) Potential applications of near infra-red hemispherical imagery in forest environments. Agric For Meteorol 143:151–156
Chen JM (1996) Optically-based methods for measuring seasonal variation in leaf area index in boreal conifer stands. Agric For Meteorol 80:135–163
Chen JM, Black TA (1992) Defining leaf area index for non-flat leaves. Plant Cell Environ 15:421–429
Chen JM, Cihlar J (1995) Plant canopy gap-size analysis theory for improving optical measurements of leaf-area index. Appl Opt 34:6211–6222
Chen JM, Black TA, Adams RS (1991) Evaluation of hemispherical photography for determining plant area index and geometry of a forest stand. Agric For Meteorol 56:129–143
Chen JM, Rich PM, Gower ST, Norman JM, Plummer S (1997) Leaf area index of boreal forests: theory, techniques and measurements. J Geophys Res 102(D24):29429–29444
Chen JM, Pavlic G, Brown L, Cihlar J, Leblanc SG, White HP, Hall RJ, Peddle DR, King DJ, Trofymow JA, Swift E, Van der Sanden J, Pellika PKE (2002) Derivation and validation of Canada-wide coarse-resolution leaf area index maps using high-resolution satellite imagery and ground measurements. Remote Sens Environ 80:165–184
Chen JM, Liu J, Leblanc SG, Lacaze R, Roujean J-L (2003) Multi-angular remote sensing for estimating carbon absorption by vegetation. Remote Sens Environ 84:516–525
Chen JM, Menges C, Leblanc SG (2005) Global mapping of foliage clumping index using multi-angular satellite data. Remote Sens Environ 97:447–457
Chen JM, Govind A, Sonnentag O, Zhang Y, Barr A, Amiro B (2006) Leaf area index measurements at Fluxnet-Canada forest sites. Agric For Meteorol 140:257–268
Cihlar J, Latifovic R, Chen J, Beaubien J, Li Z, Magnussen S (2000) Selecting representative high resolution sample images for land cover studies. Part 2: application to estimating land cover composition. Remote Sens Environ 72:127–138
Coops NC, Waring RH (2001) The use of multi-scale remote sensing imagery to derive regional estimates of forest growth capacity using 3-PGS. Remote Sens Environ 75:324–334
Disney MI, Lewis P, North PRJ (2000) Monte Carlo ray tracing in optical canopy reflectance modelling. Remote Sens Rev 18:163–196
FAO (Food and Agriculture Organization) (2017) GTOS-EVC T11 leaf area index (LAI). http://www.fao.org/gtos/ecv-t11.html. Accessed 6 March 2017
Farquhar GD, von Caemmerer S, Berry JA (1980) A biochemical model of photosynthetic CO_2 assimilation in leaves of C_3 species. Planta 149:78–90
Fournier RA, Landry R, August NM, Fedosejevs G, Gauthier RP (1996) Modelling light obstruction in three conifer forests using hemispherical photography and fine tree architecture. Agric For Meteorol 82(1–4):47–72
Fournier RA, Mailly D, Walter J-MN, Soudani K (2003) Indirect measurement of forest canopy structure from in situ optical sensors. In: Wulder MA, Franklin SE (eds) Remote sensing of forest environments: concepts and case studies. Kluwer, Dordrecht, pp 77–113
Fournier RA, Grenier M, Lavoie A, Hélie R (2007) Towards a strategy to implement the Canadian Wetland Inventory using satellite remote sensing. Can J Remote Sens 33:S1–S16
Franklin SE (2001) Remote sensing for sustainable forest management. CRC Press (Lewis), Boca Raton, FL, 407 p

Frazer GW, Canham CD, Lertzman KP (1999) Gap light analyzer (GLA), Version 2.0: imaging software to extract canopy structure and gap light transmission indices from true-colour fisheye photographs, users manual and program documentation. Simon Fraser University, Burnaby, British Columbia, and the Institute of Ecosystem Studies, Millbrook, New York. 52 p

Frazer GW, Fournier RA, Trofymow JA, Hall RJ (2001) A comparison of digital and film fisheye photography for analysis of forest canopy structure and gap light transmission. Agric For Meteorol 109:249–263

Frazer GW, Fournier RA, Leblanc SG, Walter JMN (2017) View angle-dependent clumping indices for Indirect LAI estimation. In: Fournier RA, Hall RJ Hemispherical photography in forest science: theory, methods, and applications. Springer, Heidelberg

Gillis MD, Leckie DG (1993) Forestry Canada, Petawawa National Forestry Institute. Chalk River, ON. Information Report PI-X-114

Godin C (2000) Representing and encoding plant architecture: a review. Ann For Sci 57:413–438

Gonsamo A, D'odorico P, Pellikka P (2013). Measuring fractional forest canopy element cover and openness—definitions and methodologies revisited. Oikos 122:1283–1291

Gower ST, Norman JM (1991) Rapid estimation of leaf area index in conifer and broad-leaf plantations. Ecology 72:1896–1900

Gower ST, Kucharik JK, Norman JM (1999) Direct and indirect estimation of leaf area index, f_{APAR}, and net primary production of terrestrial ecosystems. Remote Sens Environ 70:29–51

Hall RJ, Fernandes RA, Hogg EH, Brandt JP, Butson C, Case BS, Leblanc SG (2003) Relating aspen defoliation to changes in leaf area from field and satellite remote sensing perspectives. Can J Remote Sens 29:299–313

Hopkinson C, Chasmer L, Young-Pow C, Treitz P (2004) Assessing forest metrics with a ground-based scanning lidar. Can J For Res 34:573–583

Jonckheere I, Fleck S, Nackaerts K, Muys B, Coppin P, Weiss M, Baret F (2004) Methods for leaf area index determination. Part I: Theories, sensors and hemispherical photography. Agric For Meteorol 121:19–35

Jonckheere I, Muys B, Coppin P (2005) Allometry and evaluation of in situ optical LAI determination in Scots pine: a case study in Belgium. Tree Physiol 25:723–732

Jonckheere IGC, Macfarlane C, Walter J-MN (2017) Image analysis of hemispherical photographs, algorithms and calculation. In: Fournier RA, Hall RJ (eds) Hemispherical photography in forest science: theory, methods, applications. Springer, Berlin

Kira T, Shinozaki K, Hozumi K (1969) Structure of forest canopies as related to their primary productivity. Plant Cell Physiol 10:129–142

Kucharik CJ, Norman JM, Murdock LM, Gower ST (1997) Characterizing canopy nonrandomness with a multiband vegetation imager (MVI). J Geophys Res 102(D24):29455–29473

Kucharik CJ, Norman JM, Gower ST (1998) Measurements of branch area and adjusting leaf area index indirect measurements. Agric For Meteorol 91:69–88

Küßner R, Mosandl R (2000) Comparison of direct and indirect estimation of leaf area index in mature Norway spruce stands of eastern Germany. Can J For Res 30:440–447

Lacaze R, Chen JM, Roujean J-L, Leblanc SG (2002) Retrieval of vegetation clumping index using hot spot signatures measured by POLDER instrument. Remote Sens Environ 79:84–95

Lang ARG (1987) Simplified estimate of leaf area index from transmittance of the sun's beam. Agric For Meteorol 41:179–186

Lang ARG (1991) Application of some Cauchy's theorems to estimation of surface areas of leaves, needles and branches of plants, and light transmittance. Agric For Meteorol 55:191–212

Lang ARG, Xiang Y (1986) Estimation of leaf area index from transmission of direct sunlight in discontinuous canopies. Agric For Meteorol 35:229–243

Layton PA, Guynn ST, Guynn DC (2004) Biodiversity metrics in Sustainable Certification Programs. J of Forestry 102(3):46–52

Leblanc SG (2002) Correction to the plant canopy gap-size analysis theory used by the tracing radiation and architecture of canopies (TRAC) instrument. Appl Opt 31:7667–7670

Leblanc SG (2006) Digital hemispherical photography (DHP) manual, version 2.3b. Natural Resources Canada, Canada Centre for Remote Sensing, Ottawa
Leblanc SG, Chen JM (2000) A windows graphic user interface (GUI) for the five-scale model for fast BRDF simulations. Remote Sens Rev 19:293–305
Leblanc SG, Chen JM (2001) A practical scheme for correcting multiple scattering effects on optical LAI measurements. Agric For Meteorol 110:125–139
Leblanc SG, Fournier RA (2014) Hemispherical photography simulations with an architectural model to assess retrieval of leaf area index. Agric For Meteorol 194:64–76
Leblanc SG, Bicheron P, Chen JM, Leroy M, Cihlar J (1999) Investigation of directional reflectance in boreal forests using an improved 4-Scale model and airborne POLDER data. IEEE Trans Geosci Remote 37:1396–1414
Leblanc SG, Chen JM, White HP, Latifovic R Fernandes R, Roujean J-L, Lacaze R (2002) Mapping leaf area index heterogeneity over Canada using directional reflectance and anisotropy canopy reflectance models. In: IGARSS 2002, Proceedings of the IEEE international geoscience and remote sensing symposium, 24–28 June, Toronto, Ontario. CD-ROM. IEEE, Piscataway, New Jersey
Leblanc SG, Chen JM, Fernandes R, Deering DW, Conley A (2005a) Methodology comparison for canopy structure parameters extraction from digital hemispherical photography in boreal forests. Agric For Meteorol 129:187–207
Leblanc SG, Chen JM, White HP, Latifovic R, Lacaze R, Roujean J-L (2005b) Canada-wide foliage clumping index mapping from multiangular POLDER measurements. Can J Remote Sens 31:364–376
Li X, Strahler A (1985) Geometric-optical modeling of a conifer forest canopy. IEEE Trans Geosci Remote 23:705–721
LI-COR (1991). LAI-2000 plant canopy analyzer operating manual. LI-COR Inc., Lincoln, Nebraska
Liou KN (1980) An introduction to atmospheric radiation. International geophysics series, vol 26. Academic Press, San Diego
Liu J, Chen JM, Cihlar J, Park WM (1997) A process-based boreal ecosystem productivity simulator using remote sensing inputs. Remote Sens Environ 62:158–175
Lowell JL, Jupp DLB, Culvenor DS, Coops NC (2003) Using airborne and ground-based ranging lidar to measure canopy structure in Australian forests. Can J Remote Sens 5:607–622
Macfarlane C, Hoffman M, Eamus D, Kerp N, Higginson S, McMurtrie R, Adams M (2007a) Estimation of leaf area index in eucalypt forest using digital photography. Agric For Meteorol 143:176–188
Macfarlane C, Grigg A, Evangelista C (2007b) Estimating forest leaf area using cover and fullframe fisheye photography: thinking outside the circle. Agric For Meteorol 146:1–12
Manninen T, Stenberg P, Rautiainen M, Voipio P, Smolander H (2005) Leaf area index estimation of boreal forest using ENVISAT ASAR. IEEE Trans Geosci Remote 43:2627–2635
Miller JB (1967) A formula for average foliage density. Aust J Bot 15:141–144
Monsi M, Saeki T (1953) Über den Lichtfaktor in den Pflanzengesellschaften und seine Bedeutung für die Stoffproduktion. Jpn J Bot 14:22–52
Monteith JL (1973) Principles of environmental physics, 1st edn. Edward Arnold, London
Myneni RB, Nemani RR, Running SW (1997) Estimation of global leaf area index and absorbed PAR using radiative transfer models. IEEE Trans Geosci Remote 35:1380–1393
Neumann HH, den Hartog G, Shaw RH (1989) Leaf area measurements based on hemispheric photographs and leaf-litter collection in a deciduous forest during autumn leaf-fall. Agric For Meteorol 45:325–345
Nilson T (1971) A theoretical analysis of the frequency of gaps in plant stands. Agric For Meteorol 8:25–38
Nilson T (1999) Inversion of gap frequency data in forest stands. Agric For Meteorol 98–99: 437–448
Nilson T, Peterson U (1991) A forest canopy reflectance model and a test case. Remote Sens Environ 37:131–142

Nobis M, Hunziker U (2005) Automatic thresholding for hemispherical canopy-photographs based on edge detection. Agric For Meteorol 128:243–250
Norman JM (1993) Scaling processes between leaf and canopy levels. In: Ehleringer JR, Field CB (eds) Scaling physiological processes: leaf to globe. Academic Press, San Diego, pp 41–66
Norman JM, Campbell GS (1989) Canopy structure. In: Pearcy RW, Ehleringer J, Mooney HA, Rundel PW (eds) Physiological plant ecology: field methods and instrumentation. Chapman & Hall, London, pp 301–325
Oker-Blom P, Smolander H (1988) The ratio of shoot silhouette area to total needle area in Scots pine. For Sci 34:894–906
Parresol BR (1999) Assessing tree and stand biomass: a review with examples and critical comparisons. For Sci 45:573–593
Pisek J, Lang M, Nilson T, Korhonen L, Karu H (2011) Comparison of methods for measuring gap size distribution and canopy nonrandomness at Järvselja RAMI (Radiation Transfer Model Intercomparison) test sites. Agric For Meteorl 51:365–377
Privette JL, Myneni RB, Knyazikhin Y, Mukelabai M, Roberts G, Tian Y, Wang Y, Leblanc SG (2002) Early spatial and temporal validation of MODIS LAI product in the Southern Africa Kalahari. Remote Sens Environ 83:232–243
Privette JL, Tian Y, Roberts G, Scholes RJ, Wang Y, Caylor KK, Frost P, Mukelabai M (2004) Vegetation structure characteristics and relationships of Kalahari woodlands and savannas. Global Change Biol 10:281–291
Pypker TG, Bond BJ, Link TE, Marks D, Unsworth MH (2005) The importance of canopy structure controlling the interception loss of rainfall: examples from a young and an old-growth Douglas-fir forest. Agric For Meteorol 130:113–129
Rich PM (1990) Characterizing plant canopies with hemispherical photographs. Remote Sens Rev 5:13–29
Ross J (1981) The radiation regime and architecture of plant stands. Dr. W. Junk Bublishers, The Hague
Running SW, Nemani RR, Peterson DL, Band LE, Potts DF, Pierce LL, Spanner MA (1989) Mapping regional forest evapotranspiration and photosynthesis by coupling satellite data with ecosystem simulation. Ecology 70:1090–1101
Ryu Y, Nilson T, Kobayashi H, Sonnentag O, Law BE, Baldocchi DD (2010) On the correct estimation of effective leaf area index: does it reveal information on clumping effects? Agric For Meteorol 150:463–472
Schleppi P, Paquette A (2017) Solar radiation in forests: theory for hemispherical photography. In: Fournier RA, Hall RJ (eds) Hemispherical photography in forest science: theory, methods, applications. Springer, Berlin
Skowronski N, Clark K, Nelson R, Hom J, Patterson M (2007) Remotely sensed measurements of forest structure and fuel loads in the Pinelands of New Jersey. Remote Sens Environ 108:123–129
Stenberg P (1996) Simulations of the effects of shoot structure and orientation on vertical gradients in intercepted light by conifer canopies. Tree Physiol 16:99–108
Stenberg P (2006) A note on the G-function for needle leaf canopies. Agric For Meteorol 136:76–79
Stenberg P (2007) Simple analytical formula for calculating average photon recollision probability in vegetation canopies. Remote Sens Environ 109:221–224
St-Onge B, Véga C, Fournier RA, Hu Y (2008) Mapping canopy height using a combination of digital stereo-photogrammetry and lidar. Int J Remote Sens 29:3343–3364
Sun H, Zhang J, Duan A He C (2007) A review of stand basal area growth models. Forestry Stud China 9:85–94
Thomas SC, Winner WE (2000) A rotated ellipsoidal angle density function improves estimation of foliage inclination distributions in forest canopies. Agric For Meteorol 100:19–24
Treuhaft RN, Asner GP, Law BE, Van Tuyl S (2002) Forest leaf area density profiles from the quantitative fusion of radar and hyperspectral data. J Geophys Res 107(D21):4568. doi:10.1029/2001JD000646

Treuhaft RN, Asner GP, Law BE (2003) Structure-based forest biomass from fusion of radar and hyperspectral observations. Geophys Res Lett 30:1472–1476

Wagner S (1998) Calibration of grey values of hemispherical photographs for image analysis. Agric For Meteorol 90:103–117

Wagner S (2001) Relative radiance measurements and zenith angle dependent segmentation in hemispherical photography. Agric For Meteorol 10:103–115

Wagner S, Hagemeier M (2006) Method of segmentation affects leaf inclination angle estimation in hemispherical photography. Agric For Meteorol 139:12–24

Walter J-MN, Fournier RA, Soudani K, Meyer E (2003) Integrating clumping effects in forest canopy structure: an assessment through hemispherical photographs. Can J Remote Sens 29:388–410

Wang W-M, Li Z-L, Su H-B (2006) Comparison of leaf angle distribution functions: effects on extinction coefficient and fraction of sunlit foliage. Agric For Meteorol 143:106–122

Warren Wilson J (1960) Inclined point quadrats. New Phytol 59:1–8

Warren Wilson J, Reeve JE (1959) Analysis of the spatial distribution of foliage by two-dimensional point quadrats. New Phytol 58:92–101

Weiss M, Baret F, Smith GJ, Jonckheere I, Coppin P (2004) Review of methods for in situ leaf area index (LAI) determination: Part II. Estimation of LAI, errors and sampling. Agric For Meteorol 121:37–53

Welles JM, Norman JM (1991) Instrument for indirect measurement of canopy architecture. Agron J 83:818–825

Zhang Y, Chen JM, Miller JR (2005) Determining digital hemispherical photograph exposure for leaf area index estimation. Agric For Meteorol 133:166–181

Chapter 4
Acquiring Hemispherical Photographs in Forest Environments: From Planning to Archiving Photographs

Richard A. Fournier, Daniel Mailly, Jean-Michel N. Walter and Inge G.C. Jonckheere

Abstract This chapter deals with the practical aspects of hemispherical photography that pertain to equipment, field preparation and data acquisition. Despite the large number of scientific articles that report use of hemispherical photographs, many questions remain for those wanting to determine how to use this technology in forestry applications. For example, commonly asked questions include what equipment should be used and what are their technical requirements? This chapter was designed to guide those interested in acquiring hemispherical photography by providing procedures about what equipment to select, how to reduce the uncertainties of data collection, how to prepare data acquisition plans, and what to collect and archive, especially with respect to forestry applications. In short, it summarises the informal learning required to progress from the planning of data acquisition to the collection of a suitable data set of hemispherical photographs. While the capabilities of digital photography are continually advancing, limitations vary among different systems that are seldom known. To help improve knowledge about how to acquire hemispherical photos, this chapter includes information for preparing field sites, presents examples of field data sheets, and describes procedures in support of acquiring photography. Steps for successful photo acquisition require knowledge of lighting and environmental conditions. Following data acquisition in the field, the raw photographs need to be normalized in preparation

R.A. Fournier (✉)
Département de géomatique appliquée, Centre de Recherche et d'Applications en Télédétection (CARTEL), Université de Sherbrooke, Sherbrooke, QC J1K 2R1, Canada
e-mail: Richard.Fournier@USherbrooke.ca

D. Mailly
Decizio, 929 ave Sir Adolphe-Routhier, Québec G1S 3P4, QC, Canada
e-mail: daniel.mailly.1@ulaval.ca

J.-M.N. Walter
Département de botanique, Université Louis-Pasteur, Strasbourg, France
e-mail: Jean-jmnw@free.fr

I.G.C. Jonckheere
FAO of the United Nations, viale delle Terme di Caracalla, 00153 Rome, Italy
e-mail: inge.jonckheere@fao.org

R.A. Fournier and R.J. Hall (eds.), *Hemispherical Photography in Forest Science: Theory, Methods, Applications*, Managing Forest Ecosystems 28,
DOI 10.1007/978-94-024-1098-3_4

for analysis. This includes removing geometric and radiometric distortions. In the case of analog photography, the digitizing processes are discussed. In addition, defining the outer ring (90° zenith angle) of the viewing perspective for each hemispherical photograph is critical for the analysis. Procedures to define this outer ring are proposed and examples are given. Lastly, the archival of hemispherical photographs as a data set is discussed with emphasis on the description fields included in a suggested metadata.

Keywords Acquisition of hemispherical photographs · Camera · Fisheye lens · Exposure setting · Lighting conditions

4.1 Introduction

The development of an ultra-wide-angle (180° field of view—FOV) or "whole-sky" photographic lens for use in cloud surveys was first reported by Robin Hill, Biochemical Laboratory, University of Cambridge, in 1924 (Hill 1924). In 1951, two Cambridge botanists, G.C. Evans and D.E. Coombe, learned about Hill's ingenious 'fisheye' camera and set about developing the first formal technique to estimate direct sunlight and diffuse skylight transmission through woodland canopies using hemispherical photographs (Evans and Coombe 1959; Coombe and Evans 1960). In these early applications of 'hemispherical' or 'fisheye' canopy photography, solar tracks marking the path of the sun over the course of a day and for specific days of the year were superimposed directly on top of the photograph, so that temporally integrated estimates of direct and diffuse light penetration could be obtained. This work was followed by further applications to estimate the light regime in forest environments by Brown (1962), Brown and Worley (1965), and Magdwick and Brumfield (1969). The method proposed by Anderson (1964a, b, c, 1966a, b, 1970, 1971), through manual analyses, built upon this work to develop most of the theoretical and practical foundations of light regime estimation using HP. In turn, other authors proposed the theoretical basis to assess forest structural attributes like Becker (1971), Bonhomme et al. (1974), Hutchison and Matt (1977), and Nilson and Ross (1979), to name a few. The difficulty of taking direct measurements in tropical forests led to the development of further new analytical techniques to retrieve canopy parameters (e.g., Bonhomme and Chartier 1972; Chazdon and Field 1987). The interest in using HP in fieldwork has continued to expand to this day. This is largely due to ongoing developments in photographic and computer products, which have facilitated the use of HP, but also have brought about significant advances in the specialized HP techniques. Most HP applications fall under three main themes: (1) measuring solar radiation in the forest; (2) measuring the quantity of forest cover components; and (3) providing visual archives.

The development of computer and digital camera technologies facilitate access to low-cost, high-quality photo products, but the implementation of HP depends heavily upon the availability of proven analytical techniques. Furthermore, the decision to use HP is usually based on three criteria: (1) the quality and suitability of the data to be gathered; (2) the cost of equipment and processing; and (3) the ease of data acquisition. There is an extensive body of work on techniques for using and analyzing HP, but most of this information is scattered throughout a number of fairly technical articles. The growing interest in using HP in forest sciences demonstrates the very real need for a practical manual (instead of case studies, of which there are many). Those interested in using HP for forestry applications will instead find syntheses on current techniques and reviews of publications. To date, the most useful syntheses include those of Pearcy (1989), Rich (1989, 1990), Fournier et al. (2003), and Jonckheere et al. (2004).

This chapter covers practical aspects that are rarely described in scientific reviews or articles. Therefore, it aims at providing practical advice for purchasing equipment, guidelines for taking photographs, and selecting appropriate options for data acquisition, but also describes how to transform the acquired images into images that can be archived and processed.

4.2 Equipment

An HP system consists of a fisheye lens, a film-based or digital camera, and possibly other equipment such as a tripod (or self-leveling mount), and light-emitting diodes (LED) to help level the camera-lens system and to identify North on the photograph. This section describes and summarizes the equipment needed to obtain high-quality HP images in the field. It first describes the relevant information regarding camera selection. The many rapid advances in digital camera technology have resulted in a wide range of choices, and have complicated the decisions that must be made relative to the quality of photographs required for the intended application and the funding resources that might be available. A description of the lens follows, a critically important component of the system as it influences image quality, FOV, projection geometry, and angular transparency. Lastly, accessories and other equipment are introduced as components, which would complete a system for acquiring HPs.

4.2.1 Camera

Both traditional film-based and digital cameras now come with a wide range of features, including exposure settings, image resolution, choice of automatic settings, exposure compensation, white balance, amongst others. It is also available in a wide range of build quality and price. Choosing the degree of sophistication of the camera often comes down to the cost and degree of practical functionality that the camera features offer.

4.2.1.1 Film-Based Cameras

Most film-based systems that are used to acquire HP are based on a standard camera body that are also used for regular photography, most often allowing manual settings. For example, the F2 camera body from Nikon Corporation (Tokyo, Japan; 1971–1980) is a standard camera that was used extensively before the use of digital cameras became prevalent. It combined ease of use, simplicity, efficiency and sturdiness. Although the basic features have remained unchanged, top-of-the-line cameras now include a number of features that make it easier to adjust the settings that can completely automate the picture-taking process. However, sound judgement is required in choosing a system for HP, since such high-end features may un-necessarily complicate the acquisition of HP. For example, automatic film advance is not needed in most cases, and adds extra weight. In contrast, a databack might be useful. Databack is a detachable unit that can be fitted in the back of the camera and it is connected to the electronic circuitry to imprint information on the film. While not strictly necessary, a databack is useful because it prints the time of day and date directly onto the film. The permanent visual record that is encoded on the photo makes it easier to orient, classify and store photographs and, in many cases, may prevent identification problems. Other options, such as active-matrix light setting or a remote release, may make setting up the camera easier. The user should be judicious in choosing camera features, however, since often all one really needs is a basic camera with manual controls and a databack.

4.2.1.2 Digital Cameras

The greatest technological advance in photography in recent years is the increasing availability of digital cameras. Nowadays for the large majority of HP users, digital cameras have replaced film photography in most all recent studies that have been published. Their spatial resolution has reached, and even exceeded in some cases, those available from film-based cameras, without having to deal with the film processing steps. When the camera-lens system is chosen carefully, the digital format can be an attractive solution from technical and cost perspectives. The ability to produce a digital image that is almost immediately usable significantly reduces the number of steps that are required for a final HP, resulting in substantial time savings for film development and scanning purposes (Frazer et al. 2001). Relative to analog cameras, digital sensors have the potential for real-time processing. At the same time, they offer the opportunity to preview digital images in the field to determine optimal exposure settings, with further advantages for image processing and data extraction in the field, thereby creating a more streamlined process (Frazer et al. 2001).

A primary concern in the selection of a digital camera system for HP is the dynamic range, which is the range of discrete brightness (intensity) levels an imaging system can discern. Normal photographic film may not provide a dynamic range exceeding a 6-bit depth, or 64 discrete brightness levels (Hinz et al. 2001).

In contrast, most digital cameras on the market usually have a radiometric range of 8 bits per pixel (256 gray levels per color), such as the Nikon Coolpix 950. Newer and more expensive systems offer a CCD matrix easily exceeding 12-bit (4096 gray levels) and even 16-bit (65,536 gray levels) pixels. Having a low dynamic range causes difficulties in distinguishing sunlit leaves from relatively small, underexposed gaps in the canopy. Conversely, a high-dynamic range digital camera has the potential for improving the separability between vegetation elements and sky (Jonckheere et al. 2004, 2005b). Therefore, a high dynamic range offers a significant advantage in viewing subtle radiometric textures.

Significant advancements are anticipated both in radiometric accuracy and in the number of pixels in the CCD matrix covering the active field. For example, immediate viewing of HP when using digital camera is a great advantage, since adjusting the light setting correctly is not only difficult but critical. The potential to separate woody branches and boles from photosynthetically active leaves using digital photography (Chapman 2007) is another improvement with some digital cameras. Digital cameras, therefore, have key advantages over standard cameras, which will no doubt quickly multiply interesting features for HP acquisition in the years to come. This being said, a standard film camera, even one with manual settings, is sufficient for many applications of HP. However, advances in digital cameras, support equipment and image processing has helped to facilitate the use of HP for a large range of applications. Overall, an analysis of environmental conditions, intended applications, and operating requirements is necessary in defining the most suitable system (i.e., the camera-lens combination) for HP acquisition.

4.2.2 Fisheye Lens

A lens with a very wide FOV is required in HP and, as a result, the lens is the most important piece of equipment in an HP package. The most common type that is used is a fisheye lens, with a FOV of 180°. In other words, when pointed upwards, a fisheye lens provides an image of the entire sky hemisphere. A lens that was commonly used in the past was the second generation of the Nikkor 8 mm f/2.8 fisheye (Rich 1989, 1990; Rich et al. 1993; Fournier et al. 1996, 1997) (Fig. 4.1a). Unfortunately, this high-quality lens was discontinued, resulting in the need to consider alternative manufacturers of fisheye lenses, such as Sigma Corporation (Tokyo, Japan), Tamron Co., Limited (Saitama, Japan), and Coastal Optics (JENOPTIK Optical Systems, Inc., Jupiter, FL, USA). An example of these products is shown in Fig. 4.1c. The quality of lenses varies widely. Moreover, the list of available fisheye lenses will likely change with time, assuming that choice and quality also will improve with time. However, it requires a careful search at the time of purchase for the suitable product. For forestry applications, the choice is not always that crucial, since the FOV at very large zenith angles (in general >75°) is almost always obscured by tree foliage or other obstacles such as trunks, woody structure, or shrubs. However, the minimum FOV is also dictated by the software that will ultimately be used to undertake the analysis. For

Fig. 4.1 Examples of key pieces of equipment for HP: **a** 8 mm f/2.8 Nikkor fisheye lens mounted on a Nikkormat manual film camera, **b** Fisheye converter (FC-E9) mounted on a Nikon CoolPix 8800 using an adaptor (UR-E18), **c** Coastal Optic 185° fisheye lenses, and **d** Camera with a self-leveling mount SLM7 from Delta-T

example, the GLA software will only analyze images taken with a circular 180° fisheye (Frazer et al. 1999). Therefore, the user must consider all steps of the process before deciding on equipment needs.

A wide range of fisheye lenses are available, including 8 mm and 15 mm focal length fisheye lenses, and fisheye converter. For example, a 15 mm lens from Canon (Tokyo, Japan) will provide a FOV of 180°, diagonally from corner to corner, 142° horizontally, and 92° vertically, when passing through the center of the photo. The wide range of different lenses that have been used in the literature reflects the diversity of lenses that are available on the market. Full-frame fisheye lenses with a full, 180° circular FOV include the Minolta (Konica Minolta, Inc., Tokyo, Japan) 7.5 mm lens (Barrie et al. 1990; Mailly and Kimmins 1997), Canon 7.5 mm lens (Chan et al. 1986), and Olympus Corporation (Tokyo, Japan) 8 mm lens (Whitmore et al. 1993). The 10.5 mm lens from Nikon Corporation (Tokyo, Japan) that is made for their digital cameras provides the 180° ring but only on the image corners. Other researchers have used fisheye adapters that can be screwed onto a regular (50 mm) lens (Chen et al. 1991; Clearwater and Gould 1995) (e.g. Fig. 4.1b). Anderson (1981) used a Zeiss (Carl Zeiss AG, Jena, Germany)

30 mm lens with a 154° FOV. The 8 mm, f/4 lens from Sigma Corporation, with a 179° FOV has been used by Hale and Edwards (2002) and Jonckheere et al. (2005a, 2006). A FOV of 180° or greater allow having a full-frame fisheye whereas smaller FOV may not allow viewing the whole 180° ring on the image, therefore with some part of that ring out of the imaging frame. Some HP software will only process images with full-frame fisheye. While very important, FOV is not the only consideration in selecting the appropriate lens.

In addition to the FOV, the following criteria are important to selecting a lens for HP: aperture range and optical quality (geometric and radiometric distortions), which are both dependent on manufacturing; and design quality, which in turn will influence purchase price. First, a large aperture (or small f-stop) is important because photographs must often be taken under low light conditions. Second, geometric distortion is particularly relevant to viewing errors, a problem that has been discussed in detail by Herbert (1986, 1987), Clark and Follin (1988), Rich (1990), and Frazer et al. (1997). Geometric distortion problems can be assumed to be more prevalent to less expensive, poorer quality lenses, or for those configurations that employ a lens adaptor. Grossly simplified, the deviation between the image that is obtained and a perfect polar projection must be known in order to reduce errors in positioning and in the surface area on the image that is used in quantitative analysis. Most fisheye lenses employ polar coordinates, but it may be useful to know that other types of projection exist, such as the orthographic or the stereographic projections. Before adopting any other projection system than polar, it is important to know if it can be accommodated by the HP processing software to be employed. Another important consideration is the resolution of the lens system. This is given by the modulation transfer function (MTF), which provides the dimension of the smallest distance that can be resolved. In addition to geometric distortions, radiometric distortions or aberrations are also encountered. The radiometric distortions can include (i) a decrease in illumination from the center of the image to the edges (called vignetting), (ii) decreased transmission of certain wavelengths, and (iii) multiple reflections in the infrared band. Geometric and radiometric distortions are discussed in greater detail later in this chapter (Sects. 4.2.2.2 and 4.2.2.4).

Apart from optical considerations, it should be kept in mind that fisheye lenses are designed to fit the camera body of a specific brand and not necessarily others. In some cases, it might be possible to make the lens physically fit the camera body of another manufacturer, but doing so usually affects image quality. This may result in some blurring or unsuitable results. For example, when the distance between the focal point of the lens and the image plane is not far enough, the full circle of the full-fisheye image can be truncated on the edge, which may limit the FOV, and consequently, some analysis. A practical solution for such a situation would be to add lens rings to increase the distance between the focal point and the image plane.

Selecting a lens with suitable specifications is particularly important if image processing is based on quantitative measurements. The criteria listed below will assist users in determining if a particular lens is satisfactory for the intended application. Particular attention must be paid to the geometric and radiometric accuracy that is required for the user's needs. In addition, lens manufacturing and design quality is very important because forestry applications often result in their use under very challenging working environments. Some of the signs of a well-constructed lens configuration include a high-quality lens that is capable of attaining a wide range of apertures with an f-stop up to f/2.8, a sturdy lens mount which provides greater longevity, and light settings (f-stop, shutter speed) that are easily adjusted. Purchasing a lens with known characteristics is an advantage in obtaining images with good geometric and radiometric quality. Any type of photography involves geometric and radiometric distortions. A description of the geometric factors involved in HP first requires an understanding of hemispherical projections.

4.2.2.1 Geometric Projections

HP links or map 3D points in the hemispherical object region to points on a 2D image plane according to some theoretical projection model (i.e., polar, stereographic, orthographic, etc.) following the geometry illustrated in Fig. 4.2. Two vectors are useful in understanding the geometry of HP: the viewing direction vector (V) and radius vector (R), which goes from the center point to a ring corresponding to a zenith angle between 0° and 90°. V represents the reverse path that the light ray travels to the point on the photograph. V_θ and V_ϕ, therefore, represent the zenith angle (θ) and azimuth angle (ϕ) of the viewing direction to the point on the photograph. All the points on the photograph with the same V_θ form a circle with radius R. V_θ ranges from 0° in the center to $V_{\theta max}$ = 90° at the edges of the photograph (at R_{max}). The distances between the circles for V_θ at regular intervals of θ characterize the projection. The vector R represents the distance between the central point of the photograph, which corresponds to the projection of the vector towards the zenith (V_θ = 0°), and the ring on the image that corresponds to a constant viewing angle. The projection provides the mathematical relationship between R and V for values of θ between 0° and 90°. The most common projection used in HP is called the equiangular, equipolar, polar or Hill projection, depending on the author (e.g., Herbert 1987; Mitchell and Whitmore 1993). In this projection, which we will call an equiangular projection, the increase in length ΔR is linearly proportional to ΔV_θ and corresponds to the functions:

$$R/R_{max} = V_q/90 \quad \text{or} \quad R = C_1 V_\theta \tag{4.1}$$

where the coefficient C_1 is specific for each lens. Equiangular projections are popular since the linear increase of R with V_θ simplifies the analysis. Frequently, if the lens projection and geometrical distortion are unknown, the equiangular projection is assumed (which is the projection of a perfect lens). However, this assumption may lead to significant errors since, in practice, all lenses introduce geometric distortions (Frazer et al. 1997, Sect. 2.1), with the surface area subtended by the hemispherical projection differing from that subtended in the actual forest canopy. Since the

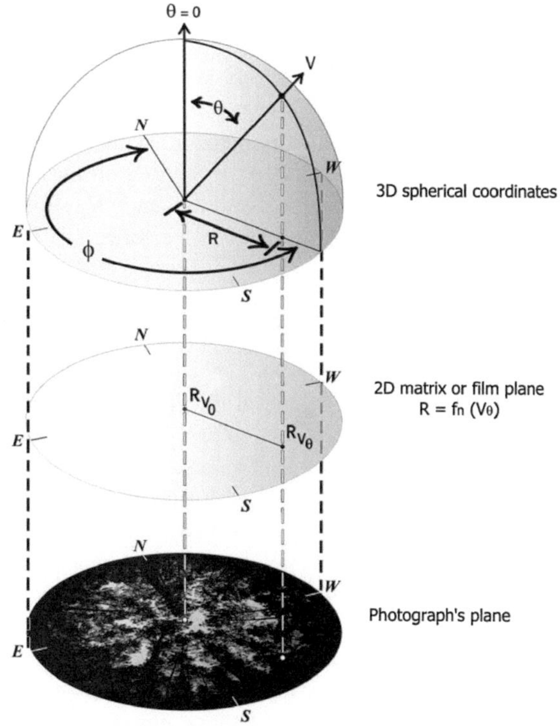

Fig. 4.2 Geometry in spherical coordinate associated with generation of the hemispherical image on a film or on a CCD matrix of a camera. The viewing vector V corresponds to a position on the 2D plane of the film or matrix at position R. Therefore all distance on the image can be associated with its corresponding vector V associated with an azimuth and zenith angle of the 3D spherical coordinates (this figure is adapted from Rich 1989)

distance between the camera and the crowns of the trees is smaller at lower zenith angles, the area subtended in the forest cover increases with the zenith angle. This aspect is sometimes neglected in analyses of HP. Some lenses have an equal area projection; however, converting from an equal area projection to an equiangular projection is complicated. This aspect is discussed in greater detail in the work of Clark and Follin (1988), Nomura et al. (1992), Mitchell and Whitmore (1993), and Frazer et al. (1997). Despite its limitations, the equiangular projection is by far the most commonly used since it allows the simple calculation of viewing angles, which is important in producing sun tracks or locating openings in the forest cover.

4.2.2.2 Geometrical Distortions Introduced by the Lens

An equation can usually be used to compensate for the geometric distortion caused by the lens in relation to an equiangular projection, usually a polynomial of the form R = fn (V_θ). To illustrate how this polynomial is calculated, let us imagine a situation in which a camera with a fisheye lens is installed horizontally to point directly upward to the zenith. For two objects at infinity (treated as point sources), the first of which (P_1) is placed directly in the central optical axis running through the zenith and the second (P_2), at varying zenith angles ranging between 0° and 90°. R corresponds to the distance between P_1 and P_2 on the film. In the lab, one can measure the exact relation between R and the angle V_θ separating points P_1 and P_2 to express it as a polynomial regression in the form:

$$\mathrm{R} = \sum_{i=1}^{n} a V_\theta^n \tag{4.2}$$

where n is the order of the polynomial fit. Fitting the polynomial to the lens correction function forces the polynomial to pass through zero and the maximum angle of the lens (Jonckheere et al. 2005b). As an example, the more recent (yet discontinued) Nikkor 8 mm fisheye lens has a maximum R_{max} value of 11.5 mm at $V_\theta = 90°$. The coefficients calculated by Nikon and Herbert are given in Table 4.1. The maximum angular distortions measured were around 3.05° at $V_\theta = 54°$ (Herbert 1986, 1987). Although the adjustment was quite small, according to Frazer et al. (1997), neglecting these aberrations may result in errors of up to 9% of the projected area, 19% of the sky proportion and 7% of LAI values. Such a polynomial is generally calculated from data provided by the fisheye constructor, or the outcome of a particular in-house calibration of the fisheye lens (e.g. Clark and Follin 1988; Inoue et al. 2004). In addition, the correction factors of some lenses are already available in some specialized HP software. Generally, a polynomial of degree 3 or 4 gives good approximation for any particular fisheye lens. It must not be forgotten that some fisheye lenses have a FOV beyond 180° ($V_\theta > 90°$), this in practice, creates difficulty in discerning the end of the image ($V_\theta = 90°$) and sometimes greater errors can be introduced than those caused by geometric distortion.

Table 4.1 Regression coefficients to compute viewing angle V_θ of the Nikkor 8 mm fisheye lens and the radial displacement (R) from the center point of the hemispherical photograph

Coeff.	Nikkor 8 mm lens (R_{max} = 1 at $V_\theta = 90°$) (Ref. Rich 1989)	Nikkor 8 mm lens (R_{max} = 11.5 mm at $V_\theta = 90°$) (Ref. Nikon 1993)	FC-E8 (Nikon) (Ref. Inoue 2004)	Sigma 8 mm f/4 Fisheye (Sigma) (Ref. Jonckheere et al. 2005a, b)
C_0	1.256138×10^{-3}	0	0	−0.0009
C_1	1.090603	0.1392	1.483×10^{-1}	0.1367
C_2	3.432221×10^{-4}	5.011×10^{-5}	2.067×10^{-5}	-1.545×10^{-5}
C_3	-4.067174×10^{-5}	-3.733×10^{-6}	-9.367×10^{-7}	-1.549×10^{-6}
C_4	9.732894×10^{-7}	3.733×10^{-8}	1.969×10^{-8}	–
C_5	-1.745721×10^{-8}	-0.1771×10^{-9}	-2.185×10^{-10}	–
C_6	1.666496×10^{-10}	–	1.484×10^{-12}	–
C_7	$-6.403817 \times 10^{-13}$	–	-6.0505×10^{-15}	–

Overall, geometrical distortion of a lens can be corrected if the coefficient of the correction function are known and can be applied. Many of the HP software allow using predefined or deriving correction functions. Some software (like GLA) use a binary search and linear interpolation algorithm to implement lens specific projections assuming a minimum of 20 calibration points (Frazer et al. 1999). With all these options, correcting lens distortion depends primarily on having a function already available for that lens or having calibration points, and using software that will apply the proper interpolation calculations. If the HP software does not allow such calculations or if no suitable data or functions are available, the user must resort to using the equiangular or polar projection by default since most fisheye lenses use that projection. If another projection than polar is used, there are several programs that will transform from one projection to another which facilitates image transformation between projections.

4.2.2.3 Resolution of Optical System

The resolution of the optical system determines the level of detail that can be perceived in the photograph. For forestry applications, this means the ability to distinguish the fine texture of the foliage and other components of the forest cover such as branches from trunks. It is characterized by the Modulation Transfer Function (MTF), which is measured in the laboratory and which describes how faithfully the optical system reproduces (or transfers) detail from the forest cover to the image that is produced by the lens. This subject is thoroughly discussed in optics manuals, particularly in those dealing with optical instrumentation (e.g., Ray 2002; Kasunic 2011). Fundamentally, if MTF is not known, then a calibration image containing lines of various thickness and spacing can be used to determine the spatial resolution of the system. Although this information is useful, however, it is often ignored by users.

Another aspect related to the resolution of an optical system, combining the specifications of the lens and camera, is the varying footprint at different distances from the focal point of the image. For fisheye, the footprint of a pixel of a hemispherical photograph is directly proportional with the distance between the point in the canopy and the camera. However, considering that the path through the canopy can double or triple with the viewing angle, the footprint within the canopy for each pixel may vary considerably. Footprint radius can easily vary 10 times from lower branches just above the camera with the top of the canopy at a viewing angle of 45°. This effect, although always present, may affect to a lesser degree, the results if the footprint resolution is much smaller than canopy object (leaf, branches, stem) dimension. Digital cameras are continuously increasing their number of pixels, therefore decreasing the footprint size, and consequently reducing the impact of this problem.

4.2.2.4 Radiometric Distortions

Radiometric distortions are less well-documented than lens-generated geometric distortions in the literature on HP, despite the fact that they are well known in optical engineering. Since radiometric data derived from images is often used in quantitative applications, here we focus on three types of potential radiometric distortions in lenses: (1) the decrease in sky brightness from the center to the edges of the photograph (vignetting); (2) the difference in transmission for different portions of the spectrum; and (3) multiple reflections.

The first problem, vignetting, refers to the fact that in a lens illuminated by an isotropic (uniform) light source, the lighting on the edges of the image will be of lower intensity (from around 50–60%) compared to the center of the image. This reduction in light transmission, which is caused by the lens, is combined with an accompanying decrease in sky brightness on the photograph as the viewing angle goes from (center) to (edges of the image). The second problem is tied to the fact that transmission capacities of the lens' optical elements are wavelength dependent; in general, better transmission is obtained in the blue and green range of the spectrum, but this gradually decreases towards the red and near infrared (NIR) end. The third problem frequently encountered is that of multiple reflections in the NIR (>800 nm) and ultraviolet (360–400 nm) ranges. These multiple reflections create ring-shaped areas of variable (and unpredictable) light intensity, due to the lack of transparency of the multicoating on the lens' internal components at that particular wavelength. For applications involving images in these ranges, we strongly recommend that you verify the resulting images prior to field work. To do this, take a photograph of a uniform white surface under diffused light and check the uniformity of the resulting image.

There are several other documented problems related to radiometric distortions that may occur. For example, images from the CCD of a digital camera can be altered by electron migration or blooming, color interpolation errors or CCD anomalies (e.g., striping). Therefore, it is best to test new equipment thoroughly just after purchase to avoid encountering manufacturing flaws during field data collection. These problems are documented in technical engineering documents on the CCD, and they can generally only be resolved by the camera manufacturer.

4.2.3 Other Photographic Equipment

While the lens and camera are fundamental components of an HP system, other data acquisition equipment are also necessary. These components can be divided into two groups that entail (1) the preparations for, and (2) the actual taking of, hemispherical photographs (Table 4.2). The use of this equipment is discussed in detail in Sect. 4.3.5 on taking the photographs. The support for the camera and lens is a particularly important accessory, because it provides the means for orienting the optical axis of the camera system towards the zenith. For this purpose, some

Table 4.2 Equipment and supplies required for HP acquisition in the forest

Equipment and supplies required in preparation for taking hemispherical photographs
Cleaning kit for camera and lens
Extra battery pack (for camera, display and lamps)
Padded carrying case for all equipment
Umbrella and/or impermeable sheet (for protection in case of sudden rain)
Indelible pen/markers
Duct tape
Flagging tape in several different colors
Allen key (if required for self-leveling mount)
Repair tools and multipurpose pocket knife, multitool, etc.
Tape measure and/or other measuring device
Flags with metal rods
Pre-cut wooden stakes and sledgehammer
Equipment and supplies required for taking photographs
Lenses for HP (8, 16 mm, …)
Standard lens: 35 or 50 mm
Filters (if applicable)
Spare batteries and several charged batteries (for digital camera)
Camera (a backup camera, if possible)
Databack (for film camera)
Compass and clinometer
Measuring tape
GPS
Remote control for delayed acquisition (if not available on camera)
Self-leveling mount for camera and lens or
Tripod or monopod with a level (e.g., bubble level)
Waterproof log and field sheets for entering data
Pencils
Accurate watch
List of procedures for preparation and taking photographs
A small backpack
Solar angle chart (including time of sunrise and sunset)
Small flashlight or headlamp
Extra digital memory card, or film (if film camera is used)
Portable computer for immediate data transfer
A two-direction bubble-level (attachable to the camera frame)

manufacturers (e.g., Delta-T Devices, Burwell, Cambridge, UK) offer a self-leveling mount on a telescopic pole similar to that originally used by Rich (1990) (Fig. 4.2d). This is a practical idea since it allows quick settings, adaptable photo height, and greater mobility than a conventional tripod. A practical alternative to the self-leveling mount is a sturdy monopod, with two spirit-bubble levels (see Fig. 4.2b). These levels are widely available and can be affixed to the top of the camera frame. Another alternative is the use of a tripod that is the only sure way to provide a stable, level, and motion-free support for a camera. Considering that most acquisition procedures use remote or delay camera release and are often taken at a slow shutter speed, a stable camera platform is necessary to avoid blurry photographs. With a tripod, however, more time is required to adjust the viewing angle for each photograph than it would from a self-leveling mount or a monopod. Selecting a camera support device is thus a compromise between operational requirements, instrument manipulation carefulness at the time of acquisition and the resulting quality of the photographs.

When exact image orientation is required on the photographs, this imposes a new set of procedures and equipment. Most applications designed to assess forest structure (e.g., gap fraction and LAI), do not need to be spatially explicit or orientation specific. Conversely, most applications designed to estimate the light regime in the canopy, need to tie each photograph to a specific orientation. With the ability to identify magnetic or true North, solar tracks can be precisely positioned on the hemispherical photographs to allow estimation of variables such as light penetration and sunfleck duration. For these applications, it is useful for field operations to have a system of lights to orient the images with respect to a geographic reference, generally true North. Pearcy (1989) suggested using a field assistant standing in the North direction of the azimuth, at a certain distance from the photographer, and shine a flashlight toward the camera. LEDs that are installed on the top of the self-leveling mount have been used to help identify the location of North (Rich 1990). These positioning devices not only provide the azimuth direction of the reference point, but they also help to determine the edge of the image where the viewing angle is 90° ($V_\theta = 90°$). Some systems are sold with two LEDs that are installed symmetrically on either side of the lens, which may make it difficult to determine which diode shows the proper azimuth. To avoid confusion, the operator should ensure that there are an odd number of LEDs to identify a direction. When no equipment is used to define a chosen direction on the HP, it is conveniently assumed that the bottom of the camera is placed towards the magnetic or true North. The operator should always consider this important aspect in the acquisition procedure. One benefit of marking the photographs is to broaden the choice of analysis, now or in the future, which is an asset for retrospective studies.

Although the basic system components are crucial, additional accessory equipment is also important for facilitating the taking of photographs for subsequent image analysis. The most complex and expensive elements in HP systems are the (i) camera, (ii) lens and (iii) self-leveling mount, tripod or monopod. Additional items necessary to complete a HP acquisition system are listed in Table 4.2.

4.3 Canopy Photographs

Acquiring HP of the forest canopy requires paying attention to specific steps in the preparation process. This chapter attempted to regroup the important considerations for a successful field campaign using HP. This first involves establishing the survey plots and dealing with two important aspects precluding any HP acquisition: selecting suitable sky conditions and having a good strategy for the exposure settings. These procedures encompass assembling the necessary equipment, and applying a set of procedures for photo collection. Despite the differences from experiment to experiment, there are some basic principles that apply to all HP acquisition projects. The keys to successful data gathering are high-quality equipment, careful preparation, and attention to detail during the actual operations.

4.3.1 Establishment of Survey Plots

Having inventory measurements at plot locations results in data that may be associated with attributes derived from HP. This activity is somewhat similar to traditional forest inventories or ecological surveys (e.g., Avery and Burkhart 2001).

Establishing a survey plot requires taking into account the measurement strategy and the efficiency of field operations. First of all, the plots that are chosen must be as representative as possible. Deciding upon the location, size and the number of study plots requires a thorough knowledge of the ecological components under study as well as statistical sampling principles, subjects that go beyond the scope of this chapter. Here, readers are reminded that choices must be made carefully and justifiable from a specialist's point-of-view.

Timing of photo acquisition must take into account plant phenology, constraints on site access, weather conditions, and the availability of personnel. In hardwood forests, for example, adequately monitoring plant phenology requires more frequent photography sessions during the spring (from before budbreak until the termination of leaf elongation), a few sessions in the summer (which is a stable period), and one or two sessions during autumnal leaf senescence. In comparison, the phenology of softwoods is less variable, and requires less follow-up in terms of the number of site visits.

The size, shape and location of the plots must reflect the physical structure of the components under study. Plots may be circular or rectangular, depending upon the positioning of the photographic points. Attention must be given to avoid significant anomalies near the plot that can alter the results. For instance, plots should be located far enough from a road or any man-made alteration to avoid accounting for artificial openings in the analysis. Although a small plot will suffice if there is just a single point in the center, it is always better to establish plots that are sufficiently large to contain several photographic points. In addition, a greater number of photographs must often be taken to compensate for forest structure, the spatial distribution of which may be variable. In the case of multiple photo points, three strategies are often employed with respect to arranging points: along a transect or line, in a grid, or in a star shape. When inventory data are crucial for the analysis, studies can often benefit from the establishment of plots that are large enough to represent the variability observed in HP. In this case, we would propose a simple rule: in boreal or temperate forest, the boundaries of the plot must include trees with crowns that can be seen from a 45° zenith angle in the image. For example, Fournier et al. (1997) used rectangular (50 m × 60 m) plots, with the points located 20 m from the edge, to model the spatial distribution of the foliage; a plot of 160 m × 160 m was simulated, based on data gathered in the original plot to produce a complete model of the hemispherical images. Determining the location, number and size of survey plots, therefore, requires good judgment to balance field data and image analysis requirements. Fortunately, HP is often used to complete an existing study. In this case, the user just needs to ensure that criteria for the establishment of plots and field measurements correspond to image analysis requirements.

Several studies have investigated the number of hemispherical measurements that are needed to represent the natural variability of a forest stand. In a study on canopy structure, Nackaerts et al. (2000) concluded after applying Monte Carlo simulation techniques that around 16 hemispherical measurements were needed to represent natural forest stand variability within a 1 ha area. Some field protocols (e.g. Fernandes et al. 2014) recommended ~8–10 HPs. Leblanc (2014) and Leblanc and Fournier (2014) also addressed this question by using a modeling framework. They suggested that 8 HPs may be sufficient for relatively homogeneous sites, but this number should increase for heterogeneous sites with 16 HP being sufficient in most cases, except for very heterogeneous canopies.

Lastly, by following certain rules for establishing survey plots, errors will be minimized, ensuring successful follow-up of the experiment. In the case of permanent plots, all HP points must be well-known, identified, and documented. Permanent plots require permanent markers that can withstand the elements for a number of years; plastic (PVC) or wooden markers that are painted in bright colors and labeled in indelible ink are almost essential. Furthermore, we suggest that a distance-based system be used for labeling the markers to avoid a double system of co-ordinates (numbered in meters or with cardinal points). Reference to at least one point with a known location, which is determined with a GPS, is often required. In this way, the site map that is produced will correspond directly to the position of the HPs, eliminating possible errors. In addition, the points in the plot can be identified with brightly colored flagging tape, which aids in the photographic process since photographs are often taken under difficult conditions. In analyzing the photographs, the operator may have to deal with patterns of spatial texture that are very difficult to interpret. Consequently, having plots and field data with characteristics that anticipate requirements of the analysis would permit more efficient picture talking, and allow the user to extract the maximum information from HP.

4.3.2 Lighting Conditions

The ideal condition for HP is diffuse irradiance (skylight) that provides contrasting lighting to the foliage in the forest cover. These conditions are encountered when the entire sky provides an approximate isotropic (uniform) irradiance field. When the solar disc is directly visible, two main anomalies may prevent the user from getting a good contrast between the foliage and the sky. First, the extreme brightness on and around the solar corolla results in foliage obstructions being under-evaluated; this part of the image is overexposed so that you cannot distinguish the exact texture of the foliage or other elements that obstruct the skylight. Secondly, when the solar disc is in the sky, it causes direct reflections on part of the foliage, which often makes it difficult to distinguish between the sky and the reflections in the photograph. To prevent these two problems, hemispherical photographs are usually taken under two types of lighting conditions: (1) a uniformly (or pseudo-uniformly) overcast sky or (2) when the solar disc is just below the horizon and there is a clear, cloudless sky.

A uniformly overcast sky provides the best conditions for taking photos, usually providing sufficient irradiance so that a contrasting filter (such as red) can be used. In addition, the excellent light conditions mean that a finer grain film (e.g., ISO 100) can be used, which increases the inherent resolution of the textures of the obstructions. Furthermore, with an overcast sky, pictures can be taken at any time of the day (instead of just at dawn or dusk), for extended periods. The lighting does not have to be completely uniform and a certain amount of variability is acceptable as long as the sun does not break through the clouds and there are no significant variations in lighting (dark or light portions) or large openings in the clouds.

Appropriate conditions for taking photos under a clear, cloudless sky are (1) neither a portion of the foliage at the treetops nor the entire treetops are illuminated by the solar disc at dawn or dusk and (2) there is enough light in the entire sky to provide a good contrast between the light-colored sky and the dark foliage. At dawn or dusk, the light is generally poor and requires light-colored filters and sensitive (and therefore larger grained) films. In addition, at dawn or dusk, the period of time in which lighting conditions are acceptable is very limited, around 45 min, although this depends on the latitude, relief and the presence of clouds on the horizon. For dawn or dusk photo sessions, excellent planning is required to deal with the small window of opportunity for taking the HP and the time of the day in which it has to be done.

Lighting conditions are often a significant limiting factor when planning photography sessions. The number of days with favorable conditions is often a problem for certain areas in the middle latitudes. Good lighting conditions are crucial for high-quality hemispherical photographs. Only with experience is it possible to develop reliable judgement as to whether lighting conditions are adequate.

4.3.3 Exposure Setting

The tonal range of an HP is critical to produce images suitable for analysis. The tonal range is defined by the range of the gray values on the histogram of an HP, for which extremes represent the deep shadow associated with the minimum values and the bright sky pixels for the maximum values. The tonal range of an HP image should not be confused with the dynamic range. The tonal range is image dependent and the dynamic range is system dependent. Tonal range varies for each of the RGB channels of an image. An image offering its best potential for analysis maximizes the tonal range without having a significant number of pixels at the saturation value (the maximum value). A good tonal range is important when using systems with 8-bit per channel and less critical with systems offering higher dynamic range. The importance of tonal range on estimates of gap fraction has been reported by several authors (Chen et al. 1991; Wagner 1998; Mcfarlane et al. 2000). This raises the importance of a suitable procedure to determine the correct exposure (aperture-speed setup) in order to maximize the tonal range on the output image. An overexposed photograph will result in the overestimation of the actual gap fraction that cannot be corrected with any subsequent image processing. Conversely, a photograph that is underestimated will restrict analysis options due to its low dynamic range in radiometric values.

The exposure made by the camera operator will define the tonal range of the resulting image. The lens' shutter aperture (referred to as the f-stop), and camera speed in fractions of a second (e.g., 1/250 s) will determine image exposure. It is important to keep in mind that selecting a f-stop (inversely proportional to aperture) impacts the "depth of field"; i.e., a large f-stop (small aperture e.g., f/8) will keep all canopy elements, near and far, in focus in HP applications. Consequently, an operator should favor larger f-stop to make sure it has a large depth of field resulting on having clear (on focus) canopy elements. Conversely, selecting a small f-stop (e.g., f/1.4) may limit the depth of field and consequently render parts of the canopy off focus on the image. Aside from addressing the f-stop value, the process for determining the correct exposure first depends upon metering available light to suggest a potential setting. Built-in camera light meters measure the luminance of the subject that is being photographed and the camera calculates 'automatic' exposure settings, assuming that the light comes from a mid-gray surface (18% visible reflectivity) by converting it to photographic exposure. The extent to which the photographs should be overexposed depends upon the relative contribution of the sky and the canopy to the solid angle of the hemisphere and on the internal light meter of the camera. The pixels of the sky need to be towards the highest values of the dynamic range ($\sim$255 for a 8 bit system) to allow a more accurate thresholding to produce a binary image (0: sky and 1: canopy). Selection of a suitable exposure setting has been dealt with in detail by Zhang et al. (2005). Exposure setup implies first metering the light intensity from either outside or below the canopy by different operators (e.g., Canham et al. 1990). Chen et al. (1991) and Zhang et al. (2005) have suggested overexposing the sky pixels by two aperture stops, compared with the light measurement that is taken outside of the canopy. This approach may not always be convenient, as an operator has to deal with extended stands and quickly changing light conditions. As a result, Rich (1990), Fournier et al. (1997), and Frazer et al. (2001) suggest using the light meter under the canopy but taking photographs through several gradations of overexposure (e.g. 2 and 4 f-stops). Fortunately, with the advancement of digital cameras, there are new adaptable light metering programs, but more importantly, the operator is able to assess in situ the tonal range of the photographs. Proper settings, therefore, can be taken with confidence and reassessed as light intensity changes.

4.3.4 *Assembling the Necessary Equipment and Data Sheets*

4.3.4.1 Preparations for Taking Photographs

Two separate sets of recommended procedures are required in preparing to take photographs: one for the procedures to be followed prior to going to the field, and a second to be implemented just prior to taking the photographs (Table 4.3). In general, the purpose of the pre-fieldwork procedures is to ensure that all of the equipment that is required is available and in good working order. It should be checked before going into the field or prepared the day before, especially when the user is taking HP at dawn. It also represents a chance to collect any spare equipment or supplies that the user will need for contingencies.

The procedures to be followed just before taking measurements have been designed with ergonomics in mind, using the minimum amount of equipment. This is why it is best taking a small backpack into the field, containing only what is absolutely needed, so that the user is unencumbered and free to move around when taking photographs. These are the last preparations that should be taken before actually taking the actual photographs and can often be done in the vehicle at the edge of the site. One of the most important elements in these preparations is choosing the appropriate camera light settings (or film, in the case of analog photography).

For analog photography, black and white (B&W) film is most popular, however, color film provides more flexibility in image processing, and is therefore recommended. The choice of film speed is always a compromise: using a high-speed film (such as ISO 400) is preferable under low light conditions, but results in greater image graininess than when a lower film speed (for example, ISO 100) is used. Choosing between standard and professional film is also a compromise between the higher development quality of professional film, and the ease of purchasing and developing the lower cost of standard film. Standard film is often used because it is a good compromise in most situations. When loading the film, most cameras will adjust their film speed setting automatically. However for the basic or old cameras frames the user should always check the film speed indicator and correct it, if necessary. This step is easy to forget and may result in a significant and sometimes irrecoverable error in the light setting.

Most projects now employ digital cameras since they now provide images with spatial resolutions approaching those of film emulsions (Hale and Edwards 2002), and they also benefit from many other advantages, such as increased dynamic range, in situ visualization of images and histograms, and direct access to digital images. The quality of the selected camera may depend upon the project. However, there may places and circumstances more suited to the use of a film-based camera.

When using a digital camera, the possibilities for adjusting the camera settings greatly depend upon the different camera types. Normally, the ISO range of a good camera allows the set-up to be made between at least ISO 100 and ISO 800. For more sophisticated cameras such as the Nikon D500, for example, the possible

Table 4.3 Procedures for preparing to take HP: (**a**) before going out in the field and (**b**) before taking the actual photographs

(a) Pre-fieldwork procedures
Ensure that all the equipment is working properly
Clean the lens with lens paper (make sure not to leave streaks)
Prepare the self-leveling mount or tripod so it is ready to receive the camera and lens
Make sure you bring enough digital memory cards (or films in case of analog photography)
For analog photography: Label the film canisters and cassettes with indelible ink
Set your compass to true North or magnetic North according to the reference system of lights (using true North is recommended)
Check your equipment list to make sure you have everything
Put in a backpack everything you will need when photographing (camera, lens, flashlight, spare memory card (or film), spare batteries, repair tools, knife and bug spray)
(b) Pre-measurement procedures
Set the camera settings (or load the film) best suited to the lighting and other conditions
In case of a camera with film Make sure the film speed setting on the camera corresponds to the speed of the film actually inside the camera (e.g., 400 ASA) Write the film (roll) number on the data sheet and on the film canister Ensure that the databack shows the correct information, generally either the photo number and time or the date and time Set the film speed on the databack so that it corresponds to the actual film speed (L, M, H) Make sure the film properly loaded so that it unwinds correctly (to check, watch the circular movement of the left spool while you are advancing the film)
In case of a digital camera Make sure the memory card is properly inserted with sufficient available memory remaining Verify if the date, time, file format and size, as well as all imaging options are properly set (e.g. when available the option of flashing saturated pixels is particularly useful for quick camera set-up during field operation)
Make a final check of the self-leveling support for the camera
Turn on the LEDs on the lens mount or prepare the reference system of lights and write the direction of the azimuth used (magnetic or true North) on the data sheet
Set the focus on the lens to infinity and secure it with tape
Make sure the filter on the lens is the one you want to use (transparent, red, etc.) and record it on the data sheet

(continued)

Table 4.3 (continued)

Complete the top part of the data sheet in order to prepare yourself to take photographs. This involves filling out the fields for the date, site, weather conditions, remarks and other relevant information
Make sure that the lighting conditions are adequate

range may go from ISO 100 to ISO 51,200, thereby allowing for use under subtle light environments gradients. Once these preparations have been completed, remains to wait for the right lighting conditions to proceed with acquiring HP.

The use of a lens filter is another option to consider. Unfortunately this option is not available on all camera-lens systems. An example of a filter is the deep-blue KB 12 (80B) 77 mm coated filter [B&W Color Cast-Correction Filters (CC Filters), Schneider, Kreuznach, Germany], which solely transmits the blue and absorbs the green and red components. It can be placed in front of the hemispherical lens to convert images from color to monochrome (gray scale). The filter reduces the amount of light through the system, which requires adjustment of the exposure setting to compensate for the light that is absorbed by the filter. It must be noted that newer, more expensive digital cameras have separate CCD sensors for the three color channels (Red, Green, Blue—RGB), which reduces chromatic aberration (i.e., light bending differently in different wavelengths due to the imperfection of the lens systems), when compared to older digital models. Separate sensors also eliminate the need for color interpolation achieved by the digital cameras firmware, which induce spectral anomalies. Therefore separate RGB channels have the advantage of making use of spectral combinations of techniques during analysis (e.g. Fernandes et al. 2003; Jonckheere et al. 2005a, b). As a result, filters are used only if multi-channel processing is not an option.

4.3.4.2 Field Data Sheets

The field data sheet is the link between the photographs that are taken and field information. The usefulness of a data sheet as a work tool will depend upon its clarity, conciseness, and accuracy. Given the nature of fieldwork, a compact sheet that can be inserted into a field notebook is preferable to a letter-size (or A4) sheet. Often, there is only a very small window of opportunity in terms of the time when light conditions for a photograph are appropriate. Maximizing the ergonomics of data entry is therefore essential, particularly if only a single person is in the field both to take photographs and to record the data. The use of a small voice-activated tape recorder can be an alternative to writing notes on a data sheet in cases of a one person operation when the time period for data collection is limited. Similarly, the voice-recording capability of some digital cameras could be used, while taking the photographs. If the tape recorder were to malfunction or your voice were inaudible for whatever reason, it is unlikely that you would be able to link the photographs with the associated data. As a result, it is strongly recommended that field notes be taken in written form in hard-copy or entered directly on a portable field computer or data logger. The shortest time possible between the measurements and data recording would ensure greater rigor and accuracy in the follow-up. That is why a two person configuration is best to have one person concentrating of taking good quality photographs and the other person dedicated to filling the data sheets rigorously and double checking the data collection procedures.

In the data sheet that we have used for our research, we had to be careful about entering certain information. Some of the fields were potentially confusing, such as the determination of the azimuth (magnetic or true North), the time that the photograph was taken (local mean time or daylight savings time), and the date format (DD/MM/YY or MM/DD/YY). Other elements that are critical for the analysis include the study site or plot number, and the type of filter, shutter speed and light aperture that is used for all photographs. In addition, any comments that could facilitate the analysis should be recorded, such as the light conditions, the site layout, topography, under-growth conditions or problems that were encountered when taking photographs (for example, clouds appearing in a visible portion of the sky are important when processing the image). One of the most important elements of information to be recorded is the paired photograph number and position. For film photographs, it is important to note if the photo number recorded corresponds with what is printed on the photograph, or to the order in which the photograph was developed (i.e., photograph #1 corresponds with the first photograph to be developed, and so on). The system of reference that is used for the plots must also be readily available when establishing a link between the photograph and its location in the plot.

4.3.5 Taking Photographs

The quality of data will depend upon, among other things, how carefully the data acquisition procedures are followed. This is especially critical in performing HP, since a poor-quality photograph will limit the user's ability to discern desired features during analysis. Taking photographs is a matter of following the same sequence of steps in every session. Due to the repetitive nature of the work, a step is frequently omitted, which may compromise the quality of the data. Some procedures may not be necessary but they are designed to avert problems. This subsection describes the procedures that should be followed before and during the process of taking hemispherical photographs.

A suggested set of procedures for taking HP is provided in Table 4.4, where operations are given in a sequence to favor complete archiving of all the information that is required, obtaining quality photographs, and minimizing operator time. Although all of the steps are important, determining the light setting by adjusting the aperture opening and shutter speed is among the most critical in obtaining high-quality photographs. After setting the exposure, the operator takes several photographs at the same setting until the light intensity changes. At every

Table 4.4 Procedures for taking HP in the forest

Procedures for Taking Photographs
Determine whether or not a filter is to be used (record this information on the data sheet)
Set the exposure by adjusting the shutter speed and aperture opening (this operation should be done at the outset and every ~ 15 min). If the sky is overcast, with no significant opening in the clouds, the setting is done on the largest gap in the forest cover. Do not forget that the light meter on the camera is zone-sensitive (the reading is often taken from the central portion of the image). See the note on setting the exposure. In case a comparison is made between different study sites: take a picture of a Kodak gray card (Kodak Inc., USA) in order to set the white balance of your camera. As such, all different pictures should be taken with the same standard light setting, so inter-comparisons are possible
Position the camera on the point that has been identified for taking the photograph
Position it 1.3 m above the ground surface and a minimum of 2 m away from any large tree (record tree diameter at breast height of the surrounding trees in the field notebook). Branches and other foliage within 1 m of the lens must be pushed away before taking the photographs as they may create disproportioned obstruction
Determine the magnetic North and the azimuth of the mount
Set the self-timer (>5 s) to allow the self-leveling support to become stabilized
For an analog camera: advance the film
Take the photograph, making sure that no individual or object in the site is present in the photograph (i.e., above the horizontal line at the lens level)
Record the information about the photograph on the data sheet, including the time, f-stop, shutter speed, photograph number, location and height
When photographs are taken on a station with significant slope, the maximum slope and its direction should be recorded
For the next photograph, begin at step three of the procedure (positioning of the camera)

station, the bottom of the camera must be set in the direction of either to true geographical North or to magnetic North. One or the other is a matter of preference but it must be clearly noted. A self-leveling mount equipped with LEDs for printing marks on the image facilitates image processing. At every image acquisition, we suggest using the timer to delay image acquisition to a few seconds to give the operator a chance to ensure the camera is steady, immobile and well leveled, and to give enough time to position him/her-self below the lens level. Lastly, make sure you take the time to note the information correctly on the data sheet. This is particularly important in linking the photograph number with the correct location. If two people are present, one should be in charge of recording the data carefully on the data sheet, as well as helping the operator with the camera to move quickly from one marker to the next. Since the window of opportunity for photographs is usually short, especially at dawn or dusk, and the work particularly intense, errors are easy to make. Using the same set of procedures regardless of field and exposure conditions will speed the picture taking process and ensure consistent information is recorded on the data sheet.

4.4 Photos Digitization and Image Archive

Archiving HP entails transforming the photographs from their original format into a normalized format ready for future use. For photographs taken with conventional cameras, this involves digitizing the film product so that computerized analysis methods can be used. If required, the digitized image must also be corrected for geometric and radiometric distortions. The resulting product is a digital image file that can be archived for future processing.

4.4.1 Digitizing Photographs

Interpreting an HP requires an image in digital form, while conventional cameras produce images on photographic film, which consists of celluloid coated with various emulsions. Film images come in three formats (negatives, slides and paper) and all three can be digitized by using a scanner or video system.

Scanners often come with various options, but four are particularly important for the quality of the resulting digital image produced from the original photograph: (1) the ability to adjust the background brightness, (2) the ability to select the pixel size or resolution of the digitized image, (3) the ability to select a range of gray levels (dynamic range), and (4) the ability to adjust the contrast, brightness and light intensities. For negatives and slides, the decision whether to use a normal or bright background depends on how much the dark details need to be enhanced. A bright background tends to bring out subtle textures but will also result in the saturation of bright or large open areas in the canopy. Therefore, a compromise must be reached between overexposing parts of the image and obtaining the required level of detail. In terms of the resolution, this will depend on the pixel resolution. The image spatial resolution is sometimes described in terms of the number of line pairs that can be resolved (Russ 2016). Digital cameras now have a spatial resolution close to that of film emulsions (Hale and Edwards 2002) and professional digital cameras with extremely high-resolution detector arrays (>15 million pixels) that are producing images that are approaching those from traditional 35 mm and larger format film cameras (Russ 2016). A Kodak 35 mm color film, for example, can be scanned at 1080 or 2000 lines. Likewise, a cropped 15 megapixel image using a film scanner such as a Nikon Coolscan V ED, 5000, will produce a resolution approximating the original film. Indeed, such resolutions offer crisp images with fine details. In addition, high spatial resolution greatly helps the thresholding operation to transform the HP into a binary image because the proportion of mixed pixels is lower compared with those from low-resolution images (Blennow 1995; Walter et al. 2003; Jonckheere et al. 2006).

The range of radiometric intensities (or dynamic range) identifies the radiometric resolution of each color in the digitized photograph. Increasing the image bit depth from 8-bits to 12- or 16-bits results in a marked difference in terms of being able to distinguish subtle radiometric changes. High dynamic range may prove essential in distinguishing trunks from foliage or in discerning patterns of light transmission through the foliage. There must be a balance between the file size and the level of detail desired, based on the available analyzing tools and application. Lastly, tools for adjusting image contrast through look-up tables (or radiometric conversion tables) allow enhancing specific radiometric arrangements to facilitate image interpretation or analysis. Also, these look-up tables applied to HP intensity values can also be used to reduce the size of images without compromising analysis possibilities. Therefore, there is a range of options for transforming the HP into digital images. Making an optimum choice requires some preliminary testing to evaluate the most suited options.

Optical scanners are available in many different forms, providing a wide range of options for digitizing photographs. In descending order of accuracy, first comes slide or negative scanners. Their main advantage is that they scan directly from the film; in addition, the top-of-the-line products come with a wide range of options. Another possibility is to use the services of companies specializing in the development of photographs. Video cameras for digitizing negatives were a popular option that was often used in studies on image analysis methods (e.g., Becker et al. 1989; Rich 1989, 1990). Digital cameras and high-quality scanners have made this technique obsolete.

Digital cameras save images using different image file formats. Most cameras allow image storage in a proprietary or manufacturer specific format usually called "raw image format". Some cameras allow storing pixel values (grey-level intensities) in a non-standard (even if well documented), uncompressed, and even graphics formats. In both cases it would be prudent to convert these so called "raw files" into some standardized, uncompressed graphics format prior to archive, which would enable post-processing and conversion to all image formats. Widely used photographic storage formats include .tif or .tiff ("Tagged Image Format"), .jpeg ("Joint Photographic Experts Group"), and .bmp ("Bitmap") files. These formats can be imported directly for processing in most image-processing programs. In general, raw and tiff data formats require more space than the jpeg format for which a compression algorithm is applied. We suggest the use of an uncompressed format to avoid impact on the extraction of gap fraction. Usually a short meta-data file is automatically saved with the digital photograph and is very useful since most necessary information regarding data acquisition time and the camera settings would have been recorded. Supplemental recording of these details would then be redundant during field work which is a significant help to improving the efficiency of data acquisition.

4.4.2 *Archiving*

Planning an HP project also requires devising a strategy for archiving the images and the products of the analyses, ideally before taking the actual photographs. The process involves first determining if study sites will be compared with one another or if the same site will be compared at different times of the year. The number of digital images to be generated and the purpose of the analysis will define the requirements for documentation. There are a number of existing standards (e.g., Federal Geographic Data Committee—FGDC and International Organization for Standardization—ISO) on the descriptive metafiles required for spatial databases but their application is usually cumbersome. However, a project related metafile can be created containing essential information on data acquisition to ensure that HP can be used in a later analysis. Without going into detail, a minimum structure would consist of a main directory containing one or more files with all the relevant information on the project, such as (1) a general description of the project and its objectives and methodology; (2) a list of the measurement procedures used; (3) a description of the database directory and file structure; and (4) a description of the fields used in the database records. We strongly recommend that you provide users with a quick-reference chart to view the data available quickly and easily. The archiving of the actual photographs is often done by site, date and type of product (e.g., HP, binary image, LAI table, site index, etc.). A number of software or shareware packages are now available to facilitate the archiving of images, either by making it possible to view the images, by facilitating the transfer from one format to another, or by providing cataloguing tools. In short, formulating an archiving strategy is an essential part of planning an HP program, without which the chances of losing data or not having the essential information increases dramatically.

4.5 Summary and Conclusions

HP has already proven to be a powerful indirect method for measuring various components of canopy structure and understory light regimes throughout a wide range of forestry studies that have undertaken in boreal, temperate and tropical forests. Numerous advances in HP analysis, which have taken place over the last decade, are directly related to evolving computer, photographic, and digital technologies and scientific modeling methods. Hemispherical photographs can be archived, reprocessed as improved models become available, and used to perform other measurements, providing a major asset to be a visual archive for forestry applications.

For *purchasing the equipment*, the most important factors to be considered are the quality of the lens, the needed spatial resolution of the image, the compatibility of the lens-camera combination and the support material to ensure that high quality photographs are taken. A high dynamic range (12-bit) camera is preferable in order to discriminate between shadowed and illuminated elements in the image, and to allow for measurements being undertaken under direct and diffuse conditions.

For *image preparation*, a rigorous procedure following the preliminary steps for preparation, including preparing data sheets and verifying the availability and operating condition of all equipment has to be done in order to generate high quality data. The impact of camera settings (aperture, shutter speed, ISO setting) and lighting conditions have a significant influence on the acquired data. These influencing factors have to be taken into account when making comparisons with other optical devices for forestry applications and between forest stands.

During data acquisition, special attention is needed to ensure proper exposure of the HP. This is especially true since different sky brightness conditions can lead to varying estimates for measurements with the same cameras. For instance, photographs, taken under different conditions with inexpensive digital cameras with automatic settings, might be not comparable due to different (automatic) settings used. In general, obtaining several photographs that provide a gradation of overexposures in addition to the automatic exposure metered outside the canopy is recommended.

The final steps include normalization of the image to remove lens geometrical distortion and documenting all data sets acquired. Although this last part has become redundant due to the metadata in digital photograph storage, with analog HP and in general for clarity reasons, this step is still recommended.

Acknowledgements We wish to thank Dr. Ron Hall for providing helpful suggestions to improve the manuscript. We are also grateful to the two reviewers that provided insightful suggestions to the original manuscript.

References

Anderson MC (1964a) Studies of the woodland light climate. I. The photographic computation of light conditions. J Ecol 52:27–41

Anderson MC (1964b) Studies of the woodland light climate. II. Seasonal variation in the light climate. J Ecol 52:643–663

Anderson MC (1964c) Light relations of terrestrial plant communities and their measurement. Biol Rev 39:425–486

Anderson MC (1966a) Some problems of simple characterization of the light climate in plant communities. In: Bainbridge R, Evans GC, Rackham O (eds) Light as an ecological factor. Blackwell, Oxford, pp 77–90

Anderson MC (1966b) Stand structure and light penetration II. A theoretical analysis. J Appl Ecol 3:41–54

Anderson MC (1970) Interpreting the fraction of solar radiation available in forest. Agric Meteorol 7:19–28

Anderson MC (1971) Radiation and crop structure. In: Sestak Z, Catsky J, Jarvis PG (eds) Plant photosynthetic production manual of methods. Dr. W. Junk N.V., The Hague, pp 77–90
Anderson MC (1981) The geometry of leaf distribution in some South-Eastern Australian forests. Agric Meteorol 25:195–205
Avery TE, Burkhart HE (2001) Forest measurements, 5th edn. McGraw-Hill, New York
Barrie J, Greatorex-Davies JN, Parsell RJ, Marrs RH (1990) A semi-automated method for analysing hemispherical photographs for the assessment of woodland shade. Biol Conserv 54:327–334
Becker M (1971) Une technique nouvelle d'utilisation des photographies hémisphériques pour la mesure du climat lumineux en forêt (A new technique using hemispherical photographs for measuring the light climate in a forest). Ann Sci For 28:425–442
Becker P, Erhart DW, Smith AP (1989) Analysis of forest light environments. I. Computerized estimation of solar radiation from hemispherical canopy photographs. Agric For Meteorol 44:3–4
Blennow K (1995) Sky view factors from high-resolution scanned fisheye lens photographic negatives. J Atmos Oceanic Technol 12:1357–1362
Bonhomme R, Chartier P (1972) The interpretation and automatic measurement of hemispherical photographs to obtain sunlit foliage area and gap frequency. Israel J Agric Res 22:53–61
Bonhomme R, Varlet Granger C, Chartier P (1974) The use of hemispherical photographs for determining the leaf area index of young crops. Photosynthetica 8:299–301
Brown HE (1962) The canopy camera. USDA Forest Service, Rocky Mountain Forest and Range Experiment Station, Ft. Collins, Colorado. Station Paper No. 72. 22 pp
Brown HE, Worley DP (1965) Some applications of the canopy camera in forestry. J Forest 63:674–680
Canham CD, Denslow JS, Platt WJ, Runkle JR, Spies TA, White PS (1990) Light regimes beneath closed canopies and tree-fall gaps in temperate and tropical forests. Can J For Res 20:620–631
Chan SS, McCreight RW, Walstad JD, Spies TA (1986) Evaluating forest vegetative cover with computerized analysis of fisheye photographs. For Sci 32:1085–1091
Chapman L (2007) Potential applications of near infra-red hemispherical imagery in forest environments. Agric For Meteorol 143:151–156
Chazdon RL, Field CB (1987) Photographic estimation of photosynthetically active radiation: evaluation of a computerized technique. Oecologia 73:525–532
Chen JM, Black TA, Adams RS (1991) Evaluation of hemispherical photography for determining plant area index and geometry of a forest stand. Agric For Meteorol 56:129–143
Clark JA, Follin GM (1988) A simple "equal area" calibration for fisheye photography. Agric For Meteorol 44:19–25
Clearwater MJ, Gould KS (1995) Leaf orientation and light interception by juvenile *Pseudopanax crassifolius* (Cunn.) C. Koch in a partially shaded forest environment. Oecologia 104:363–371
Coombe DE, Evans GC (1960) Hemispherical photography in studies of plants. Med Biol Illustrations 10:68–75
Evans GC, Coombe DE (1959) Hemispherical and woodland canopy photography and the light climate. J Ecol 47:103–113
Fernandes RA, Miller JR, Chen JM, Rubinstein IG (2003) Evaluating image-based estimates of leaf area index in boreal conifer stands over a range of scales using high-resolution CASI imagery. Remote Sens Environ 89:200–216
Fernandes R, Plummer S, Nightingale J, Baret F, Camacho F, Fang H, Garrigues S, Gobron N, Lang M, Lacaze R, LeBlanc S, Meroni M, Martinez B, Nilson T, Pinty B, Pisek J, Sonnentag O, Verger A, Welles J, Weiss M, Widlowski JL (2014) Global leaf area index product validation good practices. (Schaepman-Strub G, Román M, Nickeson J, (eds) version 2.0. Land Product Validation Subgroup (WGCV/CEOS)
Fournier RA, Landry R, August NM, Fedosejevs G, Gauthier RP (1996) Modelling light obstruction in three conifer forests using hemispherical photography and fine tree architecture. Agric For Meteorol 82:47–72

Fournier RA, Rich PM, Landry R (1997) Hierarchical characterization of canopy architecture for boreal forest. J Geophys Res 102(D24):29445–29454

Fournier RA, Mailly D, Walter J-MN, Soudani K (2003) Indirect measurement of forest structure from *in situ* optical sensors. In: Wulder M, Franklin S (eds) Methods and applications for remote sensing of forests: Concepts and case studies. Kluwer, Dordrecht, pp 77–113

Frazer GW, Trofymow JA, Lertzman KP (1997) A method for estimating canopy openness, effective leaf area index, and photosynthetically active photon flux density using hemispherical photography and computerized image analysis techniques. Natural Resources Canada, Canadian Forest Service, Pacific Forestry Centre, Victoria, BC. Information Report BC-X-373, 73 pp

Frazer GW, Canham CD, Lertzman KP (1999) Gap Light Analyzer (GLA), Version 2.0: Imaging software to extract canopy structure and gap light transmission indices from true-colour fisheye photographs, users manual and program documentation. Simon Fraser University, Burnaby, British Columbia, and the Institute of Ecosystem Studies, Millbrook, New York. 52 pp

Frazer GW, Fournier RA, Trofymow JA, Hall RJ (2001) A comparison of digital and film fisheye photography for analysis of forest canopy structure and gap light transmission. Agric For Meteorol 109:249–263

Hale SE, Edwards C (2002) Comparison of film and digital hemispherical photography across a wide range of canopy densities. Agric For Meteorol 112:51–56

Herbert TJ (1986) Calibration of fisheye lenses by inversion of area projections. Appl Opt 25:1875–1876

Herbert TJ (1987) Area projections of fisheye photographic lenses. Agric For Meteorol 39:215–223

Hill R (1924) A lens for whole sky photographs. Q J Royal Meteorol Soc 50:227–235

Hinz A, Dörstel C, Heier H (2001) DMC2001-The Z/I imaging digital camera system. In Proceedings of ASPRS Conference: Gateway to the new millennium, April 2001, St. Louis, Missouri, pp 23–27

Hutchison BA, Matt DR (1977) The distribution of solar radiation within a deciduous forest. Ecol Monogr 47(2):185–207

Inoue A, Yamamoto K, Mizoue N, Kawahara Y (2004) Calibrating view angle and lens distortion of the Nikon fish-eye converter FC-E8. J For Res 9:177–181

Jonckheere I, Fleck S, Nackaerts K, Muys B, Coppin P, Weiss M, Baret F (2004) Methods for leaf area index determination. part I: Theories, sensors and hemispherical photography. Agric For Meteorol 121:19–35

Jonckheere I, Muys B, Coppin P (2005a) Allometry and evaluation of in situ optical LAI determination in Scots pine: a case-study in Belgium. Tree Physiol 25:723–732

Jonckheere I, Muys B, Coppin P (2005b) Derivative analysis for in situ high-dynamic range hemispherical photography and its application in forest stands. IEEE Geosci Remote Sens Lett 2:296–300

Jonckheere I, Nackaerts K, Muys B, van Aardt J, Coppin P (2006) A fractal dimension-based modelling approach for studying the effect of leaf distribution on LAI retrieval in forest canopies. Ecol Model 197:179–195

Kasunic K (2011) Optical systems engineering. 1st edn. McGraw-Hill Education. Press Monograph Series. 464 pp. ISBN 0071754407/9780071754408

Leblanc SG (2014) Étude de l'estimation de paramètres structuraux de la forêt àl'aide d'un cadre de modélisation de la photographie hémisphérique et du LiDAR. PhD thesis. Département de géomatique appliquée, Université de Sherbrooke

Leblanc SG, Fournier RA (2014) Hemispherical photography simulations with an architectural model to assess retrieval of leaf area index. Agric For Meteorol 194:64–76

MacFarlane C, Coote M, White DA, Adams MA (2000) Photographic exposure affects indirect estimation of leaf area index in plantations of *Eucalyptus globulus* Labill. Agric For Meteorol 100:155–168

Magdwick HAI, Brumfield GL (1969) The use of hemispherical photographs to assess light climate in the forest. J Ecol 57:537–542

Mailly D, Kimmins JP (1997) Growth of *Pseudotsuga menziesii* and *Tsuga heterophylla* seedlings along a light gradient: resource allocation and morphological acclimation. Can J Bot 75:1424–1435

Mitchell PL, Whitmore TC (1993) Use of hemispherical photographs in forest ecology. Oxford Forestry Institute, Department of Plant Sciences, University of Oxford. OFI Occasional Papers No. 44. 39 pp

Nackaerts K, Coppin P, Muys B, Hermy M (2000) Sampling methodology for LAI measurements with LAI-2000 in small forest stands. Agric For Meteorol 101:247–250

Nilson T, Ross V (1979) Characterization of the transparency of a forest canopy by fisheye photographs [Vooremaa Forest Ecology Station]. Estonian contributions to the International Biological Programme (progress report) 2:117–130

Nomura Y, Sagara M, Naruse H, Ide A (1992) Simple calibration algorithm for high-distortion-lens camera. IEEE Trans Pattern Anal Mach Intell 14:1095–1099

Pearcy RW (1989) Radiation and light measurements. In: Pearcy RW, Ehleringer J, Mooney HA, Rundel PW (eds) Plant physiological ecology: field methods and instrumentation. Chapman & Hall, London, pp 95–116

Ray SF (2002) Applied photographic optics: lenses and optical systems for photography, film, video, electronic and digital imaging. 3rd edn. Focal Press, 656 pp. ISBN 0240515404/ 9780240515403

Rich PM (1989) A manual for analysis of hemispherical canopy photography. Los Alamos National Laboratory Report LA-11733-M, Los Alamos, New Mexico. http://www.creeksidescience.com/files/rich_et_al_1989_canopy_manual.pdf

Rich PM (1990) Characterizing plant canopies with hemispherical photographs. Remote Sens Rev 5:13–29

Rich PM, Clark DB, Clark DA, Oberbauer SF (1993) Long-term study of solar radiation regimes in a tropical wet forest using quantum sensors and hemispherical photography. Agric For Meteorol 65:107–127

Russ J (2016) The image processing handbook, 7th edn. CRC Press, Boca Raton, Florida

Wagner S (1998) Calibration of grey values of hemispherical photographs for image analysis. Agric For Meteorol 90:103–117

Walter J-MN, Fournier RA, Soudani K, Meyer E (2003) Integrating clumping effects in forest canopy structure: an assessment through hemispherical photographs. Can J Remote Sen 29:388–410

Whitmore TC, Brown ND, Swaine MD, Kennedy D, Goodwin-Bailey CI, Gong W-K (1993) Use of hemispherical photographs in forest ecology: measurement of gap size and radiation totals in a Bornean tropical rain forest. J Trop Ecol 9:131–151

Zhang Y, Chen JM, Miller JR (2005) Determining digital hemispherical photography exposure for leaf area index estimation. Agric For Meteorol 133:166–181

Chapter 5
Image Analysis of Hemispherical Photographs, Algorithms and Calculations

Inge G.C. Jonckheere, Craig Macfarlane and Jean-Michel N. Walter

Abstract This chapter is a logical continuation of the chapter on *acquisition* of hemispherical photographs (HP) in forest environments (Fournier et al. in Hemispherical photography in forest science: theory, methods, applications, Springer, Berlin, 2017), and describes the *analysis* of hemispherical or 'fisheye' canopy photographs, including the calculation of canopy solar radiation and structure. The workflow for digital images proceeds through several steps, from image input to the calculation of canopy structure and solar radiation variables:

Image *input* includes image selection and quality control, while image *processing* consists of editing and contrast enhancement, registration with coordinate systems, and the creation of configuration or parameter files.

Image *classification* is a critical step in the HP workflow. In the past, pixels were classified as either canopy or sky using manual selection of a threshold grayscale value. More recently, automated methods have become available for classifying pixels and have greatly increased the efficiency and productivity of HP workflow. We discuss some of these new methods for producing a binary data set from the grayscale information in HP.

Calculations and *output* of results from HP requires the definition of sampling grids or circular transects, along which gap fractions and gap sizes are measured as a function of zenith and azimuth directions. Quantitative solar and canopy structure indices, such as site factors and leaf area index (LAI), are output to spreadsheets or text files.

I.G.C. Jonckheere (✉)
Forestry Department, FAO of the United Nations, viale delle Terme di Caracalla, 00153 Rome, Italy
e-mail: inge.jonckheere@fao.org

C. Macfarlane
CSIRO Ecosystem Sciences, Private Bag No. 5, 6913 Wembley, WA, Australia
e-mail: Craig.Macfarlane@csiro.au

J.-M.N. Walter
Département de botanique, Université Louis-Pasteur, Strasbourg, France
e-mail: jmnw@free.fr

R.A. Fournier and R.J. Hall (eds.), *Hemispherical Photography in Forest Science: Theory, Methods, Applications*, Managing Forest Ecosystems 28,
DOI 10.1007/978-94-024-1098-3_5

Conclusions. This final step includes the interpretation of the results, which are generally output to spreadsheets, databases and models.

In addition to discussing the steps in the workflow, this chapter also addresses some issues for modelling solar radiation that typically have been neglected, such as steep slopes, and briefly discusses alternative parameters and approaches for quantifying canopy structure.

Keywords Biophysical data extraction · Hemispherical photography algorithms · Image classification · Sun maps · Vegetation canopy structure

5.1 Introduction

This chapter describes the analysis of hemispherical or 'fisheye' canopy photographs, including the calculation of canopy solar radiation and structure. Hemispherical photographs (HP) have some unique features that can affect the analysis of images and subsequent calculations. Hemispherical lenses provide an extremely wide (generally 180°) field of view (FOV). They project the hemisphere onto a plane, but the exact nature of the projection varies according to the lens that is used (Rich 1990). The simplest and most commonly assumed hemispherical lens geometry is the polar or equiangular projection, for which the resulting circular image shows a complete view of all sky directions, with the zenith in the center of the image and the horizon at the edges. However, few lenses exhibit truly equiangular projection (Blennow 1995; Frazer et al. 2001; Pekin and Macfarlane 2009), such that even the relationship between zenith angle and pixel location on the photograph requires careful evaluation before image analysis can begin.

What is image analysis? In this chapter, we use the term *analysis* to refer collectively to general methods of processing, classification and calculation. *Processing* consists of image registration, editing and enhancement; *classification* involves distinguishing between canopy and sky (or another class, e.g. soil background); *calculation* is the extraction of useful quantitative variables or indices concerning eco-physiological processes or structure. Therefore, image analysis of HP is a workflow whereby canopy and open sky are distinguished and quantitative variables or indices of solar radiation and canopy structure are calculated. Analysis of HP is not a recent method but, with the advent of fast personal computers, manual processing by sampling grids on printed photographs has been superseded by automated digital processing using integrated HP-specific software. Credit should be given to the pioneering work of Anderson (1964), Bonhomme and Chartier (1972), Chazdon and Field (1987), Becker et al. (1989) and Rich (1989), who were very likely the first to automate HP analysis with computer images and who inspired a great deal of later research.

In the past, HP was mostly acquired in analog format using orthochromatic (Wagner 1998) and panchromatic film (Rich 1989) and by using 35 mm single lens reflex (SLR) cameras, with color films becoming more prevalent in later years (Blennow 1995). Analog photographs needed to be converted to a digital format using negative or slide scanners, or a video digitizer. Such CCD-scanners (Charged Coupled Device) typically produced digital images on an 8-bit (0–255) radiometric scale (Fraser 1997; Fournier et al. 2017). Digitization of HP is now seldom necessary because most HPs are acquired with digital cameras (Frazer et al. 2001; Hale and Edwards 2002; Jonckheere et al. 2004). Furthermore, existing and emerging digital cameras are equipped with higher dynamic range (32 bit to 64-bit) sensors and high-resolution detector arrays (up to 40 million pixels and more) that provide new opportunities for improving image quality and contrast, which in turn may improve the quality of the final data. Although some studies have compared images with different pixel resolutions (Englund et al. 2000; Inoue et al. 2004; Pekin and Macfarlane 2009), the potential of higher-dynamic range images remains largely unexplored to date.

The software tools that are available for image analysis are discussed in Hall et al. (2017). Briefly, the new end-user is faced with two options: (1) use HP-specific commercial software or (2) use non-commercial scripts and programs possibly in combination with commercial software. In the first case, the entire workflow from image input, processing, analysis and calculation through to results output is usually performed within a single software package with a Graphical User Interface (GUI). The main advantage is a well-integrated and convenient workflow, but this may occur at the expense of flexibility. In the second case, the user must choose from a variety of options. These range from well-integrated open source software or commercial packages to customized software solutions. The latter may involve pre-processing of images with commercial software, followed by extraction of gap fractions and sizes, and calculations using one or more custom-developed scripts or command-line programs; this approach requires an in-depth knowledge of the HP analysis workflow by the user. As digital photography has greatly increased the rate at which HP can be acquired compared to film photography, batch processing options (the processing of many data files in an automated single run) should be important selection criteria for choosing between any commercial, open source or custom software solutions. The primary purpose of obtaining HP may be to characterize either solar radiation regimes beneath the forest canopy or forest structure. While many processing and analysis steps are common to both, there are also steps that are specific to one or the other and this will also affect the choice of software.

Gap fraction and gap size are key concepts. *Gap fraction* is the proportion of openness for a given sky direction, typically either along a sun path for calculation of direct radiation or for an entire sky region for calculation of diffuse radiation, sky view factor, and canopy structural characteristics. *Gap size* is the length or area of contiguous 'sky pixels' that are either along circular transects or within defined zenithal and azimuthal regions, and which is commonly used for deriving clumping factors or indices. The remainder of the chapter is divided into four sections that follow the typical Hemispherical photography Processing (HP) workflow, as is illustrated in Fig. 5.1:

Fig. 5.1 Flowchart

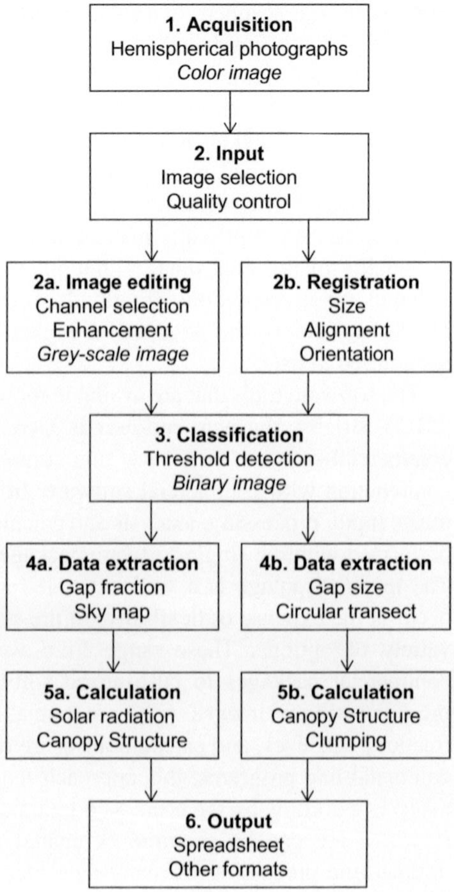

1. Image input and processing. This step includes image selection and quality control, any editing or contrast enhancement, and registration within a coordinate system (image size, translation, and rotation).
2. Image classification. This term originates from remote sensing analysis, and involves distinguishing canopy from sky using automatic classification techniques that are based on local and global algorithms. The end result is usually a binary (black-and-white) dataset or image. It is also referred to as 'image segmentation' (e.g., Wagner and Hagemeier 2006).
3. Calculations and result output. This step involves the extraction of gap fractions and gap sizes from the binary image, often as a function of sky direction (zenith and azimuth) or position along the sun track. Models and algorithms are used to calculate canopy solar radiation and vegetation structural variables and indices. Results are generally output to spreadsheets or to other file formats.
4. Conclusions.

5.2 Image Input

5.2.1 Image Selection and Quality Control

Good quality images are needed to ensure consistency in subsequent analytical steps and maximum accuracy of results. Image quality can be described using criteria such as visual aspect, exposure and histogram features, background sky conditions, image spatial resolution and dynamic range. First and foremost, the images should 'look right' (Was there an obvious error during image acquisition?). Images should be correctly exposed such that histograms have good dynamic range and display a bimodal form, viz., two well-separated peaks: one on the left side (black pixels, canopy elements), and the other on the right side (white pixels, sky). Uneven sky radiance and clouds should be avoided where possible because this can complicate image analysis and lead to less accurate results (e.g., Macfarlane 2011). The pixel resolution of the image can also affect subsequent processing (Walter et al. 2003), as discussed in Sect. 5.3.2.

Much of the aforementioned information is determined at the time of *acquisition* of HP, which has already been described by Fournier et al. (2017). Briefly, the relevant factors that are required to obtain good quality images are light conditions, exposure settings (shutter speed, aperture, and ISO setting), film or image resolution, and dynamic range (Rich 1989; Zhang et al. 2005; Fournier et al. 2017). Several photographs with different exposure settings (speed, aperture) can be taken at a given site, so that an optimal image can be captured that offers good contrast between sky and vegetation. Alternatively, on more recent digital cameras, the histogram can be scrutinized in the field to adjust exposure settings (see Fig. 5.2). Although black-and-white film was widely used for canopy photography in the past, modern digital cameras produce full-color images usually in the RGB (red, green and blue) color space. A camera's exposure metering is heavily weighted towards the luminance received by the green channel and this can affect the dynamic range of the red and blue channels for images that are collected under clear sky conditions (Fig. 5.2). Exposure settings should be adjusted to optimize the dynamic range of the channel (or channels) to be analyzed.

While color information is critically important for classifying foliage and background in downward-looking images (e.g., Woebbecke et al. 1995; Baret et al. 2010; Macfarlane and Ogden 2012), and some authors have suggested using all three bands to improve classification of canopy and sky (e.g., Kucharik et al. 1997), more often only a single channel is used that provides the best contrast between foliage and sky. Most image-processing software allows the decomposition of a color multi-spectral image into its three respective channels (also called planes or bands; Fig. 5.2). An advantage of using one band of the three that are available is to avoid chromatic aberration resulting from imperfections in the lens system (Frazer et al. 2001; Leblanc et al. 2002). Although the internal image processing algorithms of more modern digital Standard Lens Reflex (DSLR) cameras frequently include corrections for chromatic aberration, these are only for lenses that are produced by the same

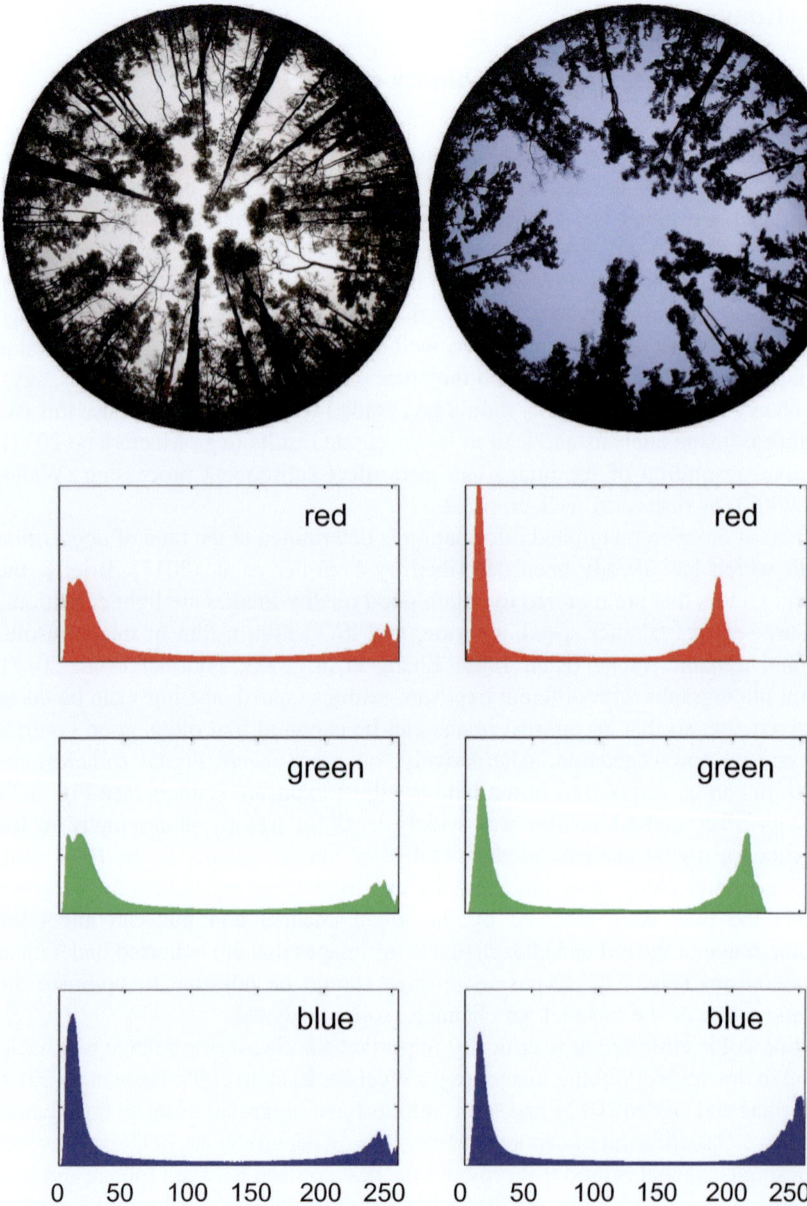

Fig. 5.2 Hemispherical photograph channels (RGB) and their respective histograms

manufacturer as the camera body and may not be readily available for hemispherical lenses that are produced by other manufacturers. The selection of a single image band is also preferable to the use of a transmission filter in front of the camera lens during image acquisition because the desired band can easily be selected after image acquisition without losing FOV. Conversely, if a physical filter is attached to the lens by means of a filter ring, this results in a reduction of the FOV (Russ 2002).

The blue channel often offers the best contrast for the analysis of HP due to reduced blue light scattering of leaves and the selective absorption of the blue portion of the electromagnetic spectrum by plant pigments, which tend to darken the foliage in HP (Rich 1989; Leblanc et al. 2005; Jonckheere et al. 2006a). For example, in the image from Fig. 5.2 that was taken under overcast conditions, the gray levels of the foliage pixels are more homogeneous and darker in the blue channel than in the other two channels. Some studies have instead used the red channel when it offered better contrast than the blue channel. The image in Fig. 5.2 was collected before dawn on a clear morning and has similar contrast between foliage and sky in all channels, although the red channel has the least dynamic range and the blue channel has the greatest saturation. The choice of channel needs to be assessed from one project to another, as it may change according to forest or illumination conditions, and the equipment that is available. The remainder of this section assumes that good quality digital, or digitized, images have been obtained in an appropriate format to be loaded into either third-party graphical software or integrated HP-specific programs.

5.2.2 Image Editing and Enhancement

HP may contain few details when acquired in dense canopies or under dim light conditions, and may benefit from some enhancement, for example, using a sharpening filter. Sharpening is an image enhancement technique that is based on edge enhancement (see Gonzalez and Woods 2008 for a detailed discussion), which may be used systematically or applied in particular conditions to increase contrast and emphasize small gaps within dense canopies. Almost all software packages with a GUI provide one or more sharpening filters. Particularly useful for very dense canopies (undergrowth, dense conifer crowns), image sharpening reveals numerous minute gaps, which will be enhanced without significantly affecting large gaps. Another effect of sharpening is decreasing the probability of encountering sky regions that are completely obstructed by foliage (i.e., zero gap fraction), which is advantageous if using the logarithmic averaging method (Lang and Xiang 1986) to calculate LAI (leaf area index, discussed in Sect. 5.4). Although a sharpening filter is optional, some authors had used it systematically (Clearwater et al. 1999; Walter et al. 2003). By emphasizing small gaps, sharpening increases the gap fraction and alters the gap size distribution. For example, Macfarlane et al. (2000) and Walter et al. (2003) found that image sharpening reduced LAI and mean leaf angle, but increased apparent foliage clumping. The most conspicuous effect of image sharpening is to *reduce the number of mixed pixels*, i.e., those containing sky and foliage components at the edge of gaps. In turn, this facilitates the detection of threshold gray values for separating sky and foliage. The interaction between image resolution, sharpening and the number of mixed pixels is discussed in Sect. 5.3.2.

5.2.3 Registration Protocols

Image registration is the process of identifying the boundary of the circular image and defining the hemispherical coordinates of the system. The *registration protocol* consists of two steps: (1) Identifying the circular boundary corresponding to a 90° zenith angle (image edge, or horizon circle), and (2) Defining the direction of magnetic or true north on the HP. Registration generally follows image editing and enhancement, and precedes classification.

Registration requirements differ somewhat for characterizing solar radiation regimes inside and beneath canopies or for characterizing canopy structure (see Table 5.1 for details). Some HP software integrates canopy structure and radiation regimes, while others calculate either one or the other. Some basic requirements are shared: image identification and geographic location, geometric lens distortion, and definition of the image boundary and lens projection function. Slope, aspect and use of a topographic mask may be important in particular topographic conditions (steep slopes). Other requirements, such as definition of image orientation and magnetic declination, are specific to radiation modeling. Software settings, which contain necessary input data for the models in the software, are usually saved in a configuration file, or a parameter file, which can be edited, saved and reloaded. Creation of separate configuration files for different batches of HP allows for archiving of analytical settings, along with images and results, and is also necessary for automatic processing of a set of HP that is taken under the same conditions.

The limit of the image is often difficult to separate from the background in canopy images, especially in dense canopies. 'Brightness/contrast', 'levels' and 'curves' adjustment tools that are available in graphical software and visible LED (Light Emitting Diode) marks on the image may be helpful in identifying the image edge. Acquiring a non-canopy image in a wide-open area is also useful for defining the boundary of the circular image for a given camera and lens. Depending upon the software, the boundary may be defined by the coordinates of the image center and radius values, by the coordinates of two opposite points (north and south), or by three points that define the image edge. An alignment overlay, which is available in most HP software, is helpful, especially if images are automatically centered for a given combination of camera and lens (e.g., Leblanc 2006).

Precise orientation of the image with respect to true or magnetic north is needed for characterization of canopy radiation regimes. It is good practice to consistently align the top of the camera with north during image acquisition to simplify image analysis. Otherwise, image orientation requires either manual or automated adjustment by dragging and rotating the image to align it with, for example, any LED marks that are on the image. Once oriented correctly, magnetic north and true north can be visualized along with other geographic coordinates.

Table 5.1 Overview of main input data required to run HP programs

Parameters	Canopy structure	Radiation regimes
General data		
Image ID	+	+
Geometric correction	+	+
Channel (RGB)	+	+
Number zenith rings	+	+
Number azimuth sectors	+	+
Zenith view angle	+	+
Topographic data		
Slope	+	+
Aspect	+	+
Topographic mask	–	+
Altitude	–	+
Geographic data		
Orientation, North	+	+
Magnetic declination	–	+
Latitude	–	+
Longitude	–	+
Time zone	–	+
Biological data		
Leaf reflectance	–	+
Leaf transmittance	–	+
Soil reflectance	–	+
Seasonal period	–	+
Atmospheric data		
Cloudiness index	–	+
Spectral fraction	–	+
Beam fraction	–	+
Condensable water	–	+
Air turbidity	–	+
Astronomic data		
Solar constant	–	+
Year (Leap)	–	+
Days of year	–	+
Solar time step	–	+
Time integration	–	+
Calculation options		
Sky radiance models	–	+
Radiation units	–	+

5.3 Classification of Sky and Foliage

5.3.1 Classification Methods

An important prerequisite for extracting gap fractions and gap sizes from HP is to select and apply an appropriate classification technique that separates foliage from sky. After quality control, editing, enhancement, channel selection and registration of the color photograph, which result in a grayscale image, binary datasets need to be created that separate vegetation and sky, in order to analyze solar radiation regimes and canopy structure. Image classification consists of selecting an optimal brightness threshold (termed *thresholding*) to produce a binary (black-and-white) image from the grayscale image and is a critical step in the image analysis workflow (Blennow 1995; Jonckheere et al. 2004). Owing to their importance, thresholding and pixel classification have been widely studied over the last two decades (Olsson et al. 1982; Chen et al. 1991; Baret et al. 1993; Blennow 1995; Wagner 1998, 2001; Ishida 2004; Leblanc et al. 2002, 2005; Jonckheere et al. 2005, 2006a; Nobis and Hunziker 2005; Cescatti 2007; Inoue et al. 2011; Macfarlane 2011). Alternative approaches exist for obtaining gap fractions and gap sizes directly from grayscale images (e.g., Wagner 1998; Leblanc et al. 2005; Cescatti 2007), but these approaches are less common and require back-correction of the gamma-function that is applied to the image during in-camera processing. A concise history of the development of canopy image classificationmethods is given in Wagner and Hagemeier (2006).

Accurate thresholding relies upon the gray values of pixels that represent vegetation being substantially different from the gray values of pixels that belong to the sky (see discussion in Sect. 5.2.1 on image quality) and is of the utmost importance because it will have a significant impact on subsequent analyses. Note that the gray values are also variously referred to as ‘brightness values,’ ‘luminance values’ and ‘digital numbers’ in the literature–these terms are synonymous. Traditionally, an image threshold is subjectively selected by the operator through visual comparison of grayscale and binary images, a process that is made simple by using graphics software or integrated programs for HP. However, subjective thresholding has proven to be a source of inconsistency and error (Jonckheere et al. 2005; Inoue et al. 2011) and may not account for local errors in the image, resulting in over- or underestimation of the real gap fraction in some regions of the image and under- or overestimation of canopy structural and radiation parameters (Nackaerts et al. 1999). Numerous commercial and free software packages have been developed specifically for processing HP with 1- to 24-bit imagery (Hall et al. 2017), but many of these rely upon the subjective choice of a threshold for the whole image. More efficient and robust classification techniques are needed, both to reduce subjectivity and to automate the thresholding process for more efficient and consistent post-processing of HP.

Jonckheere et al. (2005) have presented a review of 35 automatic general-purpose algorithms that were designed for thresholding images. The algorithms were divided into five categories:

1. Histogram shape-based methods: Analysis is based on the histogram of the gray values in the image (peaks, valleys, etc.). The methods tested in this group were those of Rosenfeld, Sezan, Olivo, and Ramesh (e.g., Olivo 1994);
2. Clustering-based thresholding methods: The gray values are clustered into two classes as foliage and sky. Tested methods were those of Ridler, Yanni, Otsu, Lloyd, Kittler, and Jawahar (e.g., Ridler and Calvard 1978);
3. Entropy-based methods: Algorithms based on the entropy of foreground and background regions, and the cross-entropy between the original and binary image. The methods of Kapur, Sahoo, Pun, Li, Brink, and Shanbag were investigated (e.g., Kapur et al. 1985);
4. Object attribute-based methods: Algorithms based on a measurement of similarity between the gray value and binary images, such as edge coincidence. The Tsai method, Hertz, OGorman, Huang, and Pikaz method were tested (e.g., Tsai 1985);
5. Spatial methods: High-order probability distribution and correlation between pixels are used for segmentation. The methods of Pal, Abutaleb, and Beghdadi were tested (e.g., Abutaleb 1989).

The algorithms were tested on a digital RGB hemispherical canopy photograph and compared to both the reference ground-truth image (Fig. 5.3, top left) and an image that was manually thresholded (Fig. 5.3, top right) to determine the worst to best methods based on visual inspection. The authors most often recommended the iterative method of Ridler and Calvard (1978) for thresholding of HP. The method works well for very closed canopies, where it emphasizes minute canopy openings without significantly affecting large openings. Conversely, the 'edge detection' algorithm of Nobis and Hunziker (2005; SideLook software) preserves well-lit foliage in more open canopies at the expense of small openings. The user should be aware that a 'perfect' method for image thresholding does not exist, owing to factors such as: (i) sky luminance heterogeneity, even under deeply overcast conditions, and (ii) lens vignetting, i.e., the gradual light fall-off of the lens, particularly at wide aperture. While some graphical software may correct vignetting, the effect of sky luminance variation is more difficult to avoid, even in correctly exposed images (Wagner 1998, 2001; Schwalbe et al. 2006). This is illustrated in Fig. 5.4; the ground truth image (Fig. 5.4a) contains brightly lit foliage in gaps that have gray values that are similar to those of dimly lit gaps in dense foliage. Application of the Ridler method preserved gaps in dense foliage but misclassified foliage in large gaps as sky (Fig. 5.4b). Conversely, the Niblack (1986) method successfully detected foliage in gaps but misclassified gaps in dense foliage (Fig. 5.4c).

Classification of HP could be improved by applying different thresholds to different regions within an image. A single threshold that is applied to an entire image is termed a *global* threshold, while a threshold that is applied to parts of an

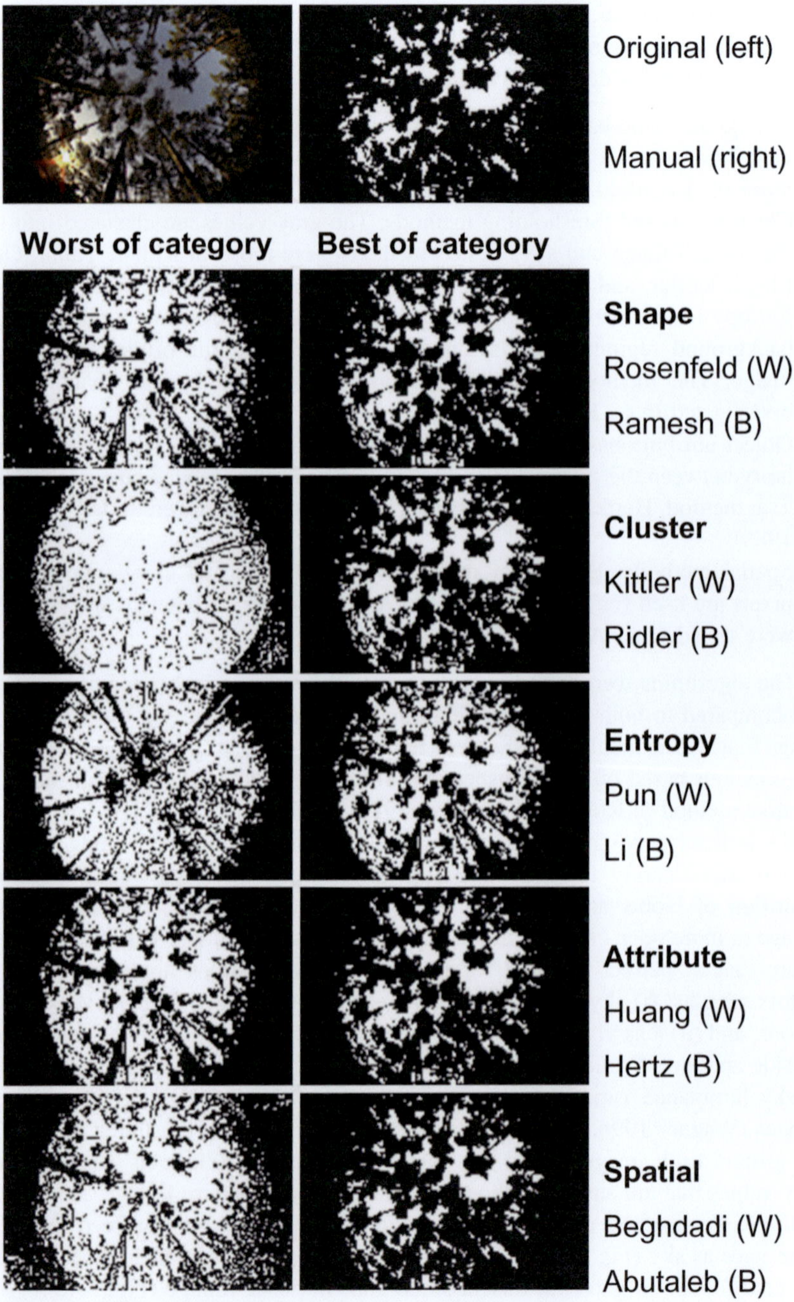

Fig. 5.3 Classification algorithms

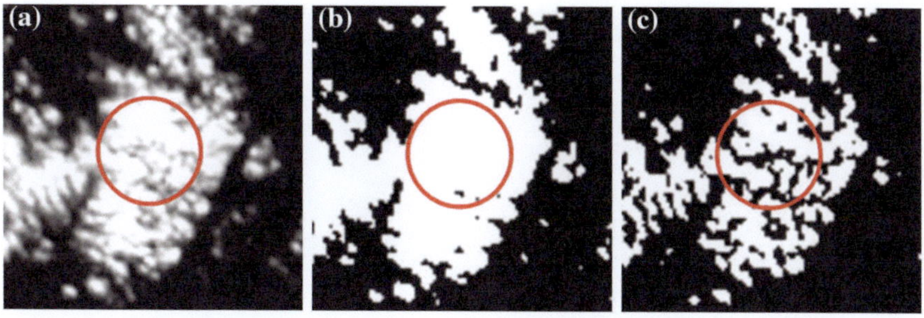

Fig. 5.4 Sky divisions

image is termed a *regional* threshold. At the extreme a threshold may be determined on a pixel-by-pixel basis, termed a *local* threshold. While most cited thresholding procedures are global, many others are regional, usually proceeding stepwise from zenith to horizon across annuli (Wagner 1998; Leblanc et al. 2005; Leblanc 2008). Alternatively, Macfarlane (2011) applied different thresholds to canopy and sky regions of the image to emphasize gaps in dense canopy and foliage in bright sky. Local threshold methods have typically been applied either to only mixed pixels within HP (Schwalbe et al. 2006) or to all pixels within a narrow zenith annulus (Jonckheere et al. 2006a). The software package DHP (Leblanc 2008) allows the user to specify different thresholds for different zenith regions (annuli) of an image. In contrast, Schwalbe et al. (2006) first identified 'pure' sky and canopy pixels, and then classified mixed pixels as sky or foliage based on a comparison of each mixed pixel's gray value with those of nearby pure pixels.

Another example of a local threshold method is the application of an edge-detection algorithm to a narrow zenith view angle (e.g., 57°). An alternative to the analysis of entire images or image regions is to define narrow annuli, e.g., 1° in width, over the desired range of zenith angles. Such *circular transects* are suitable for characterizing canopy structure. Circular transects (i) create profiles of pixel gray values that are similar to those provided by the Tracing radiation and Architecture of Canopies (TRAC) instrument (Leblanc et al. 2005; Leblanc 2006), or (ii) identify sequences of black and white pixels that are needed to measure gap size for the estimation of clumping factors and LAI (Walter et al. 2003; Leblanc et al. 2005; Gonsamo and Pellikka 2009; Frazer et al. 2017). Circular transects can be extended over the whole range of useful zenith angles in 1° increments,, such as between 10°–80°, or other appropriate zenith view angles (e.g., 57°; Fig. 5.5). In principle, any of the threshold methods that are discussed in Sect. 5.3.1 could be applied to circular transects, but they are particularly well suited to locally applied edge-detection algorithms such as those that are described by Jonckheere et al. (2006a).

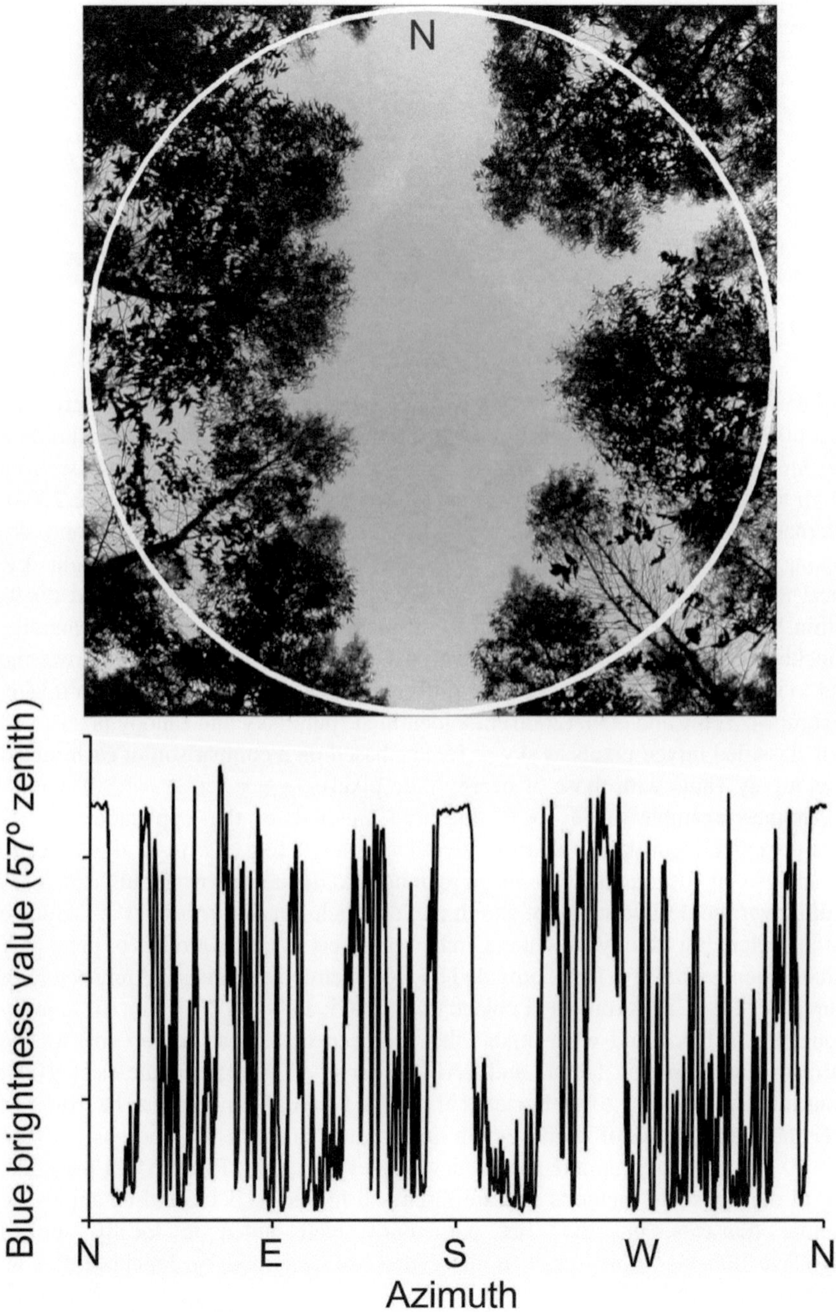

Fig. 5.5 Circular transects

5.3.2 Mixed Pixels

The problem of classifying mixed pixels has been regularly noted (Blennow 1995; Walter et al. 2003; Jonckheere et al. 2005; Leblanc et al. 2005; Schwalbe et al. 2006), but it has been rarely defined. Schwalbe et al. (2006) have referred to 'pure' and 'mixed' pixels, implying that mixed pixels are those that are not pure foliage or sky. Leblanc et al. (2005) and Macfarlane (2011) adopted a similar approach by first identifying homogeneous regions of canopy and sky, and then defined the remaining pixels as 'mixed.' There have also been a few studies that have *quantified* the number of mixed pixels in HP. Zhang et al. (2005) calculated that ~5% of pixels were mixed in an image of dense canopy, and that the proportion of mixed pixels increased with increasing exposure. In this section, we examine the impact of image acquisition and processing on the proportion of mixed pixels in an image, based on published studies and new analyses.

The single most important determinant of the proportion of mixed pixels is the gap fraction of the image; this applies to both HP and narrow FOV images (Macfarlane 2011). The number of mixed pixels increases as the number of canopy pixels increases; more canopy in the images leads to more edges and more mixed pixels. The greatest percentage of mixed pixels that were found in a single HP by Macfarlane (2011) was 15%. It is possible that images of a denser canopy might contain even greater numbers of mixed pixels, despite Zhang et al.'s (2005) previous estimate. Photographic exposure also has a strong influence on the proportion of mixed pixels in an image. In a single image, Zhang et al. (2005) found that increasing exposure by one f-stop increased the proportion of mixed pixels by 1–2%, while Macfarlane (2011) found an increase of 2–3% across many images. Although it has frequently been suggested that image sharpening, image resolution (pixel number) and dynamic range will influence the proportion of mixed pixels, these effects have yet to be tested.

To evaluate the effect of image resolution and sharpening on both the proportion of mixed pixels and the gap fraction, we reanalyzed 48 HP from the study of Macfarlane (2011) to test the effects of image sharpening and image resolution. The original 10 megapixel JPEG images were resampled using scaling factors of 0.7, 0.5, 0.35, 0.2 and 0.1 (see Table 5.2) to create lower resolution copies of the originals. Seven of those JPEG images had corresponding NEF images, which we analyzed to test the effects of dynamic range on the proportion of mixed pixels. NEF ("Nikon Electronic Format") is Nikon's proprietary RAW file format, and images were first converted to 16-bit TIF files using Nikon's ViewNX software. RAW image files, sometimes referred to as digital negatives, contain all the image information captured by the camera's sensor, along with the image's metadata (the camera's identification and its settings, the lens used and other information). The NEF file is written to the memory card in either an uncompressed or "lossless" compressed form. In 16-bit format, the image histogram contains a large number of empty bins; to analyze the histogram shape, it was necessary to perform 1-D interpolation (viz., a cubic spline) of the histogram frequency values and apply a smoothing average filter to the result. This was unnecessary for the 8-bit JPEG images. Sharpened and unsharpened versions of both the 8-bit JPEG images and 16-bit TIF images were

Table 5.2 Matrix of a low-grade sharpening filter

0	0	0	0	0
0	**−1**	**−1**	**−1**	0
0	**−1**	**12**	**−1**	0
0	**−1**	**−1**	**−1**	0
0	0	0	0	0

Table 5.3 Effect of image dynamic range and sharpening on % mixed pixels and total gap fraction of the hemisphere (n = 7). All images were 3872 by 2592 pixels

	sharpened		unsharpened	
Image type	% mixed	GF	% mixed	GF
16-bit TIF	7.8	0.66	10.5	0.65
8-bit JPG	7.9	0.66	10.6	0.64

classified using the dual threshold method of Macfarlane (2011), and their gap fractions and the proportions of mixed pixels were then calculated.

The proportion of mixed pixels was very strongly correlated with the HP gap fraction, as has been demonstrated in previous studies, but we have focused on the effects of sharpening and image resolution, as shown in Table 5.2. Reducing the resolution of the JPEG images increased the proportion of mixed pixels by 2–3%, when going from 10-megapixel to 0.1-megapixel resolution; however, most of the increase was in images with less than 1-megapixel resolution. Images with 1- to 10-megapixel resolution had a very similar proportion of mixed pixels and a very similar gap fraction. Image sharpening decreased the proportion of mixed pixels by 2–3%, an effect that was similar magnitude to altering the exposure by one stop. The effect of the sharpening filter on the gap fraction was greatest in very low resolution images (0.1 megapixel) and least in images with 5–10-megapixel resolution. There was no significant difference between the 16-bit TIF images and 8-bit JPEG images in either the gap fraction or proportion of mixed pixels (Table 5.3); a similar result was recently obtained by Pueschel et al. (2012). Sharpening increased the gap fraction and reduced the proportion of mixed pixels in a similar manner for both image types.

5.3.3 Future Developments

Pixel classification is of great importance because it transforms qualitative visual information in an original grayscale image into a binary dataset that is suitable for the extraction of quantitative data. Automatic thresholding techniques, whether they be global, regional or local, are becoming increasingly widespread and readily available and should be used in preference to manual methods. Automated classification algorithms are increasingly being incorporated into the digital workflow of software packages that are specific to the analysis of HP, thereby ensuring objectivity and repeatability. Targeting of specific threshold algorithms towards specific image regions should lead to further improvements in pixel classification.

Although the proportion of mixed pixels in HP is first and foremost determined by the gap fraction, photographic exposure variation of up to one f-stop, together with image sharpening, can both alter the proportion of mixed pixels by several percent. In modern digital cameras with at least 4-megapixel resolution (i.e., Nikon Coolpix 4500 and DSLR cameras), pixel resolution is likely to have little effect upon either mixed pixels or the gap fraction. We also found little difference in the proportion of mixed pixels in 8-bit JPEG and 16-bit TIF images, but further research into the potential of wide dynamic range images (Cescatti 2007; Lang et al. 2010; Verhoeven 2010), the use of multiple exposures (Pueschel et al. 2012), and the incorporation of spectral information from multiple color channels may lead to further improvements in image analysis.

5.4 Calculation and Output

Having obtained a good quality binary image from the classification procedure, the extraction of gap fractions, gap sizes and other information requires a sampling scheme. This is not fundamentally different from other ecological sampling, whether it be from plant populations or soils, in that sampling units are identified by their spatial arrangement on grids or transects. The peculiarity of HP is the circular and polar organization of these sampling units, which require HP-specific tools and approaches to extract information (Hall et al. 2017). The user must (1) specify the lens projection function to obtain gap fraction data at each zenith angle, (2) choose the width of sun tracks for direct radiation calculations, and (3) choose the number and width of azimuth sectors and the number of zenith annuli for modeling diffuse solar radiation and canopy structure. Choices will depend upon the preferred application. For example, the preferred settings for modeling diffuse radiation may differ from those used to estimate LAI or the clumping index. Further, the whole hemisphere is generally required for deriving canopy openness, sky view factors and for modeling radiation, but the zenith and horizon regions are often excluded when estimating LAI and clumping indices, as these are prone to optical and sampling errors.

5.4.1 Data Extraction from Binary Images

5.4.1.1 Lens Projection Functions

Prior to extracting any data, a lens projection function is needed to map pixel coordinates onto zenith angles. There is a range of possible lens projection functions (see Schneider et al. 2009 for a recent summary), and fisheye lenses may not have the commonly assumed polar or equidistant projection (Herbert 1987). Even fisheye lenses with a polar projection are likely to deviate from a true polar projection (Herbert 1987; Blennow 1995; Frazer et al. 2001). Small errors in the lens projection function can result in substantial errors in the area projected on the image plane (Herbert 1987), and up to 10% error in estimates of light transmission or leaf area index (Frazer et al. 1997). Where available, specific lens calibrations should be used and many software packages allow the user to specify the lens projection function. If calibrations are not available for a specific lens, then calibration methods that range from very simple (e.g., Clark and Follin 1988; Pekin and Macfarlane 2009) to quite complex and precise (e.g., Schneider et al. 2009) are available to the user. Frazer et al. (1997) suggested that a sixth-order polynomial was required to describe the relationship between pixel coordinates and zenith angle, but a third-order polynomial has frequently been found to be sufficient (Frazer et al. 2001; Jonckheere et al. 2005; Pekin and Macfarlane 2009). At a minimum, these methods can be used to determine the generic projection function (e.g., polar or equal area). Whether or not a more accurate calibration is needed will depend upon the application.

5.4.1.2 Sun Maps

A sun map tracks the location of the sun as it moves through the sky during the day and during the year. It is used only for modeling direct radiation transmission, in conjunction with a model of sky radiance. It relies upon accurate specification of both the orientation and slope of the HP. Depending upon the software, it is frequently possible to specify both the time period (in days) for radiation modeling, and the time-step for calculations (typically in minutes or hours). The sun map is separate from the sky map and uses the whole hemisphere.

Modeling solar radiation requires that the spatial resolution of the sky map be adjusted to the time resolution (e.g., 5 min, 1 h) of the sun position (sun-tracks) between sunrise and sunset (Mitchell and Whitmore 1993). For studying sunfleck dynamics and potential sunshine beneath the canopy for any day of the year, the choice is critical if we are to consider the daily sun-track and the precise position of the sun. A suitable sky map is particularly important if a heterogeneous clear sky model is used for deriving diffuse sky radiance. Indeed, the estimation of diffuse sky radiance and diffuse irradiance beneath plant canopies depends strictly upon the position of the sun in each of the sky segments. The problem is not so critical if the sky radiance distribution is isotropic (see Schleppi and Paquette 2017).

5.4.1.3 Sky Maps

Image Acquisition

- Camera positioning
  - Horizontal/vertical position
  - Zenithal/parallel to slope

- Exposure
- Evenness of sky lighting
- Evenness of foliage lighting (reflections): direct sunlight
- Optical distortion

Image Analysis

- Image editing/enhancement
- Consideration of missing areas
- Distinguishing foliage from canopy openings
- Assumed direct sunlight distribution
- Assumed diffuse skylight distribution
- Assumed surface of interception

Violation of Model Assumptions

- Assessment of G-function variations
- Leaf angle variability
- Consideration of clumping
- Consideration of woody material
- Consideration of slope effects
- Consideration of clear sky versus overcast sky models

Studies of diffuse solar radiation regimes and canopy structural variables, such as gap fraction distribution, LAI and solar radiation regimes, require the specification of a sky map. A *sky map* consists of the intersection of a given number of zenith *annuli* and azimuth *sectors*, which define unit *elements* or *segments*, and this defines the spatial resolution of the extraction (Fig. 5.6). The term *sector* is used here in the sense of azimuthal slice. Modeling of diffuse radiation utilizes the entire hemisphere but still depends upon the number of both zenith annuli and azimuth sectors. Gap fraction inversion models for estimating LAI are most sensitive to the number of zenith annuli, except in the case of the finite-length averaging method of Lang and Xiang (1986), which is highly sensitive to both the number of zenith annuli and azimuth sectors (van Gardingen et al. 1999). When characterizing canopy structure, users are faced not only with the choice of what number of annuli to use, but over what range zenith angles should be sampled.

A variety of analytical and technical factors will influence the choice of sky map. To begin with, users are faced with a choice between sampling a single zenith direction and sampling multiple directions. The *single direction* approach usually involves sampling a single zenith annulus that is centered at about 57° (e.g., Weiss et al. 2004; Leblanc et al. 2005; Jonckheere et al. 2006a), which takes advantage of the fact that the foliage projection function ($G(\theta)$) is virtually independent of the leaf inclination distribution function (Warren Wilson 1963; Bonhomme et al. 1974), thereby simplifying the estimation of LAI. The *multiple direction* approach is applied

Fig. 5.6 Sky divisions

to more complex methods such as Miller's formula and its variants (Miller 1967), which includes graphical and finite-length averaging methods (Lang 1986; Lang and Xiang 1986), and the ellipsoidal model (Campbell 1990), among others. However, optical and sampling problems arise at extreme angles of view, near the zenith and horizon. *Near the zenith*, foliage is poorly sampled, due to the restricted view angle (Fournier et al. 1996, 1997). Foliage density and location varies strongly from place to place, such that many more images are required to account for variation near the zenith. HP is also subject to 'blooming' near the zenith as a result of image over-exposure (Leblanc et al. 2005). *Near the horizon* on horizontal terrain, openings are very small or absent, and blurring may occur due to multiple scattering effects. Together with a long path length, this blurring reduces the distinction between foliage and sky, and between small openings and sunlit foliage (Leblanc et al. 2005). Depending upon lens quality, aperture and exposure settings, vignetting (i.e. the reduction of an image's brightness or saturation at the periphery compared to the image center) also tends to obscure the horizon (Wagner and Hagemeier 2006). Thus, the horizon region is only weakly informative or not informative at all. Consequently, annuli of the sky map are sometimes discarded up to 20° above the horizon. The most informative range of angles lies between 10° and 70° for horizontal ground, but Miller's formula requires that strictly speaking, the whole hemisphere should be used (see Leblanc and Fournier 2017). The zenith angle range from 10° to 80° is sometimes used as a compromise between the optical and sampling limitations of HP at extreme view angles, and given the strict requirements of Miller's theorem (see, for example, Frazer et al. 2017). Technical issues relating to the choice of zenith angle range are also discussed by Leblanc and Fournier (2017).

As for the number of annuli, both the Hemiview manual (Rich et al. 1999) and Rich (1989) recommend that eight azimuth sectors (45° divisions) and eighteen to twenty zenith annuli (5° divisions) be used for analyses of HP. Blennow (1995) also recommended 20 annuli for radiation modeling. Further, early studies that were aimed particularly at radiation modeling (Anderson 1964, 1981; Becker et al. 1989) conformed to this standard. Some later studies (e.g., Wagner and Hagemeier 2006) used a similarly large number of zenith annuli. Since the development of the LAI-2000 instrument (Plant Canopy Analyzer, LI-COR Environmental, Lincoln, NE, USA), however, many studies that were aimed at LAI estimation, rather than radiation modeling, have compared the LAI-2000 with HP. They have used six zenith annuli, the innermost five of which correspond closely with the five rings of the LAI-2000 (e.g., Planchais and Pontailler 1999; Bréda 2003; Jonckheere et al. 2004; Macfarlane et al. 2007b). Other authors have arbitrarily opted for annuli of 9° or 10° (e.g., Chen et al. 1991; Leblanc et al. 2005; Macfarlane 2011). Leblanc et al. (2005) noted that there was little difference between using six or ten zenith annuli when applying the Chen and Cihlar (1995) clumping-correction method (discussed by Frazer et al. 2017). Systematic studies of different sky maps have primarily sought to investigate the behavior of the log-averaging method and the effects of sky map choice on canopy structure calculations (Lang and Xiang 1987; van Gardingen et al. 1999; Gonsamo et al. 2010). Hence, the choice of sky map will depend upon the purpose of the study, and whether or not a comparison is being made with the LAI-2000. More than one sky map may be desirable for studies that seek both to characterize canopy structure and to model canopy radiation. More zenith annuli are likely to be required for modeling radiation than for estimating LAI, particularly in relation to the log-averaging method, given that "the optimal combinations of zenithal annuli and azimuthal sector angular widths that suit all canopy types remains a difficult practical problem that is often overlooked (Gonsamo et al. 2010, p. 1588)."

The 'finite-length averaging' method, which is also known as 'logarithmic gap-fraction averaging' (Lang and Xiang 1986; Fournier et al. 2017; Frazer et al. 2017), deserves special mention owing to its sensitivity with respect to the sky map and the suggestion made by some authors regarding a variable number of azimuth sectors over a fixed number of zenith annuli (van Gardingen et al. 1999; Walter et al. 2003; Gonsamo et al. 2010). The method is presented in detail in the next chapter (Frazer et al. 2017). Briefly, it averages the logarithms of the gap fractions in each azimuth sector within a zenith annulus, rather than using the linear average, to obtain a more accurate, clumping-corrected estimate of LAI. However, as the logarithm of zero is undefined, a problem arises with 'empty segments' that are completely obstructed by foliage and have a null gap fraction, especially in dense canopies and near the horizon. Several solutions have been proposed to overcome this problem: (1) "adjust the gap fraction for each empty segment by assigning a gap fraction equivalent to one pixel (picture element) per segment" (van Gardingen et al. 1999, p. 250), and (2) merge each black segment with the following non-black segment to produce a bigger segment with an average value (Cescatti 2007; Gonsamo et al. 2010). The first option, which is commonly used, imposes an arbitrary gap fraction that varies with both image resolution and the sky map (Cescatti 2007), and tends to increase LAI estimates more or less in proportion to the number of empty segments within the sky map (recall that 'empty segments' are the opposite of 'empty sky'). A variant specifies a fixed gap fraction for empty segments rather than a single pixel. The second option offers a more flexible solution, to which LAI estimates are less sensitive (Gonsamo et al. 2010). The finite-length averaging method will yield larger estimates of LAI as the number of segments (either annuli or sectors) in the sky map is increased. Consequently, authors are advised to conduct their own tests of the method and compare results against destructive sampling estimates of LAI.

5.4.2 *Calculation of Solar Radiation*

Radiation exchange within canopies is the driving force for the surface energy balance as it influences canopy leaf temperature and provides the energy for photosynthesis and plant growth (Flerchinger and Yu 2007). Numerous models of radiation transfer within canopies have been proposed (for reviews, see Ross 1981; Varlet-Grancher et al. 1993; Campbell and Norman 1998; Gastellu-Etchegorry et al. 2004; Widlowski et al. 2007a, b).

Simulation models for estimating the surface radiation balance may be divided into three categories: single-layer, dual-source and multi-layer models (Flerchinger and Yu 2007). Multi-layer models require the estimation of the transmittance, the reflectance, and the scattering of radiation within each canopy layer, and the computation of downward direct diffuse radiation being transmitted, reflected, scattered and absorbed by each layer. Optical properties of leaves and bark, among other tissues, as well as their respective orientations, are the major input parameters that are required to compute the energy balance for each layer within the canopy.

In analyzing solar radiation regimes, the radiation that is scattered by foliage is mostly neglected if estimated from HP. This is often justified by the statement that scattered radiation represents, at the most, a small percentage of global radiation penetrating canopy openings, except in deep shade (up to 43% of total radiation; Mitchell and Whitmore 1993). Hence, site factors computed from HP are often for penetrated radiation only, whereas those measured by instruments include scattered radiation. Due to inherent complexities and lack of experiments and data, several other radiation components are often ignored and, thus, usually not included in radiation estimates by HP, among which are radiation scattering by leaves and other surfaces, and penumbral effects. Penumbral effects are often considered negligible compared to the other components. However, the estimation of these components is essential for a correct biophysical description of the complexity of the radiation microclimate.

The theory of radiation transmission to the Earth's surface and its interaction with plant canopies is described in detail by Schleppi and Paquette (2017). In the following sections, we deal with modeling of diffuse radiation on more complex terrain, and the choice between isotropic and anisotropic sky models. We then evaluate the use of HP to estimate the R/FR ratio, penumbral effects, and sunfleck dynamics.

5.4.2.1 Complex Terrain, Diffuse Radiation Models and Penumbra

The site factor approach is most commonly used in photographic studies of radiation regimes. Mitchell and Whitmore (1993, p. 38) define site factors as "the radiation received beneath a canopy as a fraction of that incident on the canopy, all measurement conditions being identical." The direct site factor (DSF) is for direct (beam) radiation, while the indirect site factor (ISF) is for indirect (diffuse) radiation. The weighted (for incident direct and diffuse radiation) sum of the two is the global site factor (GSF) or gap light index (GLI; see Schleppi and Paquette 2017, Eq. 2.37).

Both isotropic and anisotropic sky models, and the calculation of ISF, were discussed by Schleppi and Paquette (2017) in relation to horizontal ground; however, the horizontal surface is often not the orientation that receives the most radiation under vegetation (Widlowski et al. 2007a). In complex terrain, slope is an additional factor that should be included in the analysis of HP to correctly estimate biophysical parameters (Dubayah and Rich 1995; Frazer et al. 1997; Walter and

Torquebiau 2000; Courbaud et al. 2003; Duursma et al. 2003; Wang et al. 2005; Schleppi et al. 2007; España et al. 2008; Gonsamo and Pellikka 2008). Diffuse irradiance on slopes is determined by the fraction of the overlying hemisphere that is obstructed by topographical relief, and by additional reflected radiation from adjacent terrain. Reflected radiation may be a large fraction of the diffuse irradiance, particularly on steep slopes and under cloudless skies, when the ratio of diffuse to global radiation is small (Steven and Unsworth 1979). Under clear skies, diffuse irradiance represents about 16% of global irradiance for shortwave radiation (Grant et al. 1996), but this proportion increases with atmospheric turbidity. This component is difficult to measure accurately and is often neglected (Mitchell and Whitmore 1993), although it has been considered in some models (Dogniaux 1975; Dubayah and Rich 1995; Wang et al. 2005). Furthermore, lateral canopy openings on steep slopes are more numerous and greater downslope than upslope, resulting in greater penetration of radiation from large zenith angles compared to horizontal terrain.

Diffuse light is direct radiation that has been scattered by the atmosphere, but clear and overcast skies scatter light quite differently. The diffuse component of solar radiation under overcast conditions is usually modeled using an isotropic sky model (e.g., the Uniform Overcast [UOC] and Standard Overcast [SOC] sky models), which assumes radiation is scattered equally in all directions. In reality, most diffuse radiation originates from near the solar disk; this circumsolar radiation is confined to a circle of about 15° solid angle within 20° of solar aureole (Grant et al. 1996). The UOC model is only likely to be realistic under heavily clouded conditions, whereas the SOC model may provide reasonable estimates of diffuse radiation penetration under clear or partly cloudy conditions, over longer time periods (Moon and Spencer 1942; Grant et al. 1996).

Both isotropic models of diffuse radiation are poor estimators of instantaneous diffuse radiation, and there is evidence that even in horizontal terrain, the SOC model tends to underestimate ISF compared to more complex anisotropic clear-sky models (Hutchinson and Matt 1976). As a result anisotropic models are preferred to isotropic models for more detailed estimates of canopy radiation parameters. For example, Hutchison and Matt (1976) predicted higher penetration of Photosynthetically Active Radiation (PAR) into a deciduous forest based on an anisotropic clear sky distribution compared to the SOC or UOC. Goudrian (1988, cited by Grant et al. 1996) estimated that the sky radiance distribution influences modeled photosynthesis to a greater extent than the leaf angle distribution; hence, for modeling photosynthesis, models of diffuse radiation that are more detailed than UOC or SOC are desirable. Sunfleck dynamics—the frequency distribution and duration of sunflecks—is an important driver of photosynthesis and plant growth in seedlings, saplings, and in the understory (e.g., Ustin et al. 1984; Chazdon 1988; Pearcy 1990). HP provides a means of quantifying sunfleck dynamics through the DSF sky map, but this is complicated by the fact that the sun is not truly a 'point source' of light; rather, it has an apparent diameter of 0.5°. This causes 'penumbra' (a region in which only a portion of the light source is obscured by the occluding body), where movement of the sun's aureole across the sky causes areas of full sunlight to change to deep shade (umbra) gradually rather than abruptly (Miller and Norman 1971; Anderson and Miller 1974). A sunfleck may consist entirely of a penumbral area if the ratio between the diameter of the canopy opening and height of the canopy is 0.01 (Smith et al. 1989); thus, high spatial resolution HP are important for accurate detection of penumbral gaps in the understory. An analytical approach to estimating the amount of penumbra from HP is detailed in ter Steege (1997).

5.4.2.2 Red/Far-Red Ratio

Various publications on 'Light quality, photoperception, and plant strategy' (Smith 1982), 'The perception of shade' (Holmes et al. 1983), and 'The color of light in forests and its implications' (Endler 1993), illustrate the need for ecophysiologists and ecologists to understand the spatio-temporal variability of the canopy radiative microclimate. The red/far-red ratio is a key characteristic of light quality within the canopy, but there are few published methodologies for determining the red/far-red ratio directly from HP. Empirical relationships have been established between the red/far-red ratio and fractional transmission of PAR, by means of quantum sensors (e.g., Lee 1987; Messier et al. 1989; Capers and Chazdon 2004). Capers and Chazdon (2004) obtained the following relationship between diffuse radiation transmittance and the red/far-red ratio under overcast skies in a tropical rain forest:

$$\%T = 0.548 + \exp\left(-2.4541 + 5.6594 * (R\,/\,FR)\right) \tag{5.1}$$

This equation explained 97.2% of the variation in the percent of diffuse transmittance ($r = 0.986$). As far as we know, the first method using HP was introduced by ter Steege (1997, following Alexandre 1982), based on the Poisson theory of beam penetration into a random canopy. The photographic quality requires that the diffuse site factor not be too small because unreliable values for the red/far-red ratio are easily obtained in very closed canopies under deep shade. Both the empirical approach and modeling of the red/far-red ratio from HP require proper calibration. Statistical parameters that are derived from curve fitting may only be of local use, depending heavily upon leaf optical properties (e.g., Gates 1980; Lee and Graham 1986; Poorter et al. 1995).

5.4.3 Canopy Structure from Hemispherical Photographs

Several models are available for estimating leaf area index (LAI), together with average leaf inclination angle (ALIA) or the Leaf Inclination Distribution Function (LIDF) from HP, following the extraction of gap fractions from one or more zenith angles, based on the principle of light extinction in the canopy. The most frequently used 1-D radiation transfer models in vegetation canopies assume the Poisson model (see also Chaps. 3 and 6). One-D radiation transfer models simplify the problem of 3-D canopy radiation transfer (Pinty et al. 2004) by substituting the 'real' LAI with an 'effective' LAI, viz., L_e, which embeds foliage clumping into LAI. In 1-D representations of canopies, the canopy is assumed to be a turbid medium, and the gap fraction is considered to be equivalent to below-canopy light transmittance. In the following, we illustrate the calculation of L_e from ground-based gap fraction measurements, both in the simple case of only one zenith direction and the more complex case of measurements over a range of zenith angles. Many of these methods are used in the various commercial and freeware software packages for analyzing HP as discussed by Hall et al. (2017). While software packages themselves have occasionally been compared (e.g., Hall et al. 2017), we are aware of no formal comparisons of the various algorithms that they contain, using images that have been subjected to common processing and classification methods.

5.4.3.1 Use of a Single Zenith Direction

As described by Leblanc and Fournier (2017), Warren Wilson (1963) showed that for a zenith angle (θ) of one radian or 57.3°, the canopy extinction coefficient ($G(\theta)$) is almost independent of the leaf inclination distribution function ($G(57.3)$, $\cong 0.5$) resulting in a light extinction coefficient (k) of approximately $\cong 0.9$. Some literature cites an angle of 57.5° for convenience (e.g., Weiss et al. 2004; Jonckheere et al. 2005). The method is based on gap fraction measurements at a zenith angle of 57.3°. The gap fraction can also be averaged between about 53° and 68° or some similar range. Using the contact frequency at this angle, Warren Wilson (1963) derived L_e independently of the leaf inclination distribution function, with an accuracy of about 7%. Bonhomme et al. (1974) applied this technique using gap fraction measurements (see Eq. 2.7) and found good agreement between actual and estimated L_e values for young crops. The method has rarely been applied to forests and it has been suggested that in mature forest, tree stems may greatly influence the gap fraction at this zenith angle (Macfarlane et al. 2007b). Macfarlane et al. (2007a) suggested that the 57° method may have significant advantages over multiple-angle HP in young, short stands with a large ratio of foliage to stem wood. Although this is the simplest method to estimate L_e from HP, multiple viewing angles offer the potential for further increasing the accuracy of L_e estimation by reducing the reliance of the result on the small quantity of data that are collected at a single zenith angle, and also offer the potential for simultaneously estimating ALIA.

5.4.3.2 Use of Multiple Zenith Directions

Many methods have been developed and applied for deriving both LAI and LIDF from gap fraction measurements over multiple zenith directions (e.g., Bonhomme

1973; Campbell and Norman 1989; Jupp et al. 1980; Lang 1986, 1987; Miller 1967; Norman and Campbell 1989; Campbell 1990). In this section, we briefly describe the most commonly used methods for estimating L_e from gap fraction measurements at multiple zenith angles. In all cases, it is assumed that foliage clumping is negligible (for a detailed discussion of clumping, see Frazer et al. 2017). Generally, the analytical methods produce similar results, although van Gardingen et al. (1999) found that the constrained linear least-squares inversion method (Norman and Campbell 1989) produced more variable results than two other methods, if constraints on the inversion procedure were not adequate. Unpublished results (J-M.N. Walter, *pers. comm.*) suggest that the method of Bonhomme (1973) consistently yields larger estimates of L_e than do other algorithms, while the method of Norman and Campbell (1989) consistently yields smaller estimates of L_e.

Constrained linear least-squares inversion technique: This method estimates L_e by solving, using least-squares, simultaneous linear equations in which the gap fraction at each of several zenith angles is a function of L_e for at least two leaf angle classes. A constraint matrix is used to avoid negative or other unrealistic values. The method yields L_e and the fraction of L_e within each leaf angle class. The estimate of L_e from this method is generally more robust than the estimate of LIDF. Details of the method are presented in Perry et al. (1988) and Norman and Campbell (1989).

Constrained non-linear least squares technique: An alternative to dividing L_e between different leaf angle classes is to define the extinction coefficient using a generalized model. Campbell (1986, 1990) proposed a widely used model in which LIDF is represented by the distribution of the area on the surface of an ellipsoid of revolution. The shape of the distribution is represented by a single parameter, x, which is equal to one for a spherical LIDF. Canopies with a horizontal LIDF have $x > 1$ and canopies with a vertical LIDF have $x < 1$. Using non-linear least-squares regression techniques (see Norman and Campbell 1989 for details), L_e and x are calculated, subject to the constraint that $x > 0$. Although this method is widely used, many other models exist to represent LIDF, and recent studies have identified two-parameter LIDF models (e.g., Beta distribution of Goel and Strebel 1984) as being more robust than single parameter models (Wang et al. 2007).

Miller's theorem: Miller (1967) proposed a theorem to estimate L_e that does not require prior knowledge of the LIDF:

$$L_e = 2\int_0^{\pi/2} -\ln[P(\theta)]\cos\theta\sin\theta\, d\theta \tag{5.2}$$

Welles and Norman (1991) proposed a practical method to compute the integral of Eq. 5.2 from gap fraction measurements in several directions. This is implemented in the LAI-2000/2200 instrument to estimate L_e from measurements of the gap fraction at five zenith angle ranges between 0°–75°, although this method, strictly speaking, requires that the entire zenith angle range of gap fractions is sampled (see Sect. 4.1.2).

Lang's regression method: Lang (1987) assumed a linear relationship, with slope (A) and intercept (B) in radians, between field measurements of $-\ln P(\theta_v)\cos(\theta_v)$ and θ_v, and substituted this into Miller's theorem to calculate L_e as

$$L_e = 2\,(\mathrm{A} + \mathrm{B}). \tag{5.3}$$

ALIA was also calculated from an empirical polynomial relationship with B. Lang (1987) found that this simple approach was robust across a wide range of zenith angle ranges.

5.4.3.3 Methodological Improvements and Challenges

Accounting for foliage clumping during analysis of canopy gap fraction distribution is a key step to obtaining accurate LAI values and is dealt with in detail in the following chapter (Frazer et al. 2017). Failure to account for clumping results in violation of the assumption that foliage is randomly distributed (Lang and Xiang 1986), with underestimation of LAI by up to 40% (e.g., Wang et al. 1992; Martens et al. 1993; Eschenbach and Kappen 1996; Strachan and McCaughey 1996). Failure to account for foliage clumping will also affect the calculation of LIDF from HP; LIDF will appear more erectophile if foliage is clumped (Chen and Black 1991; Macfarlane et al. 2007b). A great deal of current research effort is directed towards the development and testing of models and methods that better quantify both foliage clumping and LIDF in forest canopies, including the application of fractal dimensions and lacunarity analyses (e.g., Plotnick et al. 1996; Jonckheere et al. 2006b; Wang et al. 2007; Frazer et al. 2017). A successful analytical method should account not only for foliage clumping, but it should also produce realistic representations of actual canopy LIDFs that are insensitive to treatments such as thinning.

A major challenge for testing of analytical methods is the lack of reliable reference measurements. Destructively measured LAI will likely achieve an accuracy of 10% at best (Kucharik et al. 1999) and it is probably unreasonable to expect indirect methods to achieve greater accuracy. Estimation of LAI from the gap fraction that is measured at 57.3° has been proposed as a reference method for LAI but has been little tested in forests. Unlike crops, forest estimates may suffer from significant bias due to woody material in sparse forests and many or large stems. Ideally, indirect methods should be evaluated not only against independent measures of LAI, but against independent measures of gap fraction, LIDF and clumping index. Independent estimates of the gap fraction can be obtained from ground-based LiDAR (Côté et al. 2011) or from telephotography at a limited range of zenith angles (Macfarlane et al. 2007b), but this is expensive in the former case and tedious in the latter. Field methods to estimate the LIDF are similarly either too expensive or tedious for routine implementation (Kucharik et al. 1998b; Sinoquet et al. 2009: Ryu et al. 2010), but such approaches may be necessary to rigorously test analytical methods that are applied to HP in research studies.

Two alternative approaches exist for obtaining highly accurate estimates of the gap fraction: LIDF and LAI in the field. The first is to obtain accurate relative changes in the gap fraction and LAI through controlled thinning studies. Macfarlane et al. (2007b) progressively thinned a stand of *Eucalyptus* trees and tested whether the field and analytical methods that were applied yielded relative decreases in LAI, relative increases in gap fraction, decreasing clumping index and constant LIDF. However, such studies are rare, owing to the cost and labor involved. Similarly, a comparison could be made between forests with contrasting LIDF (erectophile versus planophile) to test whether a proposed method detects the different leaf angles. The second alternative is to perform initial tests of methods on simulated HP images that differ in LAI, degree of foliage clumping, and LIDF. Images can be simulated using 3-D modeling and ray tracing techniques (see Frazer et al. 2017; Jonckheere et al. 2006b). Methods that perform robustly could then be tested in more expensive field studies that could involve novel methods such as ground-based LiDAR (Côté et al. 2011) to derive detailed structural information, in addition to LAI from destructive sampling.

Distinguishing woody material from foliage is another important problem that remains to be resolved for the analysis of HP. Chen (1996) introduced the α parameter, which is expressed as the ratio of woody area index to plant area index (Leblanc and Fournier 2017). Chen (1996) evaluated α from intensive destructive measurements. Barclay et al. (2000) proposed correcting L_e using a 'bole area index' that was directly measured using a map of the stand (position of the stems, diameter, and height, among other variables). Both of these methods are quite tedious, and one solution could be a classification method that can be applied to digital imagery. Using the Multiband Viewer Instrument (MVI), Kucharik et al. (1998a) used a clustering algorithm called BIRCH (Balanced Iterative Reducing And Clustering using Hierarchies). BIRCH is based on the k-means clustering algorithm (Hartigan and Wong 1979), which allows two bands to be weighted by different factors and, therefore, to identify sunlit and shaded foliage, sunlit and shaded branch area, clouds, and blue sky. One method involves the use of devices, such as digital cameras, which allow separating between green and non-green vegetation elements, thereby allowing imaging in several wavebands. Chapman (2007) recently demonstrated the adaptation of a digital camera to obtain infrared images of canopies and it has been suggested that infrared images could be used to separate stems and branches from foliage, owing to the generally greater reflection of infrared wavelengths by foliage. Another alternative is the use of ground-based LiDAR and models of tree structure to estimate the woody surface area of canopies (Côté et al. 2011). Separating woody material from green vegetation elements in HP remains an active area of research.

5.5 Conclusion

HP combines the art of photography with the science of forest micrometeorology. However, the advent of inexpensive and high-quality digital cameras and lenses, combined with continuously improving and automated image processing methods, is shifting HP ever further away from art and ever closer to an objective science. In this chapter, we have dealt in detail with image quality and image processing to create binary datasets from which measures of forest radiation and structural attributes could be modeled. Much of the control over image quality is exerted in the field during acquisition, but processing steps such as image sharpening and pixel classification have an important bearing on the final results of analyses. Automatic classification methods for a wide range of light and canopy structure conditions are becoming robust and consistent, and remove the subjectivity of the operator; a global cluster method provides a useful baseline against which more advanced image processing algorithms can be evaluated. Further research into the potential of images with wide dynamic ranges, incorporation of spectral information from multiple color channels, and targeting of specific threshold algorithms towards specific image regions should lead to further improvements in pixel classification.

HP is a versatile methodology that can be used for studying forest light climates, which is important in ecology, and for quantifying canopy structure and extent, which is important in growth and hydrological modeling. Although many algorithms for modeling radiation and canopy characteristics are well developed, there is still uncertainty surrounding the optimal sky map to use for calculating LAI, especially from the logarithmic-averaging method. The treatment of foliage clumping and woody material in canopies is under active research, and improvements in both field and analytical methods may soon result in estimates of LAI from HP that approach those obtained from destructive sampling in many circumstances. As Norman and Campbell observed in 1989 (p. 323), "indirect methods are not likely to replace direct methods but should reduce the need for many labor-intensive canopy structure measurements."

References

Abutaleb AS (1989) Automatic thresholding of gray-level pictures using two-dimensional entropy. Comput Vision Graph Image Proc 47:22–32

Alexandre DY (1982) Pénétration de la lumière au niveau du sous-bois d'une forêt dense tropicale. Ann Sci For 39:419–438

Anderson MC (1964) Studies of the woodland light climate. I. The photographic computation of light conditions. J Ecol 52:27–41

Anderson MC (1981) The geometry of leaf distribution in some South-Eastern Australian forests. Agric For Meteorol 25:195–205

Anderson MC, Miller EW (1974) Forest cover as a solar camera: penumbral effects in plant canopies. J Appl Ecol 11:691–697

Barclay HJ, Trofymow JA, Leach RI (2000) Assessing bias from boles in calculating leaf area index in immature Douglas-fir with the LI-COR canopy analyzer. Agric For Meteorol 100:255–260

Baret F, Andrieu B, Folmer JC, Hanocq JF, Sarrouy C (1993) Gap fraction measurement from hemispherical infrared photography and its use to evaluate PAR interception efficiency. In: Varlet-Grancher C, Bonhomme R, Sinoquet H (eds) Crop structure and light microclimate. Characterization and applications. INRA, Paris. p 359–372

Baret F, de Solan B, Lopez-Lozano R, Ma Kai, Weiss M (2010) GAI estimates of row crops from downward looking digital photos taken perpendicular to rows at 57.5° zenith angle: theoretical considerations based on 3D architecture models and application to wheat crops. Agric For Meteorol 150:1393–1401

Becker P, Erhart DW, Smith AP (1989) Analysis of forest light environments. Part I. Computerized estimation of solar radiation from hemispherical canopy photographs. Agric For Meteorol 44:217–232

Blennow K (1995) Sky view factors from high-resolution scanned fish-eye lens photographic negatives. J Atmos Ocean Tech 12:1357–1362

Bonhomme R (1973) Analyse de la surface des taches de soleil, de l'indice foliaire et de l'inclinaison moyenne des feuilles à l'aide de photographies hémisphériques. In: Slatyer RO (ed) Plant response to climatic factors: Proceedings of Uppsala symposium. UNESCO, Paris, pp 369–376

Bonhomme R, Chartier P (1972) The interpretation and automatic measurement of hemispherical photographs to obtain sunlit leaf area and gap frequency. Israel J Agr Res 22:53–61

Bonhomme R, Varlet-Grancher C, Chartier P (1974) The use of hemispherical photographs for determining the leaf area index of young crops. Photosynthetica 8:299–301

Bréda NJJ (2003) Ground-based measurements of leaf area index: a review of methods, instruments and current controversies. J Exp Bot 54(392):2403–2417

Campbell GS (1986) Extinction coefficients for radiation in plant canopies calculated using an ellipsoidal inclination angle distribution. Agric For Meteorol 36:317–321

Campbell GS (1990) Derivation of an angle density function for canopies with ellipsoidal leaf angle distributions. Agric For Meteorol 49:173–176

Campbell GS, Norman JM (1989) The description and measurement of plant canopy structure. In: Russell G, Marshall B, Jarvis PG (eds) Plant canopies: their growth, form and function. Cambridge University Press, Cambridge, pp 1–19

Campbell GS, Norman JM (1998) An introduction to environmental biophysics, 2nd edn. Springer, Berlin

Capers RS, Chazdon RL (2004) Rapid assessment of understory light availability in a wet tropical forest. Agric For Meteorol 123:177–185

Cescatti A (2007) Indirect estimates of canopy gap fractions based on the linear conversion of hemispherical photographs. Methodology and comparison with standard thresholding techniques. Agric For Meteorol 143:1–12

Chapman L (2007) Potential applications of near infra-red hemispherical imagery in forest environments. Agric For Meteorol 143:151–156

Chazdon RL (1988) Sunflecks and their importance to forest understory plants. Adv Ecol Res 18:1–63

Chazdon RL, Field CB (1987) Photographic estimation of photosynthetically active radiation: evaluation of a computerized technique. Oecologia 73:525–532

Chen JM (1996) Optically-based methods for measuring seasonal variation of leaf area index in boreal conifer stands. Agric For Meteorol 80:135–163

Chen JM, Black TA (1991) Measuring leaf area index of plant canopies with branch architecture. Agric For Meteorol 57:1–12

Chen JM, Black TA, Adams RS (1991) Evaluation of hemispherical photography for determining leaf area index and geometry of a forest stand. Agric For Meteorol 56:129–143

Chen JM, Cihlar J (1995) Plant canopy gap-size theory for improving optical measurements of leaf-area index. Appl Opt 34:6211–6222

Clark JA, Follin GM (1988) A simple "equal area" calibration for fisheye photography. Agric For Meteorol 44:19–25
Clearwater MJ, Nifinluri T, van Gardingen PR (1999) Forest fire smoke and a test of hemispherical photography for predicting understorey light in Bornean tropical rain forest. Agric For Meteorol 97:129–139
Côté J-F, Fournier RA, Egli R (2011) An architectural model of trees to estimate forest structural attributes using terrestrial LiDAR. Environ Modell Softw 26:761–777
Courbaud B, de Coligny F, Cordonnier T (2003) Simulating radiation distribution in a heterogeneous Norway spruce forest on a slope. Agric For Meteorol 116:1–8
Dogniaux R (1975) Variations géographiques et climatiques des expositions énergétiques solaires sur des surfaces réceptrices horizontales et verticales. Miscellanea Série B, No. 38, Institut Royal Météorologique de Belgique, 12 pp
Dubayah R, Rich PM (1995) Topographic solar radiation models for GIS. Int J Geogr Inf Syst 9:405–419
Duursma RA, Marshall JD, Robinson AP (2003) Leaf area index inferred from solar beam transmission in mixed conifer forests on complex terrain. Agric For Meteorol 118:221–236
Endler JA (1993) The color of light in forests and its implications. Ecol Monogr 63:1–27
Englund SR, O'Brien JJ, Clark DB (2000) Evaluation of digital and film hemispherical photography and spherical densitometry for measuring forest light environments. Can J For Res 30:1999–2005
Eschenbach C, Kappen L (1996) Leaf area index determination in an alder forest: a comparison of three methods. J Exp Bot 47:1457–1462
España E, Baret F, Weiss M (2008) Slope correction for LAI estimation from gap fraction measurements. Agric For Meteorol 148:1553–1562
Flerchinger GN, Qiang Yu (2007) Simplified expressions for radiation scattering in canopies with ellipsoidal leaf angle distributions. Agric For Meteorol 144:230–235
Fournier RA, Landry R, August NM, Fedosejevs G, Gauthier RP (1996) Modelling light obstruction in three conifer forests using hemispherical photography and fine tree architecture. Agric For Meteorol 82:47–72
Fournier RA, Rich PM, Landry R (1997) Hierarchical characterization of canopy architecture for boreal forest. J Geophys Res 102(D24):29445–29454
Fournier RA, Mailly D, Walter J-MN, Jonckheere IGC (2017) Acquiring hemispherical photographs in forest environments: from planning to archiving photographs. In: Fournier RA, Hall RJ (eds) Hemispherical photography in forest science: theory, methods, applications. Springer, Berlin
Fraser CS (1997) Digital camera self-calibration. ISPRS Journal. ISPRS J Photogramm 52:149–159
Frazer GW, Fournier RA, Leblanc SG, Walter J-MN (2017) View angle-dependent clumping indices for indirect LAI estimation. In: Fournier RA, Hall RJ (eds) Hemispherical photography in forest science: theory, methods, applications. Springer, Berlin
Frazer GW, Fournier RA, Trofymow JA, Hall RJ (2001) A comparison of digital and film fisheye photography for analysis of forest canopy structure and gap light transmission. Agric For Meteorol 109:249–263
Frazer GW, Trofymow JA, Lertzman KP (1997) A method for estimating canopy openness, effective leaf area index, and photosynthetically active photon flux density using hemispherical photography and computerized image analysis techniques. Natural Resources Canada, Canadian Forest Service, Pacific Forestry Centre, Victoria, BC. Information Report BC-X-373. 73 pp
Gastellu-Etchegorry J-P, Martin E, Gascon F (2004) DART: A 3D model for simulating satellite images and surface radiation budget. Int J Remote Sens 25:75–96
Gates DM (1980) Biophysical ecology. Springer, Berlin
Goel NS, Strebel DE (1984) Simple beta distribution representation of leaf orientation in vegetation canopies. Agron J 76:800–802

Gonsamo A, Pellikka P (2008) Methodology comparison for slope correction in canopy leaf area index estimation using hemispherical photography. For Ecol Manag 256:749–759

Gonsamo A, Pellikka P (2009) The computation of foliage clumping index using hemispherical photography. Agric For Meteorol 149:1781–1787

Gonsamo A, Walter J-MN, Pellikka P (2010) Sampling gap fraction and size for estimating leaf area and clumping indices from hemispherical photographs. Can J For Res 40:1588–1603

Gonzalez RC, Woods RE (2008) Digital image processing, 3rd edn. Pearson Prentice Hall, Upper Saddle River

Grant RH, Heisler GM, Gao W (1996) Photosynthetically-active radiation: ssky radiance distributions under clear and overcast skies. Agric For Meteorol 82:267–292

Hale SE, Edwards C (2002) Comparison of film and digital hemispherical photography across a wide range of canopy densities. Agric For Meteorol 112:51–56

Hall RJ, Côté J-F, Mailly D, Fournier RA (2017) Comparison of software analysis tools designed for hemispherical photography (Chapter 7). In: Fournier RA, Hall RJ (eds) Hemispherical photography in forest science: theory, methods, applications. Springer, Berlin

Hartigan JA, Wong MA (1979) Algorithm AS 136: A k-means clustering algorithm. App Stat 28 (1):100–108

Herbert TJ (1987) Area projections of fisheye photographic lenses. Agric For Meteorol 39:215–223

Holmes MG, Farmer AM, Bartley MR (1983) Perception of shade [and discussion]. Philos T R Soc B 303:503–521

Hutchison BA, Matt DR (1976) Beam enrichment of diffuse radiation in a deciduous forest. Agr Meteorol 17:93–110

Inoue A, Yamamoto K, Mizoue N (2011) Comparison of automatic and interactive thresholding of hemispherical photography. J For Sci-Jpn 57:78–87

Inoue A, Yamamoto K, Mizoue N, Kawahara Y (2004) Effects of image quality, size and camera type on forest light environment estimates using digital hemispherical photography. Agric For Meteorol 126:89–97

Ishida M (2004) Automatic thresholding for digital hemispherical photography. Can J For Res 34:2208–2216. Software: RGBFisheye.exe available at: http://www1.gifu-u.ac.jp/~ishidam/RGBFisheye.htm

Jonckheere I, Fleck S, Nackaerts K, Muys B, Coppin P, Weiss M, Baret F (2004) Review of methods for in situ leaf area index (LAI) determination. Part I. Theories, sensors and hemispherical photography. Agric For Meteorol 121:19–35

Jonckheere I, Nackaerts K, Muys B, Coppin P (2005) Assessment of automatic gap fraction estimation of forests from digital hemispherical photography. Agric For Meteorol 132:96–114

Jonckheere IGC, Muys B, Coppin B (2006a) Derivative analysis for in situ high dynamic range hemispherical photography and its application in forest stands. IEEE Geosci Remote S 2:296–300

Jonckheere I, Nackaerts K, Muys B, van Aardt J, Coppin P (2006b) A fractal dimension-based modelling approach for studying the effect of leaf distribution on LAI retrieval in forest canopies. Ecol Model 197:179–195

Jupp DLB, Anderson MC, Adomeit EM, Witts SJ (1980) PISCES—A computer program for analysing hemispherical canopy photographs. CSIRO, Institute of Earth Resources, Division of Land Use Research. Technical Memorandum 80/23. 22 pp

Kapur J, Sahoo P, Wong A (1985) A new method for gray-level picture thresholding using the entropy of the histogram. Comput Vision Graph 29:273–285

Kucharik CJ, Norman JM, Gower ST (1998a) Measurements of branch area and adjusting leaf area index indirect measurements. Agric For Meteorol 91:69–88

Kucharik CJ, Norman JM, Gower ST (1998b) Measurements of leaf orientation, light distribution and sunlit leaf area in a boreal aspen forest. Agric For Meteorol 91:127–148

Kucharik CJ, Norman JM, Gower ST (1999) Characterization of radiation regimes in nonrandom forest canopies: theory, measurements, and a simplified modeling approach. Tree Physiol 19:695–706

Kucharik CJ, Norman JM, Murdock LM, Gower ST (1997) Characterizing canopy non-randomness with a multiband vegetation imager (MVI). J Geophys Res 102(D24): 29455–29473

Lang ARG (1986) Leaf area and average leaf angle from transmission of direct sunlight. Aust J Bot 34:349–355

Lang ARG (1987) Simplified estimate of leaf area index from transmittance of the sun's beam. Agric For Meteorol 41:179–186

Lang ARG, Xiang Y (1986) Estimation of leaf area index from transmission of direct sunlight in discontinuous canopies. Agric For Meteorol 37:229–243

Lang M, Kuusk A, Mõttus M, Rautiainen M, Nilson T (2010) Canopy gap fraction estimation from digital hemispherical images using sky radiance models and a linear conversion method. Agric For Meteorol 150:20–29

Leblanc SG (2006) Digital Hemispherical Photography (DHP) Manual, version 1.0. Natural Resources Canada, Canada Centre for Remote Sensing, Ottawa

Leblanc SG (2008) DHP-TRACWin Manual. Natural Resources Canada, Saint-Hubert, Quebec, Canada

Leblanc SG, Chen JM, Fernandes R, Deering DW, Conley A (2005) Methodology comparison for canopy structure parameters extraction from digital hemispherical photography in boreal forests. Agric For Meteorol 129:187–207

Leblanc SG, Fernandes R, Chen JM (2002) Recent advancements in optical field leaf area index, foliage heterogeneity, and foliage angular distribution measurements. In: IGARSS 2002, Proceedings of the IEEE International Geoscience and Remote Sensing Symposium, 24–28 June 2002, Toronto, Ontario. CD-ROM. IEEE, Piscataway, New Jersey. 5:2902–2904

Leblanc SG, Fournier RA (2017) Measurement of forest structure with hemispherical photography. In: Fournier RA, Hall RJ (eds) Hemispherical photography in forest science: theory, methods, applications. Springer, Berlin

Lee DW (1987) The spectral distribution of radiation in two neotropical rainforests. Biotropica 19:161–166

Lee DW, Graham R (1986) Leaf optical properties of rainforest sun and extreme shade plants. Am J Bot 73:1100–1108

Macfarlane C (2011) Classification method of mixed pixels does not affect canopy metrics from digital images of forest overstorey. Agric For Meteorol 151:833–840

MacFarlane C, Coote M, White DA, Adams MA (2000) Photographic exposure affects indirect estimation of leaf area in plantations of *Eucalyptus globulus* Labill. Agric For Meteorol 100:155–168

Macfarlane C, Arndt SK, Livesley SJ, Edgar AC, White DA, Adams MA, Eamus D (2007a) Estimation of leaf area index in eucalypt forest with vertical foliage, using cover and fullframe fisheye photography. For Ecol Manag 242:756–763

Macfarlane C, Hoffman M, Eamus D, Kerp N, Higginson S, McMurtrie R, Adams M (2007b) Estimation of leaf area index in eucalypt forest using digital photography. Agric For Meteorol 143:176–188

Macfarlane C, Ogden GN (2012) Automated estimation of foliage cover in forest understorey from digital nadir images. Methods Ecol Evol 3:405–415

Martens SN, Ustin SL, Rousseau RA (1993) Estimation of tree canopy leaf area index by gap fraction analysis. For Ecol Manag 61:91–108

Messier C, Honer TW, Kimmins JP (1989) Photosynthetic photon flux density, red:far-red ratio, and minimum light requirement for survival of *Gaultheria shallon* in western red cedar-western hemlock stands in coastal British Columbia. Can J For Res 19:1470–1477

Miller EE, Norman JM (1971) A sunfleck theory for plant canopies II. Penumbra effects: Intensity distributions along sunfleck segments. Agron J 63:739–743

Miller JB (1967) A formula for average foliage density. Aust J Bot 15:141–144

Mitchell PL, Whitmore TC (1993) Use of hemispherical photographs in forest ecology. Oxford Forestry Institute, Department of Plant Sciences, University of Oxford. OFI Occasional Paper No. 44. 39 pp

Moon P, Spencer DE (1942) Illumination from a non-uniform sky. Trans Illum Engng Soc 37:707–726

Nackaerts K, Wagendorp T, Coppin P, Muys B, Gombeer R (1999) A correction of indirect LAI measurements for a non-random distribution of needles on shoots. In: Proceedings of ISSSR, Systems and sensors for the new millennium, 31 October–4 November, Las Vegas, Nevada, 4 pp

Niblack W (1986) An introduction to image processing. Prentice-Hall, Englewood Cliffs, New Jersey

Nobis M, Hunziker U (2005) Automatic thresholding for hemispherical canopy-photographs based on edge detection. Agric For Meteorol 128:243–250

Norman JM, Campbell GS (1989) Canopy structure. In: Pearcy RW, Ehleringer J, Mooney HA, Rundel PW (eds) Physiological plant ecology: field methods and instrumentation. Chapman & Hall, London, pp 301–325

Olivo JC (1994) Automatic threshold selection using the wavelet transform. Comput Vis Graph Image Process 56:205–218

Olsson L, Carlsson K, Grip H, Perttu K (1982) Evaluation of forest-canopy photographs with diode-array scanner OSIRIS. Can J For Res 12:822–828

Pearcy RW (1990) Sunflecks and photosynthesis in plant canopies. Annu Rev Plant Phys 41:421–453

Pekin BK, Macfarlane C (2009) Measurement of crown cover and leaf area index using digital cover photography and its application to remote sensing. Remote Sens 1:1298–1320

Perry SG, Fraser AB, Thomson DW, Norman JM (1988) Indirect sensing of plant canopy structure with simple radiation measurements. Agric For Meteorol 42:255–278

Pinty B, Gobron N, Widlowski J-L, Lavergne T, Verstraete MM (2004) Synergy between 1-D and 3-D radiation transfer models to retrieve vegetation canopy properties from remote sensing data. J Geophys Res 109:D21205. doi:10.1029/2004JD005214

Planchais I, Pontailler J-Y (1999) Validity of leaf areas and angles estimated in a beech forest from analysis of gap frequencies, using hemispherical photographs and a plant canopy analyzer. Ann For Sci 56:1–10

Plotnick RE, Gardner RH, Hargrove WW, Prestegaard K, Perlmutter M (1996) Lacunarity analysis: a general technique for the analysis of spatial patterns. Phys Rev E 53:5461–5468

Poorter L, Oberbauer SF, Clark DB (1995) Leaf optical properties along a vertical gradient in a tropical rain forest canopy in Costa Rica. Am J Bot 82:1257–1263

Pueschel P, Buddenbaum H, Hill J (2012) An efficient approach to standardizing the processing of hemispherical images for the estimation of forest structural attributes. Agric For Meteorol 160:1–13

Rich PM (1989) A manual for analysis of hemispherical canopy photography. Los Alamos National Laboratory Report LA-11732-M, Los Alamos, USA. (http://www.creeksidescience.com/files/rich_et_al_1989_canopy_manual.pdf)

Rich PM (1990) Characterizing plant canopies with hemispherical photographs. In: Goel NS, Norman JM (eds) Instrumentation for studying vegetation canopies for remote sensing in optical and thermal infrared regions. Remote Sens Rev 5:13–29

Rich PM, Wood J, Vieglais DA, Burek K, Webb N (1999) HemiView user manual. Delta-T Devices Ltd, Cambridge, UK, 79 pp

Ridler TW, Calvard S (1978) Picture thresholding using an iterative selection method. IEEE Trans Syst Man Cybern 8:630–632

Ross J (1981) The radiation regime and architecture of plant stands. Junk publishers, The Hague, Dr W

Russ J (2002) The image processing handbook, 4th edn. CRC Press, Boca Raton, Florida

Ryu Y, Sonnentag O, Nilson T, Vargas R, Kobayashi H, Wenk R, Baldocchi DD (2010) How to quantify tree leaf area index in an open savanna ecosystem: a multi-instrument and multi-model approach. Agric For Meteorol 150:63–76

Schleppi P, Conedera M, Sedivy I, Thimonier A (2007) Correcting non-linearity and slope effects in the estimation of the leaf area index of forests from hemispherical photographs. Agric For Meteorol 144:236–242

Schleppi P, Paquette A (2017) Solar radiation in forests: theory for hemispherical photography. In: Fournier RA, Hall RJ (eds) Hemispherical photography in forest science: theory, methods, applications. Springer, Berlin

Schneider D, Schwalbe E, Maas H-G (2009) Validation of geometric models for fisheye lenses. ISPRS J Photogramm 64:259–266

Schwalbe E, Maas H-G, Kenter M, Wagner S (2006) Profile based sub-pixel-classification of hemispherical images for solar radiation analysis in forest ecosystems. The International Archives of the Photogrammetry, Remote Sensing and Spatial Information Sciences, 36 (Part 7):(on CD-ROM)

Sinoquet H, Pincebourde S, Adam B, Donès N, Phattaralerphong J, Combes D, Ploquin S, Sangsing K, Kasemsap P. Thanisawanyangkura S, Groussier-Bout G, Casas J (2009) 3-D maps of tree canopy geometries at leaf scale. Ecology 90:283–1283

Smith H (1982) Light quality, photoperception, and plant strategy. Annu Rev Plant Phys 33:481–518

Smith WK, Knapp AK, Reiners WA (1989) Penumbral effects on sunlight penetration in plant communities. Ecology 70:1603–1609

Steven MD, Unsworth MH (1979) The diffuse solar irradiance of slopes under cloudless skies. Q J Roy Meteor Soc 105:593–602

Strachan IB, McCaughey JH (1996) Spatial and vertical leaf area index of a deciduous forest resolved using the LAI-2000 plant canopy analyzer. For Sci 42:176–181

ter Steege H (1997) Winphot 5: a programme to analyze vegetation indices, light, and light quality from hemispherical photographs. Tropenbos-Guyana Programme and Utrecht University, Utrecht, 52 pp

Tsai W (1985) Moment-preserving thresholding: a new approach. Comput Vision Graph 29:377–393

Ustin SL, Woodward RA, Barbour MG, Hatfield JL (1984) Relationships between sunfleck dynamics and red fir seedling distribution. Ecology 65:1420–1428

van Gardingen PR, Jackson GE, Hernandez-Daumas S, Russell G, Sharp L (1999) Leaf area index estimates obtained for clumped canopies using hemispherical photography. Agric For Meteorol 94:243–257

Varlet-Grancher C, Bonhomme R, Sinoquet H (eds) (1993) Crop structure and light microclimate: characterization and applications. INRA Editions, Paris

Verhoeven GJJ (2010) It's all about the format—unleashing the power of RAW aerial photography. Int J Remote Sens 31:2009–2042

Wagner S (1998) Calibration of grey values of hemispherical photographs for image analysis. Agric For Meteorol 90:103–117

Wagner S (2001) Relative radiance measurements and zenith angle dependent segmentation in hemispherical photography. Agric For Meteorol 10:103–115

Wagner S, Hagemeier M (2006) Method of segmentation affects leaf inclination angle estimation in hemispherical photography. Agric For Meteorol 139:12–24

Walter J-MN, Fournier RA, Soudani K, Meyer E (2003) Integrating clumping effects in forest canopy structure: an assessment through hemispherical photographs. Can J Remote Sens 29:388–410

Walter J-MN, Torquebiau EF (2000) The computation of forest leaf area index on slope using fish-eye sensors. C R Acad Sci III 323:801–813

Wang Q, Tenhunen J, Schmidt M, Otieno D, Kolcun O, Droesler M (2005) Diffuse PAR irradiance under clear skies in complex alpine terrain. Agric For Meteorol 128:1–15

Wang W-M, Li Z-L, Su H-B (2007) Comparison of leaf angle distribution functions: effects on extinction coefficient and fraction of sunlit foliage. Agric For Meteorol 143:106–122

Wang YS, Miller DR, Welles JM, Heisler GM (1992) Spatial variability of canopy foliage in an oak forest estimated with fisheye sensors. For Sci 38:854–865

Warren Wilson J (1963) Estimation of foliage denseness and foliage angle by inclined point quadrats. Aust J Bot 11:95–105

Weiss M, Baret F, Smith GJ, Jonckheere I, Coppin P (2004) Review of methods for in situ leaf area index (LAI) determination: Part II. Estimation of LAI, errors and sampling. Agric For Meteorol 121:37–53

Welles JM, Norman JM (1991) Instrument for indirect measurement of canopy architecture. Agron J 83:818–825

Widlowski J-L, Pinty B, Lavergne T, Verstraete MM, Gobron N (2007a) Horizontal radiation transport in 3-D forest canopies at multiple spatial resolutions: simulated impact on canopy absorption. Remote Sens Environ 103:379–397

Widlowski J-L, Robustelli M, Disney M, Gastellu-Etchegorry J-P, Lavergne T, Lewis P, North PRJ, Pinty B, Thompson R, Verstraete MM (2007b) The RAMI On-line Model Checker (ROMC): a web-based benchmarking facility for canopy reflectance models. Remote Sens Environ 112:1144–1150

Woebbecke DM, Meyer GE, Von Bargen K, Mortensen DA (1995) Color indices for weed identification under various soil, residue, and lighting conditions. T ASAE 38:259–269

Zhang Y, Chen JM, Miller JR (2005) Determining digital hemispherical photography exposure for leaf area index estimation. Agric For Meteorol 133:166–181

Chapter 6
View Angle-Dependent Clumping Indices for Indirect LAI Estimation

Gordon W. Frazer, Richard A. Fournier, Sylvain G. Leblanc and Jean-Michel N. Walter

Abstract The leaf area index (LAI) of forest canopies can be rapidly estimated by the canopy gap fraction measured using in situ optical instruments. Gap fraction analysis, however, will routinely underestimate LAI when forest canopies exhibit a strong pattern of aggregation at one or more spatial scales. Recent advances in optical-based measurement techniques and gap fraction analyses show that inclusion of a view angle-dependent clumping index $\Omega(\theta)$ may substantially improve optical-based estimates of LAI. Here we introduce four different estimates of $\Omega(\theta)$ and demonstrate how these indices can be extracted directly from hemispherical canopy photographs. Our findings suggest that estimates of $\Omega(\theta)$ can improve LAI estimation in the presence of spatial non-randomness, although their overall effectiveness was strongly dependent on the clumping index chosen, quantity of leaf area, severity and complexity of foliage clumping, and the view angles used for LAI integration.

Keywords Leaf area index (LAI) · Chen and Cihlar index · Lacunarity · Lang and Xiang index · Pielou coefficient of segregation · Foliage clumping · Gap fraction analysis · Hemispherical canopy photography

---

G.W. Frazer (✉)
GWF LiDAR Analytics, 399 Wain Road, North Saanich, BC V8L 5P9, Canada
e-mail: gfrazer@islandnet.com

R.A. Fournier
Département de Géomatique Appliquée, Centre d'Applications et de Recherches en Télédétection, Université de Sherbrooke, Sherbrooke, QC J1K 2R1, Canada
e-mail: Richard.Fournier@USherbrooke.ca

S.G. Leblanc
Canada Centre for Remote Sensing, Natural Resources Canada, 6767 route de l'Aéroport, Saint-Hubert, QC J3Y 8Y9, Canada

J.-M.N. Walter
Laboratoire de Botanique et Écologie Végétale, Faculté des Sciences de la Vie, Université Louis Pasteur, 28 rue Goethe, 67070 Strasbourg Cedex, France

R.A. Fournier and R.J. Hall (eds.), *Hemispherical Photography in Forest Science: Theory, Methods, Applications*, Managing Forest Ecosystems 28,
DOI 10.1007/978-94-024-1098-3_6

6.1 Introduction

Several direct (i.e., destructive sampling and litter fall collection), semi-direct (i.e., vertical and inclined point-quadrat sampling), and indirect (i.e., optical sensors) techniques have been developed to estimate the LAI of forest canopies (Gower et al. 1999; Bréda 2003; Jonckheere et al. 2004). Indirect methods employ linear or hemispherical optical instruments (e.g., solar radiation sensors, hemispherical canopy photography, etc.) to measure solar radiation transmittance or gap frequency data at single- or multiple-view angles beneath the canopy, and have been a popular choice for rapid stand-level LAI estimation due to their high precision, non-destructive nature, and ease of use (Welles and Cohen 1996; Fournier et al. 2003). Optically derived measurements of solar radiation transmittance or gap frequency are generally referred to as the canopy 'gap fraction' (Welles and Norman 1991).

The theoretical relationship between the canopy gap fraction and LAI has been widely described for spatially homogenous and heterogeneous forest canopies using a modified formulation of the Beer-Lambert model known as the 'Markov' model (Nilson 1971; Chen et al. 1997; Kucharik et al. 1999; Weiss et al. 2004; Jonchkeere et al. 2006; Leblanc and Fournier 2017):

$$P(\theta) = \exp[-G(\theta)\,\Omega(\theta)\,L_t/\cos(\theta)] \quad (6.1)$$

where $P(\theta)$ is the gap fraction or, more formally, the statistical probability of an infinitely narrow beam of light successfully penetrating the canopy at view (zenith) angle θ (Welles and Norman 1991); $G(\theta)$ is the mean projection coefficient of unit leaf area, i.e., the ratio of projected leaf area to one-half the total leaf surface area, on the plane oriented perpendicular to θ (Chen and Cihlar 1995; Stenberg 2006); $\Omega(\theta)$ is the element clumping index, which describes the spatial distribution of foliage elements at θ for scales greater than a conifer shoot or broad leaf (Chen et al. 1997; Kucharik et al. 1999); L_t is one-half of the total plant (including foliage, branches, and boles) surface area per unit ground surface area (Chen and Black 1992); and the multiplicative inverse (reciprocal) of $\cos(\theta)$ defines the relative path-length of a ray of light passing through the canopy at θ.

The original Beer-Lambert model (also known as the Poisson model) assumes that foliage and its support structures (i.e., twigs, branches, and boles) are spatially dispersed both randomly and independently of one another within the optical sensor's field of view (FOV) (Nilson 1999). When this assumption is violated, the canopy gap fraction will usually appear larger than what would be theoretically expected under the condition of complete spatial randomness (Lang and Xiang 1986; Lang 1991). Forest canopies, especially those of temperate and boreal coniferous forests, exhibit distinct patterns of spatial aggregation or clumping at several scales (Oker-Blom et al. 1991; Cescatti 1998). At the finest spatial scale, individual needles are clumped within shoots or twigs of varying size, shape, and needle density (Smolander et al. 1994; Stenberg 1996). In turn, shoots are spatially arranged into progressively coarser-scale structures, such as branches, whorls, crowns, and ultimately, neighborhoods of individual trees (Cooper 1961; Parker 1995; Fournier et al. 1997). Forest canopies that display marked clumping at one or more of these spatial scales will allow substantially more light to penetrate at a given LAI than a randomly dispersed canopy (Baldocchi et al. 1985; Law et al. 2001; Ryu et al. 2010). As a result, optical sensors routinely underestimate LAI in coniferous forests and most (if not all) broadleaf forests (Gower et al. 1999). The full magnitude of this underestimation, however, is lessened to some extent by the added shading that is contributed by branches and boles. It is also theoretically possible that optical methods can overestimate stand LAI when canopy elements are spatially over-dispersed (i.e., when leaves are uniformly or regularly distributed throughout the canopy volume), although such occurrences have been rarely reported in the literature (e.g., Whitford et al. 1995).

Various strategies have been proposed to improve the accuracy of optically derived estimates of LAI in spatially heterogeneous forest canopies. First, alternative gap-probability models, such as the negative binomial, positive binomial and Markov models, have been used in place of the Poisson model for LAI estimation in non-random forest canopies (Nilson 1971; Baldocchi et al. 1985; Neumann et al. 1989; Chason et al. 1991; Jonckheere et al. 2006). Second, Lang and Xiang (1986) have recommended averaging the logarithm of gap fractions sampled at finer spatial scales, where canopy elements are considered more likely to be randomly dispersed (Fassnacht et al. 1994; Planchais and Pontailler 1999; van Gardingen et al. 1999). Third, additional measurements of stand structure, such as tree height, crown cover, crown depth, stem density and indices of tree dispersion, have been incorporated into the Beer-Lambert model to explicitly account for the effects of crown size, crown porosity, site occupancy, and inter-crown spacing on the canopy gap fraction and LAI (Smith 1993; Nilson 1999; Nilson and Kuusk 2004). Fourth, a single correction factor or linear regression equation, which is derived from the empirical relationship between direct (destructive sampling and allometric regression models) and optical LAI measurements, has been applied to optical LAI data to compensate for measurement biases that are related to spatial clumping and inclusion of the woody area fraction (Gower and Norman 1991; Smith et al. 1993; Deblonde et al. 1994; Fassnacht et al. 1994; Barclay and Trofymow 2000; Jonckheere et al. 2006). Last, specialized optical sensors [e.g., Tracing Radiation and Architecture of Canopies (TRAC) and Multiband Vegetation Imager (MVI)] and supporting analytical theory have been developed to estimate $\Omega(\theta)$ for insertion into the Markov model (Chen and Cihlar 1995; Chen et al. 1997; Kucharik et al. 1997).

Recent advances in optical LAI measurement theory and methods suggest that hemispherical (fisheye) photography can be used to simultaneously retrieve estimates of gap fraction and spatial clumping for parameterization of the Markov model (van Gardingen et al. 1999; Walter et al. 2003; Leblanc et al. 2005; Jonckheere et al. 2006). Moreover, photograph-based methods offer several advantages over traditional light sensor-based technologies (e.g., TRAC) for indirect LAI estimation in heterogeneous forest canopies. First, fisheye photographs provide a permanent, two-dimensional (2D) spatial record of the size, shape, and distribution of all canopy components (i.e., leaves, woody structures, and gaps).

Second, the ability to detect small canopy gaps in fisheye photographs is constrained only by image quality and resolution, and not by the solar penumbral effect that can seriously limit the retrieval of small gaps from sunfleck data (Chen and Cihlar 1995). Third, because light-sensor-based technologies depend on the solar beam as a probe, sunfleck measurements are spatially restricted to a limited range of solar view angles. Fisheye photographs, in contrast, allow analysts to extract gap fraction and clumping indices at any angle of zenith or azimuth within the 180° hemispherical FOV. This distinction is important because these two parameters are known to be strongly view angle- and scale-dependent in spatially heterogeneous forest canopies (Black et al. 1991; Chen and Black 1991; Kucharik et al. 1999; Walter et al. 2003). Last, a single fisheye photograph can be readily used to model a broad range of canopy attributes (e.g., gap-light transmittance, sunfleck-frequency distribution, gap fraction, canopy openness, LAI, etc.), thereby eliminating the need for multiple instruments, sky conditions (clear versus overcast), sampling designs, and site visits.

Our objectives for this book chapter are threefold. First, to introduce theoretical and practical aspects of optical LAI estimation by gap fraction analysis in random and non-random forest canopies. Second, to demonstrate how the Markov model can be parameterized using view-angle-dependent estimates of element clumping and gap fraction retrieved from fisheye photographs. Third, to discuss the benefits and limitations of photograph-based element clumping indices and LAI estimation techniques in the presence of simple to complex patterns of spatial heterogeneity. We also introduce, demonstrate and compare four different view-angle-dependent element clumping indices. Two of these indices are based on established theory and methods proposed by Lang and Xiang (1986) and Chen and Cihlar (1995), while the others are less well-developed and largely experimental. The first of these two experimental methods is based on Pielou's coefficient of segregation (Pielou 1962; Walter et al. 2003), while the second is derived using the mathematical concept of 'lacunarity' first proposed by Mandelbrot (1983) (Frazer et al. 2005; Frazer 2007). We also demonstrate a simple procedure for LAI estimation by inversion of the Markov model, and through numerical integration of $P(\theta)$ and $\Omega(\theta)$ across multiple view angles θ. We have relied exclusively on the analysis of simulated fisheye photographs, so that we could retain control over the quantity and spatial distribution of canopy leaf area.

6.2 Theory of Vegetation Element Clumping Index from Gap Fraction Measurements

Most optical instruments do not distinguish between leaves and other non-foliar parts of the canopy. As a result, gap fraction analyses based on the Beer-Lambert (Poisson) and Markov models provide estimates of the plant area index (L_e or L_t) rather than LAI. Here we describe the basic theory and numerical procedures needed to parameterize and solve the Markov model. Methods required to convert L_t to LAI are discussed in detail by Leblanc and Fournier (2017).

Three computational steps are required to estimate LAI via application of the Markov model. In the first step, Eq. 6.1 is rearranged algebraically (inverted) to solve for L_t:

$$L_t = -\frac{\ln[P(\theta)]\cos\theta}{G(\theta)\,\Omega(\theta)} \tag{6.2}$$

Second, L_t is evaluated at one or more zenith view angles θ using known or estimated values of $P(\theta)$, $\Omega(\theta)$, and $G(\theta)$. $P(\theta)$ and $\Omega(\theta)$ are unknown variables to be estimated (discussed in detail below). $G(\theta)$ is wholly dependent on the foliage angle distribution and, therefore, also unknown. Miller (1967), however, demonstrated mathematically that G was independent of foliage orientation and equivalent to 0.5 when integrated between the limits $0° \leq \theta \leq 90°$ (widely referred to as Miller's Theorem):

$$G = \int_0^{\pi/2} G(\theta)\ \sin(\theta)d(\theta) = 0.5 \tag{6.3}$$

L_t can be computed following Miller's Theorem (Leblanc et al. 2005):

$$L_t = -2\int_0^{\pi/2} \frac{\ln[P(\theta)]}{\Omega(\theta)}\cos\theta\sin\theta d\theta \tag{6.4}$$

Under the assumption that leaves are randomly distributed ($\Omega(\theta) = 1$) in the horizontal domain (i.e., the Poisson model), Miller's Theorem also provides an estimate of the "effective" LAI (L_e; Welles and Norman 1991; Chen et al. 1997):

$$L_e = -2\int_0^{\pi/2} \ln[P(\theta)]\cos\theta\sin\theta d\theta \tag{6.5}$$

Numerical integration of L_t and L_e between the limits $0° \leq \theta \leq 90°$ is usually not possible due to the lack of foliage and canopy gaps, which occurs at small and large zenith angles, respectively. Furthermore, sample sizes become increasingly small as θ approaches 0°, which further complicates the estimation of $\Omega(\theta)$. To alleviate these shortcomings, L_t and L_e can be evaluated at $\theta \approx 57.3°$, where $P(\theta)$ and $\Omega(\theta)$ are readily estimated, and $G(\theta)$ is known to be near 0.5 for any foliage angle distribution (Warren Wilson 1960; Neumann et al. 1989; Weiss et al. 2004; Leblanc et al. 2005; Jonckheere et al. 2006; Leblanc and Fournier 2017). Leblanc and Fournier (2014) further showed that reliable estimates of L_e could be obtained by using a narrower range of view angles ($55° \leq \theta \leq 60°$) than what Miller (1967) had originally proposed.

Finally, L_t is converted to LAI by compensating for the clumping of needles within coniferous shoots (not required for broadleaf forests), as well as the shading effects that are contributed by the woody fraction (for details, refer to Leblanc and Fournier 2014).

In the following sections, we describe four techniques that can be used to estimate the element clumping index $\Omega(\theta)$ for binary sets or sequences of white (gap) and black (non-gap) pixels derived from fisheye photographs. In keeping with expectations of the Markov model, all four clumping indices will return values of $\Omega(\theta)$ approximately equal to one, less than one, and greater than one for random, clumped, and uniform canopies, respectively.

6.2.1 Lang and Xiang Clumping Index, $\Omega_{LX}(\theta)$

Lang and Xiang (1986) and Lang (1991) argued that the logarithmic average of $P(\theta)$ measured over several short segment-lengths (also known as the 'finite-length averaging technique') would provide a more accurate estimate of LAI than the linear average of $P(\theta)$ integrated over the entire transect length. Their reasoning behind this approach was based on two main points: (i) the inverted formulation of the Markov model (Eq. 6.2) implies that L_t is proportional to the logarithm of $P(\theta)$ rather than to $P(\theta)$ itself; (ii) the assumption of a random foliage distribution was more likely to be met over short rather than long segment-lengths in heterogeneous forest canopies.

$\Omega_{LX}(\theta)$ can be calculated for any view angle θ by partitioning a sequence of black (foliage) and white (gap) image pixels (known as a binary set) extracted at θ into several n smaller sets i based on equal-intervals of azimuth. For example, a binary set partitioned into 45° segments (sectors) of azimuth angle would yield $n = 8$ separate equal-length sets. Dividing the logarithm of the average gap fraction $P(\theta)$ calculated for the entire binary set by the average logarithm (logarithmic mean) of average gap fraction $P_i(\theta)$ computed for each angular segment i gives an estimate of $\Omega_{LX}(\theta)$ (Walter et al. 2003):

$$\Omega_{LX}(\theta) = \frac{\ln[P(\theta)]}{\frac{1}{n}\sum_{i=1}^{n} \ln[P_i(\theta)]} \tag{6.6}$$

where $P(\theta)$ and $P_i(\theta)$ are both estimated as the fraction or proportion of gap pixels found in a binary set.

6.2.2 Chen and Cihlar Clumping Index, $\Omega_{CC}(\theta)$

Chen and Cihlar (1995) developed a plant canopy gap-size theory and optical instrument known as TRAC to measure the gap fraction, element clumping and LAI of forest canopies. TRAC records instantaneous measurements of transmitted direct and diffuse photosynthetically active radiation (PAR) as it is moved along the length of a sampling transect located beneath the canopy. The recorded signal is then decomposed into an 'observed' gap-size distribution $F_m(\lambda, \theta)$ and compared to an 'expected' random distribution of gap sizes $F_r(\lambda, \theta)$ following an iterative gap removal procedure. Finally, the accumulated gap fractions associated with $F_m(\lambda, \theta)$ and the newly 'compacted' gap-size distribution $F_{mr}(\lambda, \theta)$ created by gap removal are used to quantify element clumping.

Walter et al. (2003) and Leblanc et al. (2005) both demonstrated that $F_m(\lambda, \theta)$ could also be reliably retrieved from fisheye photographs. Walter et al. (2003), for example, proposed the following iterative gap removal procedure to estimate $\Omega_{CC}(\theta)$. First, the observed gap-size distribution $F_m(\lambda, \theta)$, which is extracted from a binary set at view angle θ, is sorted by gap size in a descending order from largest to smallest. Second, a single gap, starting with the largest one, is removed, and a new 'compacted' gap-size distribution $F_{mr}(\lambda, \theta)$ is formed. Third, after each successive gap removal, the absolute difference ε between the accumulated gap fractions associated with the compacted $F_{mr}(\lambda, \theta)$ and random $F_r(\lambda, \theta)$ gap-size distributions is computed. Finally, when $\varepsilon \leq 0.005$, the gap removal process is halted [i.e., further gap removal would have very little effect on $F_{mr}(\lambda, \theta)$]. $\Omega_{CC}(\theta)$ is computed using the accumulated gap fractions associated with $F_m(\lambda, \theta)$ and $F_{mr}(\lambda, \theta)$ (Leblanc et al. 2005):

$$\Omega_{CC}(\theta) = \frac{\ln[F_m(0,\theta)]}{\ln[F_{mr}(0,\theta)]} \frac{[1 - F_{mr}(0,\theta)]}{[1 - F_m(0,\theta)]} \tag{6.7}$$

where $F_m(0, \theta)$ is the total accumulated gap fraction associated with all gap sizes λ larger than 0 for $F_m(\lambda, \theta)$, and $F_{mr}(0, \theta)$ is the total accumulated gap fraction associated with all gap sizes λ larger than 0 for $F_{mr}(\lambda, \theta)$.

6.2.3 Pielou Coefficient of Segregation Clumping Index, $\Omega_{PCS}(\theta)$

The Pielou (1962) coefficient of segregation (PCS) was initially used to detect and describe the type and degree of spatial non-randomness that occurred between two plant species found along a narrow sampling transect. Walter et al. (2003), by analogy, used PCS to quantify the spatial non-randomness found in binary sets of black (*B*, foliage) and white (*W*, canopy gaps) pixels extracted from fisheye photos at view angle θ. The spatial arrangement of *B* and *W* pixels within a set is considered (with 95% probability) to be random when (Pielou 1962):

$$\hat{b} + \hat{w} = \frac{1}{\overline{m_W}} + \frac{1}{\overline{m_B}} = 1 \pm 1.96\sqrt{s_b^2 + s_w^2} \tag{6.8}$$

where $\hat{b}$ and $\hat{w}$ are estimates of the maximum likelihoods of encountering individual *B* and *W* pixels, respectively; m_W and m_B are the average run-lengths (gap sizes) of *W* and *B* pixels, respectively; s_b^2 and s_w^2 are estimates of the sample variances that are associated with the probabilities *b* and *w* of encountering individual *B* and *W* pixels, respectively:

$$s_b^2 = \frac{1}{n_W}\frac{\overline{m_W} - 1}{\overline{m_W}^3} \quad \text{and} \quad s_w^2 = \frac{1}{n_B}\frac{\overline{m_B} - 1}{\overline{m_B}^3} \tag{6.9}$$

where n_W and n_B are the total numbers of discrete runs (sequences) of *W* and *B* pixels, respectively. Finally, $\Omega_{\mathrm{PCS}}(\theta)$ is calculated using the average run lengths of *W* and *B* pixels (Walter et al. 2003):

$$\Omega_{PCS}(\theta) = \frac{1}{\overline{m_W}} + \frac{1}{\overline{m_B}} = 1 \pm 1.96\sqrt{s_b^2 + s_w^2}. \tag{6.10}$$

6.2.4 Lacunarity-Based Clumping Index, $\Omega_{LAC}(\theta)$

Mandelbrot (1983) noted that two geometrical objects with markedly different textures (defined by the presence and pattern of structural holes or gaps—or in Latin, *lacunae*) could exhibit the same fractal dimension. Consequently, he developed the mathematical concept of 'lacunarity' to describe the scale-dependent textural properties or 'gappiness' of fractals. Lacunarity is formally defined as the scale-dependent deviation of a geometrical object or pattern from translational invariance or homogeneity (Gefen et al. 1983; Cheng 1997). Spatial patterns that exhibit 'low lacunarity' generally have small gaps, appear fine-grained, and are homogeneous with respect to the size-frequency and spatial distributions of gaps. In contrast, spatial patterns that have large gaps, appear coarse-grained, and display substantial heterogeneity in the size-frequency and spatial distributions of gaps are described as being 'highly lacunar' (Plotnick et al. 1996; Blumenfeld and Mandelbrot 1997). Lacunarity analysis is explicitly a multiscale technique, since pattern is known to be strongly dependent on the spatial scale (grain and extent) of observation (Plotnick et al. 1996; Dale 1999).

The derivation of $\Omega_{\mathrm{LAC}}(\theta)$ can be computationally expensive, because it involves repeated measurements of the lacunarity statistic $\Lambda(r)$ at different spatial scales using a potentially large number and range of box sizes *r*. For example, a binary set of $n = 1000$ pixels extracted at a single view angle θ would require 1000 iterations to estimate $\Lambda(r)$ at scales $1 \leq r \leq n$ to compute a single lacunarity curve. Estimation of $\Omega_{\mathrm{LAC}}(\theta)$ requires the computation of three separate lacunarity curves: the mean 'observed' lacunarity curve $\overline{\Lambda_{Obs}(r)}$, 'random' lacunarity curve $\Lambda_{CSR}(r)$, and 'maximal clumping' lacunarity curve $\Lambda_{Max}(r)$. Because the magnitude of $\Lambda(r)$ depends on the total length n and proportion p of occupied (black) pixels in a binary set, all curves must be normalized to a common Y intercept by dividing each value of $\Lambda(r)$ by $\Lambda(1)$. The integrated area beneath each of the three normalized lacunarity curves provides the basis for estimating $\Omega_{\mathrm{LAC}}(\theta)$ (see Appendix for a full derivation of the technique and the computational details).

6.3 Materials and Methods

6.3.1 Simulated Fisheye Canopy Photos

We utilized the three-dimensional (3D) modeling and ray-tracing techniques developed by Fournier et al. (1996) and Walter et al. (2003) to construct 12 simulated fisheye canopy photographs with known values of LAI and various degrees of foliage clumping. Five main steps were required to generate each of these simulated images. First, we constructed a 3D canopy volume V with an arbitrary vertical height of 10 m and a square horizontal base of 5 m × 5 m (25 m^2), following a method that had been similarly proposed by Castro and Fletcher (1999). V was divided into 2.5×10^6 discrete equal-sized volume elements (voxels), each with a vertical height of 1 cm, horizontal widths of 10 cm, and a total surface area of 0.024 m^2. Second, following the definition of LAI[1] that was proposed by Chen and Black (1992), we randomly populated the matrix of voxels with a specific number N of opaque or occupied cells (leaves) to create a V of pseudo-randomly distributed leaves with known LAI. To produce LAIs of about 2, 4, and 6 m^2/m^2 ($\mathrm{LAI} = N \times [0.5 \times 0.024\ m^2]/25\ m^2$), we set N equal to 4167, 8333, and 12,500 leaves, respectively. Third, we randomly located 15 cluster seeds (*CS*) or cluster centroids within each V, assigned each leaf to the closest *CS*, and computed the Euclidean distance d between each leaf and its nearest *CS*. We then allowed each leaf to migrate towards its assigned *CS* by applying a scale or clumping factor (*Fc*) of 0.2, 0.4, or 0.6 ($d' = d \times [1 - Fc]$) to each d (Fig. 6.1). Leaves were shifted along the original vectors that were used to measure d. Two or more leaves were never permitted to occupy the same voxel, so that a constant LAI was maintained at all values of *Fc*. Fourth, we generated a large, spatially contiguous (600 m 600 m) forest canopy of known LAI and *Fc* by placing multiple copies of V side by side in a manner similar to that of 'toroidal edge correction,' which is used in point-pattern analysis (see Dale 1999, p. 215). Before placement, each V was randomly rotated in the horizontal plane by 0°, 90°, 180°, or 270° to alleviate some of the repetition (uniformity) in spatial pattern that was created by this approach.

---

[1]LAI is a dimensionless metric defined as one-half the total leaf surface area (m^2) per unit of horizontal ground surface area (m^2).

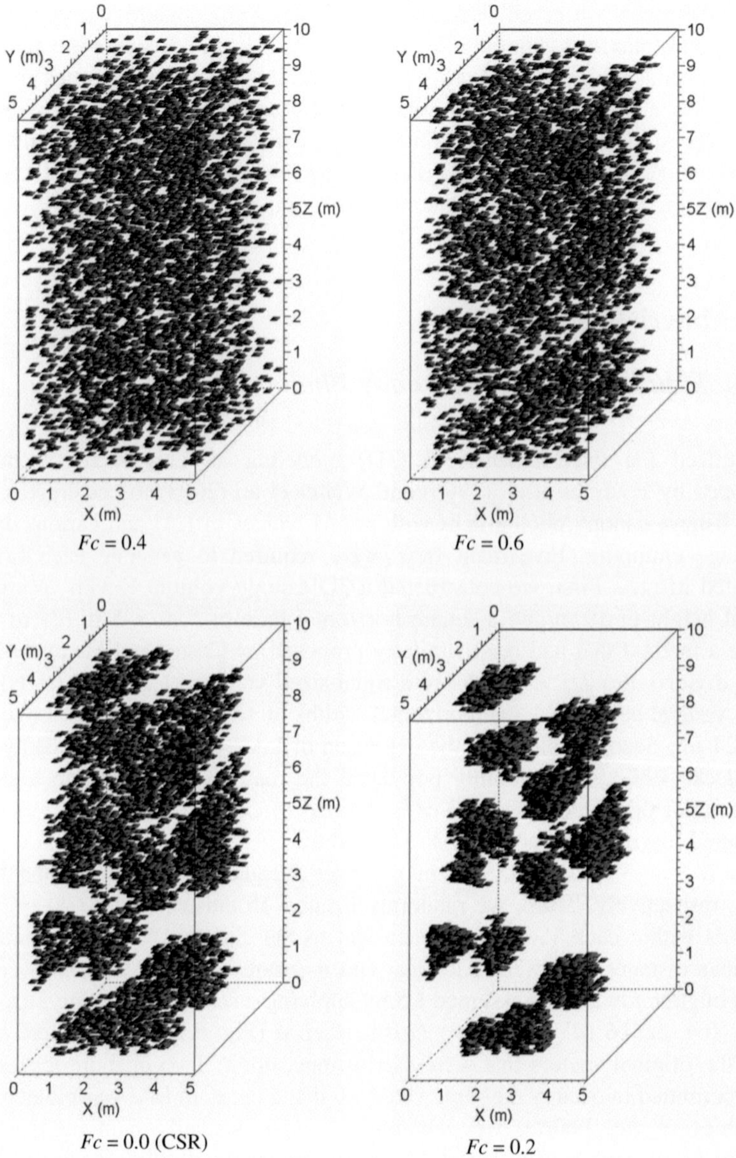

Fig. 6.1 A single simulated 5 m (width) × 5 m (length) × 10 m (height) canopy volume *V* displaying increasing amounts of spatial aggregation (*Fc* = 0.0, 0.2, 0.4, 0.6). *V* is composed of 2.5×10^6 discrete voxels, each with a vertical height of 1 cm, width and length of 10 cm, and a surface area of 0.024 m^2. *V* was randomly populated with 15 centroids that became points of attraction to which the closest surrounding leaves could migrate as *Fc* increased. We used a 120 × 120 array of randomly rotated *V*s to generate a contiguous 600 m × 600 m forest canopy

The bottom of each V was elevated to a height of 10 m, so that the vertical distribution of leaves would range from 10 to 20 m above the hypothetical ground surface. Last, we used the ray-tracing program ULooPS (Upward-Looking Photography Simulator), which was originally developed by Fournier et al. (1996), to generate skyward-looking, hemispherical (fisheye) views from the center of the spatially extended V and at 1.5 m above the ground (Fig. 6.2).

6.3.2 Retrieval of $P(\theta)$, $\Omega(\theta)$, LAI and Other Gap Statistics

We applied the same equiangular (polar) projection that was used to construct the simulated photographs to guide the spatial extraction of sequences of black (foliage) and white (gap) image pixels sampled along lines of equal zenith θ at 1° intervals between $10° \leq \theta \leq 80°$ (Fig. 6.3). Each binary set [sequence of ones and zeros, where one and zero represent occupied (foliage) and unoccupied (gap) pixels, respectively] was then spatially rearranged, so that the starting point began at the edge of the first gap that was encountered along the length of the set. We used these resampled sets to compute angle-dependent estimates of $P(\theta)$, $\Omega(\theta)$ and other gap-related statistics [i.e., total number of gaps, mean gap size, and the coefficient of variation (CV) of gap size]. $P(\theta)$ was estimated as the total number of white pixels (gaps) divided by the total number of (white plus black) pixels in a set (Frazer et al. 1997).

We employed four separate methods to estimate the element clumping index $\Omega(\theta)$ required to solve Eq. 6.4: (i) Lang and Xiang clumping index, $\Omega_{LX}(\theta)$; (ii) Chen and Cihlar clumping index, $\Omega_{CC}(\theta)$; (iii) Pielou coefficient of segregation clumping index, $\Omega_{PCS}(\theta)$; (iv) Lacunarity-based clumping index, $\Omega_{LAC}(\theta)$. Several algorithmic modifications were required to adapt these four clumping techniques to view-angle-dependent, photo-based LAI estimation procedures. To estimate $\Omega_{LX}(\theta)$, for example, we arbitrarily chose to divide each binary set into $n = 8$ equal arc-length (45°) segments i, based on the need to (i) minimize occurrences where $P_i(\theta) = 0$ (the logarithm of zero is infinite and undefined), and (ii) find a segment length (scale) at which the distribution of foliage elements was more likely to be random. In cases where $P_i(\theta) = 0$, we inserted one-half of a single gap (i.e., 0.5 divided by the total segment length measured in pixels) following recommendations by Leblanc et al. (2005).

To reduce the processing time required to estimate $\Omega_{LAC}(\theta)$, we chose to measure the lacunarity statistic $\Lambda(r)$ using a maximum of 75 equal intervals of $\ln(r)$. Also, $\Lambda(r)$ is sensitive to the position and patterning of occupied (black) and empty (gaps) pixels within a binary set, which poses a problem when there is no logical start or end to circular data. To minimize this effect, we divided each binary set sampled at view angle θ into four equal segments (sectors) of azimuth angle: 0° to <90°, 90° to <180°, 180° to <270°, and 270° to <360°, and established a new starting point at the edge of the first gap that was found within each of the four sectors. This step produced four full-length resampled sets that differed only in their

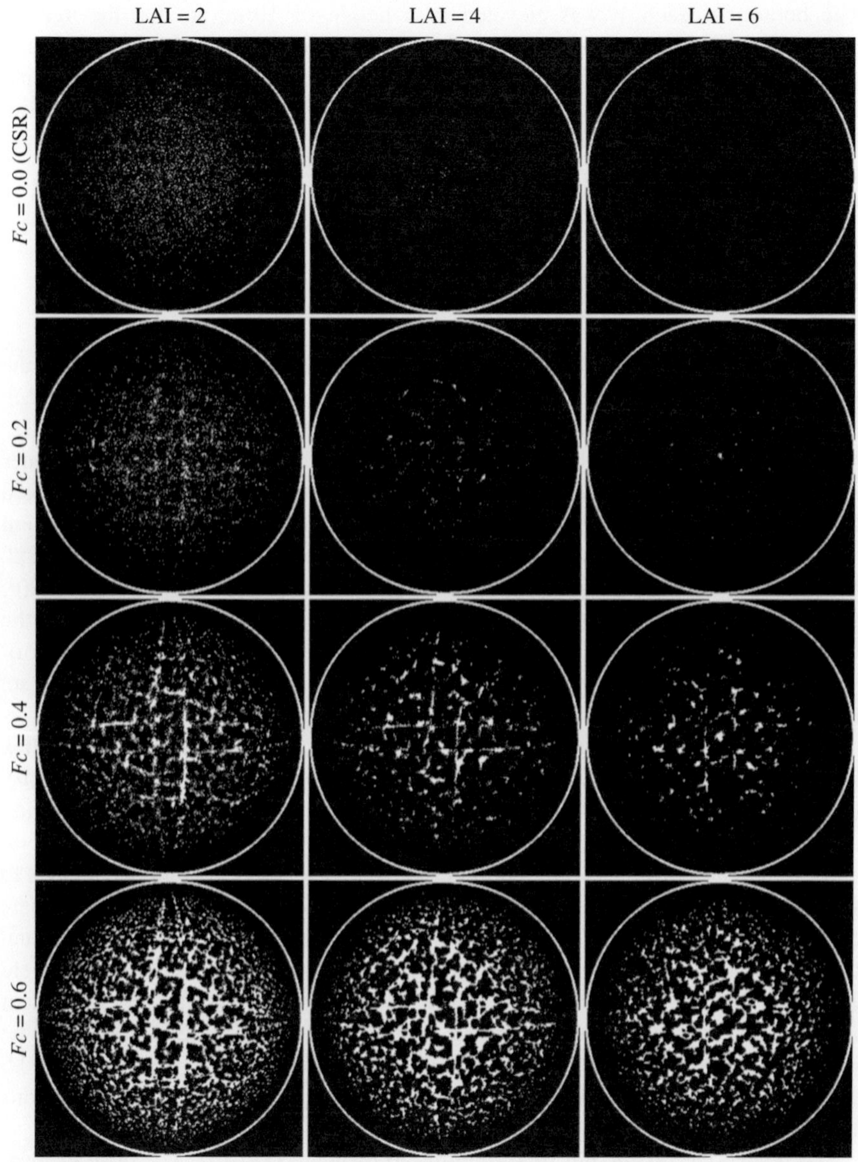

Fig. 6.2 Simulated hemispherical canopy photographs with known values of LAI (2, 4, and 6) and *Fc* (0.0 (CSR), 0.2, 0.4, and 0.6). Photographs were generated by ray-tracing through a 600 m (width) $\times$ 600 m (length) $\times$ 20 m (height) simulated volume of square, horizontal, flat leaves. Leaves were spatially restricted to vertical heights between 10 and 20 m. The simulated image resolution was 1024 $\times$ 1024 pixels

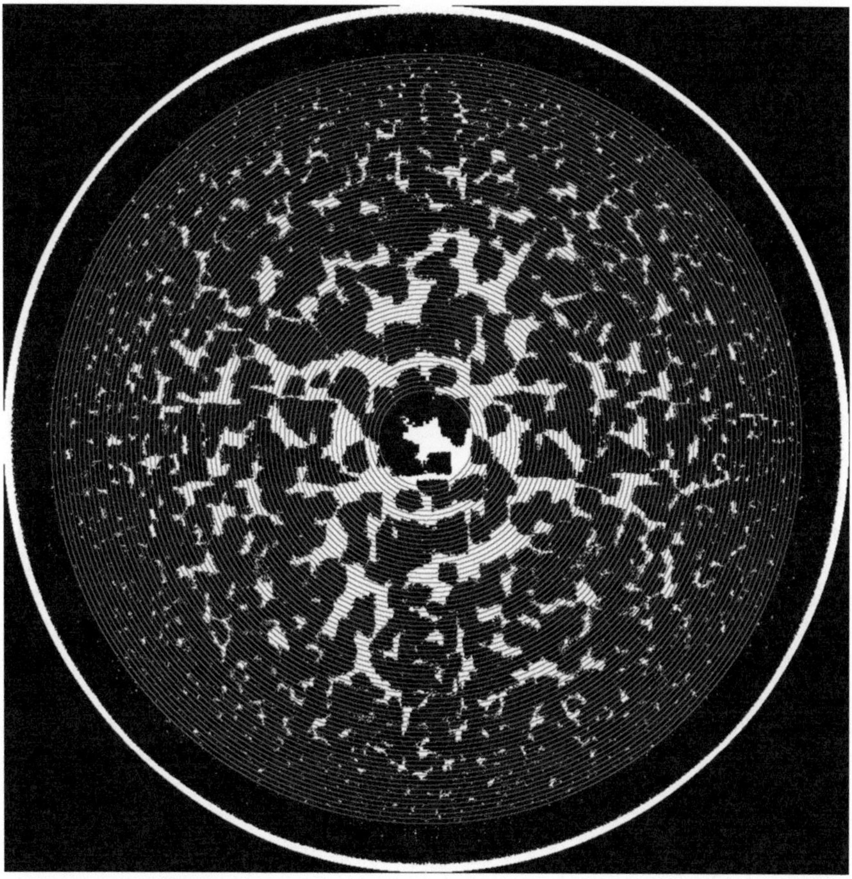

Fig. 6.3 Equiangular (polar) overlay of circular sampling transects that were placed every 1° of view angle θ between $10° \leq \theta \leq 80°$. Binary sets were created by sequentially sampling target pixels along lines of equal θ using all 360° of azimuth angle. The pixel search algorithm was designed to guard against errors of omission (missed target pixels) and duplication (target pixels selected more than once). Transect lengths become exceedingly short at $\theta < 10°$, and canopy gaps are often infrequent to nonexistent at $\theta > 80°$ in closed canopies. Therefore, estimates of $P(\theta)$ and $\Omega(\theta)$ were necessarily restricted to a reduced subset of view angles ($10° \leq \theta \leq 80°$)

starting and ending azimuth angles. We then used Eq. 6.15 to compute the mean 'observed' lacunarity statistic $\overline{\Lambda_{Obs}(r)}$ of these four resampled sets. We estimated lacunarity $\Lambda_{CSR}(r)$ under the condition of complete spatial randomness (CSR) using Eq. 6.14, while $\Lambda_{Max}(r)$ was derived from the analysis of a single binary set containing one large, spatially contiguous gap of size $n \times 0.99$ ($p = 0.01$). We calculated the total area found under each of these three normalized lacunarity curves by assuming linearity between known data points and through numerical integration, using a constant step value of 0.01 (in $\ln(r)$ units) between the limits $\ln(1)$ and $\ln(n)$. Finally, $\Omega_{LAC}(\theta)$ was estimated for each θ using Eq. 6.16.

We chose to evaluate L_t (Eq. 6.4) and L_e (Eq. 6.5) using three separate sets of limits on the integral: $10° \leq \theta \leq 80°$ (the closest approximation of Miller's Theorem); $45° \leq \theta \leq 60°$ (Leblanc and Chen 2001); and $55° \leq \theta \leq 60°$ (a generalized estimate of $\theta \approx 57.3°$; Leblanc et al. 2005; Leblanc and Fournier 2014). In this study, L_t was immediately equivalent to LAI, since our simulated plant canopies contained only square, flat leaves and no woody support structures (i.e., twigs, branches, or boles).

6.4 Results and Discussion

6.4.1 *Estimation of* L_e *and* L_t *in Random and Clumped Canopies*

Our results showed that the estimated L_e was a very good approximation of LAI when leaves were randomly and independently distributed in the horizontal domain ($Fc = 0$). The best approximation of LAI was obtained by integrating L_e between the limits $45° \leq \theta \leq 60°$ and $10° \leq \theta \leq 80°$, while the largest mean % error[2] in L_e was associated with view angles $55° \leq \theta \leq 60°$. Mean % errors for these three discrete sets of angular limits were respectively −1.9, 1.3, and −9.4% (Table 6.1a–c). However, as the spatial distribution of leaves became more clumped ($Fc > 0$), L_e increasingly underestimated LAI, especially for those canopy volumes with larger leaf areas (LAI > 2). The magnitude and rate of decline in L_e with increased foliage clumping was nonlinear and varied strongly with LAI. For example, the % error between L_e and LAI for any given Fc was always larger for LAI = 6 than for LAI = 2 or 4. When Fc reached 0.6, estimates of L_e converged toward a similar low value ($L_e < 2$) for LAI = 2, 4, and 6. Furthermore, the % error between L_e and LAI was nearly always minimized (11 of 12 photos) when L_e was integrated between the limits $45° \leq \theta \leq 60°$. Still, L_e underestimated LAI = 2, 4, and 6 by 36.2, 57.8, and 65.3%, respectively, at $Fc = 0.6$ and when measured between $45° \leq \theta \leq 60°$.

We found that estimates of L_t (Eq. 6.4), which incorporated the angle-dependent effect of $\Omega(\theta)$, generally improved optical LAI estimation in the presence of spatial non-randomness (Table 6.1a–c). The magnitude and extent of this improvement, however, was often limited and largely determined by (i) the clumping method used to compute $\Omega(\theta)$ (i.e., CC–Chen and Cihlar; LAC–lacunarity; LX–Lang and Xiang; PCS–Pielou's coefficient of segregation); (ii) the quantity and spatial arrangement of leaf area, and (iii) the angular limits (range of θ) over which the Markov model was numerically integrated. Clumping indices CC, LAC, and LX, for example,

[2]Expressed as % error = (measured – actual)/actual × 100%, where the 'measured' variable (L_e or L_t) is estimated through inversion of the canopy gap fraction, and LAI is the 'actual' or 'known' quantity.

Table 6.1 Summary of Ω, L_e, and L_t estimates computed by inversion of the Markov model and through numerical integration between the limits: (a) $10° \leq \theta \leq 80°$, (b) $45° \leq \theta \leq 60°$, and (c) $55° \leq \theta \leq 60°$

LAI	Integration		*Fc*	L_e	% Error	CC			LAC			LX			PCS		
	Min	Max				Ω	L_t	% Error	Ω	L_t	% Error	Ω	L_t	% Error	Ω	L_t	% Error
a																	
2	10°	80°	0.0	1.94	−2.8	0.99	1.98	−1.0	0.97	2.01	0.6	0.99	1.96	−1.9	0.86	2.45	22.5
			0.2	1.86	−6.9	0.98	1.92	−4.2	0.96	1.96	−2.0	0.99	1.89	−5.7	0.83	2.45	22.6
			0.4	1.64	−17.9	0.94	1.80	−10.1	0.90	1.87	−6.6	0.98	1.69	−15.6	0.72	2.69	34.4
			0.6	1.17	−41.7	0.81	1.60	−20.1	0.79	1.54	−23.1	0.97	1.21	−39.4	0.48	3.40	69.9
4	10°	80°	0.0	3.89	−2.8	0.99	3.94	−1.5	0.99	3.95	−1.2	0.98	3.99	−0.1	0.92	4.39	9.7
			0.2	3.57	−10.8	0.98	3.65	−8.8	0.97	3.72	−7.1	0.97	3.70	−7.5	0.87	4.35	8.8
			0.4	2.67	−33.3	0.92	3.00	−25.0	0.88	3.12	−22.0	0.95	2.83	−29.3	0.67	4.98	24.5
			0.6	1.55	−61.2	0.76	2.33	−41.7	0.76	2.13	−46.8	0.94	1.66	−58.5	0.36	6.39	59.8
6	10°	80°	0.0	5.99	−0.2	0.98	6.13	2.2	1.01	5.96	−0.7	1.00	6.04	0.6	0.98	6.15	2.5
			0.2	5.24	−12.7	0.98	5.38	−10.3	0.99	5.34	−10.9	0.97	5.47	−8.8	0.91	6.07	1.2
			0.4	3.62	−39.6	0.91	4.13	−31.2	0.88	4.22	−29.7	0.93	3.98	−33.7	0.65	7.32	22.0
			0.6	1.92	−68.1	0.76	2.86	−52.3	0.76	2.59	−56.8	0.94	2.07	−65.5	0.34	8.84	47.3
b																	
2	45°	60°	0.0	2.03	1.5	1.00	2.04	1.9	0.98	2.07	3.5	1.00	2.04	1.9	0.88	2.33	16.7
			0.2	1.95	−2.4	1.00	1.96	−1.9	0.97	2.02	1.1	0.99	1.97	−1.7	0.84	2.33	16.7
			0.4	1.79	−10.4	0.98	1.82	−9.0	0.92	1.96	−2.0	0.99	1.81	−9.4	0.73	2.48	24.1
			0.6	1.28	−36.2	0.86	1.49	−25.3	0.80	1.61	−19.6	0.99	1.29	−35.4	0.45	2.93	46.4
4	45°	60°	0.0	4.01	0.2	0.99	4.04	1.0	0.99	4.04	1.0	0.98	4.07	1.8	0.95	4.23	5.8
			0.2	3.66	−8.6	0.99	3.69	−7.8	0.97	3.76	−6.0	0.98	3.73	−6.7	0.89	4.15	3.8
			0.4	2.89	−27.8	0.96	3.01	−24.7	0.89	3.24	−18.9	0.97	2.97	−25.6	0.66	4.46	11.4
			0.6	1.69	−57.7	0.79	2.16	−46.0	0.75	2.27	−43.3	0.96	1.77	−55.8	0.31	5.76	44.0

(continued)

Table 6.1 (continued)

LAI	Integration		*Fc*	L_e	% Error	CC			LAC			LX			PCS		
	Min	Max				Ω	L_t	% Error	Ω	L_t	% Error	Ω	L_t	% Error	Ω	L_t	% Error
6	45°	60°	0.0	6.13	2.2	0.98	6.24	3.9	1.00	6.10	1.7	0.97	6.30	5.0	1.00	6.16	2.7
			0.2	5.52	−8.0	0.98	5.63	−6.2	0.98	5.61	−6.6	0.95	5.79	−3.5	0.92	6.07	1.1
			0.4	3.97	−33.9	0.95	4.17	−30.5	0.89	4.46	−25.7	0.94	4.22	−29.6	0.66	6.22	3.6
			0.6	2.08	−65.3	0.81	2.59	−56.8	0.75	2.78	−53.7	0.96	2.18	−63.7	0.27	7.95	32.5
c																	
2	55°	60°	0.0	1.82	−9.2	1.00	1.82	−8.8	0.99	1.84	−7.9	1.00	1.82	−8.9	0.91	1.99	−0.5
			0.2	1.81	−9.3	1.00	1.82	−9.0	0.98	1.85	−7.3	0.99	1.82	−8.9	0.90	2.02	1.2
			0.4	1.71	−14.3	0.99	1.72	−13.8	0.94	1.82	−8.9	0.99	1.73	−13.5	0.79	2.16	8.2
			0.6	1.26	−36.9	0.92	1.37	−31.7	0.83	1.52	−24.1	0.99	1.28	−36.2	0.51	2.51	25.3
4	55°	60°	0.0	3.62	−9.5	0.99	3.64	−8.9	0.99	3.65	−8.8	0.98	3.68	−8.0	0.95	3.81	−4.7
			0.2	3.32	−16.9	0.99	3.35	−16.3	0.98	3.40	−15.1	0.99	3.37	−15.7	0.92	3.61	−9.7
			0.4	2.66	−33.5	0.98	2.71	−32.3	0.91	2.92	−26.9	0.97	2.73	−31.7	0.71	3.74	−6.5
			0.6	1.62	−59.5	0.87	1.88	−53.1	0.78	2.09	−47.8	0.96	1.68	−58.0	0.36	4.56	14.0
6	55°	60°	0.0	5.43	−9.5	0.99	5.50	−8.4	1.00	5.41	−9.9	0.97	5.59	−6.9	1.01	5.40	−10.1
			0.2	5.19	−13.5	0.98	5.29	−11.8	0.99	5.23	−12.9	0.96	5.39	−10.2	0.96	5.45	−9.2
			0.4	3.66	−39.0	0.97	3.79	−36.8	0.92	3.96	−34.0	0.96	3.81	−36.4	0.75	4.96	−17.3
			0.6	1.92	−68.0	0.88	2.19	−63.5	0.77	2.48	−58.6	0.98	1.97	−67.1	0.29	6.59	9.8

L_e, L_t, and Ω were numerically integrated across the same range of zenith view angles

% Error is the percent error between 'estimated' L_e or L_t (optical) and 'actual' LAI [% Error = (Measured – Actual)/Actual × 100%]

CC Chen and Cihlar method; *LAC* Lacunarity method; *LX* Lang and Xiang method; *PCS* Pielou's coefficient of segregation method reduced the disparity between optical estimates of L_t and LAI; yet, all three indices consistently underestimated LAI, particularly at higher values of LAI and Fc. CC, LAC, and LX underestimated LAI, on average, by 17.9, 15.3, and 20%, respectively, when L_t was integrated between the limits $45° \leq \theta \leq 60°$. The magnitude of these underestimates increased with LAI and Fc, and reached a maximum ($-67.1\% \leq \%$ error $\leq -58.6\%$) at LAI = 6 and $Fc = 0.6$, for view angles $55° \leq \theta \leq 60°$. The effect of PCS was far less predictable, with the magnitude and direction (bias) of % error in L_t varying strongly with LAI, Fc, and integration method. In contrast to the other three indices, PCS consistently overestimated LAI by 22.3%, on average ($1.2\% \leq \%$ error $\leq 69.9\%$) when L_t was integrated between the limits $10° \leq \theta \leq 80°$ and $45° \leq \theta \leq 60°$. The magnitude of overestimation increased with Fc, but generally declined with a rise in LAI. Similar results for PCS have been reported by Walter et al. (2003) and by Gonsamo and Pellikka (2009). When L_t was integrated between the limits $55° \leq \theta \leq 60°$, PCS produced the lowest mean % error (9.7%) of all four clumping indices. Yet, the size and direction of these errors were strongly dependent on LAI and Fc. The next lowest mean % errors were produced by LAC (15.3%) and CC (17.4%) when L_t was integrated between the limits $45° \leq \theta \leq 60°$ and $10° \leq \theta \leq 80°$, respectively. LX consistently underestimated LAI by a similar amount ($20\% \leq \%$ error $\leq 25\%$), irrespective of the integration method that was used to compute L_t.

Our results showed that clumping indices improved optical LAI estimation in non-random canopies; however, no single index was able to compensate entirely for the increase in gap fraction that was associated with spatial clumping. Also, the efficacy of these indices was markedly idiosyncratic and tightly coupled to the range of view angles that were used to compute L_t, as well as the amount of LAI and degree of Fc. For example, CC and PCS produced the smallest % errors when L_t was integrated between the limits $10° \leq \theta \leq 80°$ and $55° \leq \theta \leq 60°$, respectively, while LAC and LX performed best when L_t was measured using view angles $45° \leq \theta \leq 60°$. Furthermore, CC, LAC, and LX seemed to work best at low LAI and low Fc, while PCS operated best at high LAI and high Fc.

6.4.2 Local Versus Global Saturation Effects

An LAI of 5–6 is considered to be the theoretical limit for optical LAI estimation in random canopies, because the presence of any more leaves would reduce the projected canopy gap fraction to zero. This is known as the global 'saturation point' in gap fraction analysis (Gower et al. 1999; Jonckheere et al. 2004). Saturation can also occur locally in hemispherical canopy photographs, and at a substantially lower LAI, due to (i) the exceedingly long path-lengths that accompany large zenith angles, and (ii) the presence of marked branch- and crown-level spatial clumping. The 'path-length' effect has been well described in the literature, and several researchers have proposed the insertion of a partial gap or a modification to the range of view angles over which LAI is integrated to alleviate the problem (Weiss et al. 2004; Leblanc et al. 2005). Results from our study suggest that the arbitrariness associated with either of these two options may contribute to further uncertainty in the optical estimation of LAI.

Our findings also indicate that individual foliage clumps can saturate at a very low LAI when forest canopies display significant spatial heterogeneity. It is well known that when local saturation occurs (often within individual branches or crowns), gap fraction analysis provides an estimate of one-half the total 'imaginary' surface area of clumps and not the surface area of individual shoots or leaves (Chen et al. 1997; Nilson 1999). Practical solutions to resolve the local saturation issue are not entirely apparent, although possibilities might include the use of (i) higher spatial and spectral resolution digital cameras (Hwang et al. 2016), (ii) 'thresholding' algorithms that operate locally rather than globally (Jonckheere et al. 2005; Leblanc et al. 2005; Macfarlane 2011; Macfarlane et al. 2014) and (iii) non-fisheye lenses (i.e., longer focal lengths) that would capture more spatial detail over a smaller portion of the canopy (Kucharik et al. 1999; Macfarlane et al. 2007a, b).

6.4.3 Angular Dependence of $P(\theta)$, $\Omega(\theta)$, and Other Statistics

The size-frequency distribution, spatial arrangement, and magnitude of the canopy gap fraction vary strongly with view angle θ, because θ alters the linear distance (path-length) and line-of-sight (perspective view) through the canopy, as well as the spatial extent or footprint (i.e., FOV) of the optical sensor (Kucharik et al. 1999; Walter et al. 2003; Leblanc et al. 2005). At small θ angles, for example, path-lengths are relatively short, the perspective view is near vertical, and the spatial footprint of the optical sensor comprises only one to a few canopy trees. Large θ angles, in contrast, produce longer path-lengths, oblique views of the canopy, and include many more trees in the sensor's FOV. Path-length and line-of-sight directly affect the size, number, and spatial arrangement of canopy gaps (Black et al. 1991; Kucharik et al. 1999; Leblanc et al. 2005), while instrument FOV and image spatial resolution determine the potential range of spatial scales over which attributes of canopy structure can be measured (Walter et al. 2003).

Scatterplots of selected canopy gap statistics, which were extracted at multiple view angles θ from the 12 simulated fisheye photographs, reveal the strong angular dependence of these metrics (Fig. 6.4). $P(\theta)$, for example, generally declined with increasing θ (Fig. 6.4a–c), because longer path-lengths through a horizontally contiguous forest canopy dramatically reduce the probability of light transmittance (Welles and Norman 1991). The total number (frequency) of discrete canopy gaps (Fig. 6.4d–f) tended to rise with increasing θ in response to the longer length (circumference) of circular sampling transects and the fragmentation of large gaps into smaller ones as the line-of-sight was obscured by more and more foliage layers. Gap frequency, however, reached a maximum at some larger value of θ (approx. 60°–80°)

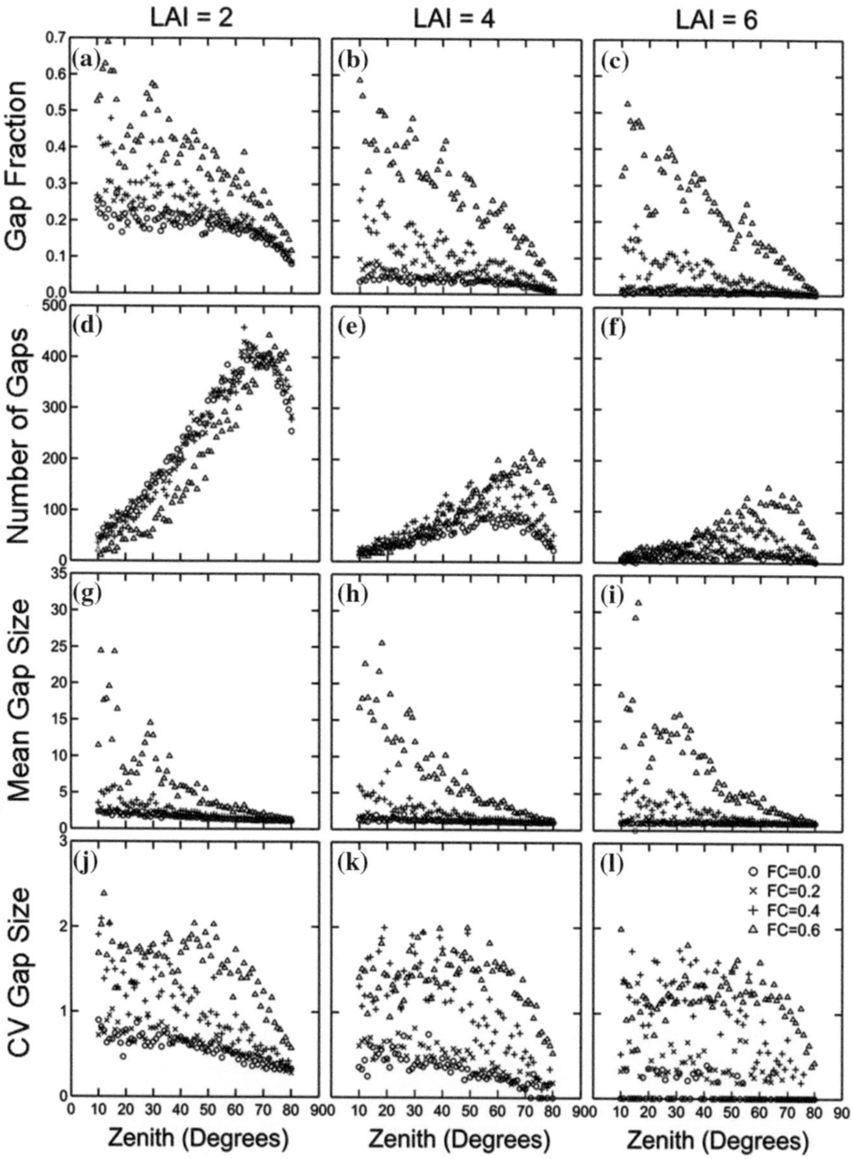

Fig. 6.4 Angular variation of **a–c** the canopy gap fraction, **d–f** gap frequency, **g–i** mean gap size, and **j–l** coefficient of variation (CV) for gap size data that were derived from the simulated fisheye photoset

and then rapidly declined towards zero once the path-length became long enough to eliminate gaps altogether. Both the mean (Fig. 6.4g–i) and coefficient of variation (CV) (Fig. 6.4j–l) of the canopy gap-size distribution generally declined with increasing θ due to the combined influences of path-length, line-of-sight, and FOV.

For example, both large (between-clump) and small (within-clump) gaps tended to be preserved at about $\theta < 50°$, because near vertical views from beneath the canopy resulted in fewer foliage layers to fragment and obscure the sky view. More oblique views ($\theta > 50°$), in contrast, eliminated many of the larger between-clump gaps, thereby producing gap-size distributions that were dominated by small openings. The pattern of angular variation in $P(\theta)$ and its associated gap-size distribution will ultimately be determined by all aspects of forest canopy structure (Kucharik et al. 1999).

Quantitative estimates of $\Omega(\theta)$ are directly linked to the size-frequency and spatial distribution of canopy gaps and, therefore, are also strongly influenced by θ. Results derived from our simulated canopies generally showed that $\Omega(\theta)$ increased nonlinearly from a minimum near the zenith towards a maximum value of approximately unity closer to the horizon (Fig. 6.5). Moreover, the range and pattern of angular variation in $\Omega(\theta)$ were markedly different for each of the four clumping indices. Lower values of $\Omega(\theta)$ always occurred at small values of θ in response to the patchy distribution of larger between-clump gaps, while larger $\Omega(\theta)$ estimates were associated with smaller, less frequent, and more randomly dispersed canopy gaps near the horizon (Fig. 6.3). A similar pattern of angular variation in $\Omega(\theta)$ has been previously reported by Kucharik et al. (1999), Walter et al. (2003), and Leblanc et al. (2005).

Differences in the magnitude and angular pattern of $\Omega(\theta)$ among each of the four clumping methods (CC, LAC, LX and PCS) can be explained by their inherent mathematical differences. PCS, for example, was derived from the sum of multiplicative inverses (reciprocals) of average run-lengths of white (gap) and black (canopy) pixels (Eq. 6.10) and, therefore, was strongly and nonlinearly correlated ($r = 0.99$) with the mean size of canopy gaps (Fig. 6.6). As a result, PCS estimates of $\Omega(\theta)$ declined with increasing mean gap size more rapidly and to substantially lower values compared to all other methods. LX, in contrast, generally produced the largest $\Omega(\theta)$ estimates and was the least responsive to changes in spatial clumping and mean gap size ($r = 0.61$). The limited sensitivity of LX to changes in the gap-size distribution was the direct consequence of the relatively long segment-length (45°) used to sample $P_i(\theta)$ (van Gardingen et al. 1999). Segment-length is a critical parameter because it determines (i) the measurement scale for which canopy elements (i.e., foliage, branches, and crowns) are aggregated and $P_i(\theta)$ is averaged, and (ii) the magnitude of the between-segment variance in $P_i(\theta)$, with shorter segment lengths producing a higher variance in gap fraction than longer segments, especially in the presence of spatial heterogeneity. We found that LX estimates of $\Omega(\theta)$ approached unity when $P_i(\theta)$ varied little among discrete segment-lengths i, while large among-segment differences in $P_i(\theta)$ produced $\Omega(\theta)$ values substantially less than unity (Welles and Norman 1991). Insertion of a partial gap (0.5 of a pixel) into segments where $P_i(\theta) = 0$ resulted in LX estimates of $\Omega(\theta) > 1$.

Estimates of $\Omega(\theta)$ derived by LAC and CC were generally lower than LX and higher than PCS at $Fc > 0$ (Fig. 6.5). LAC and CC, however, could be distinguished from one another by their distinct angular patterns of response to changes in the size-frequency and spatial distribution of canopy gaps. LAC, for instance,

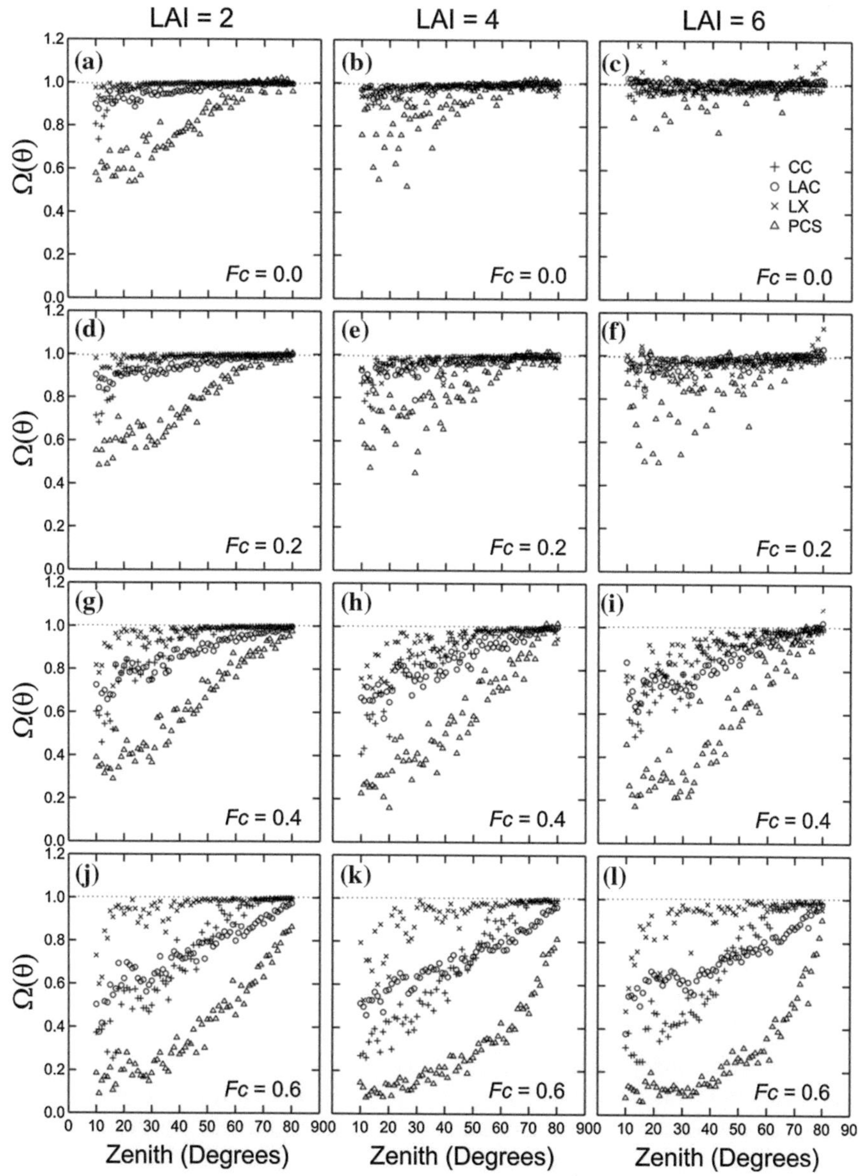

Fig. 6.5 Scatterplots displaying the angular response of $\Omega(\theta)$, which was derived from the simulated photoset using four different clumping methods (CC, LAC, LX, PCS). LAI influences leaf density, the opacity of discrete foliage clumps, and interacts strongly with *Fc* to determine the angular pattern of $\Omega(\theta)$

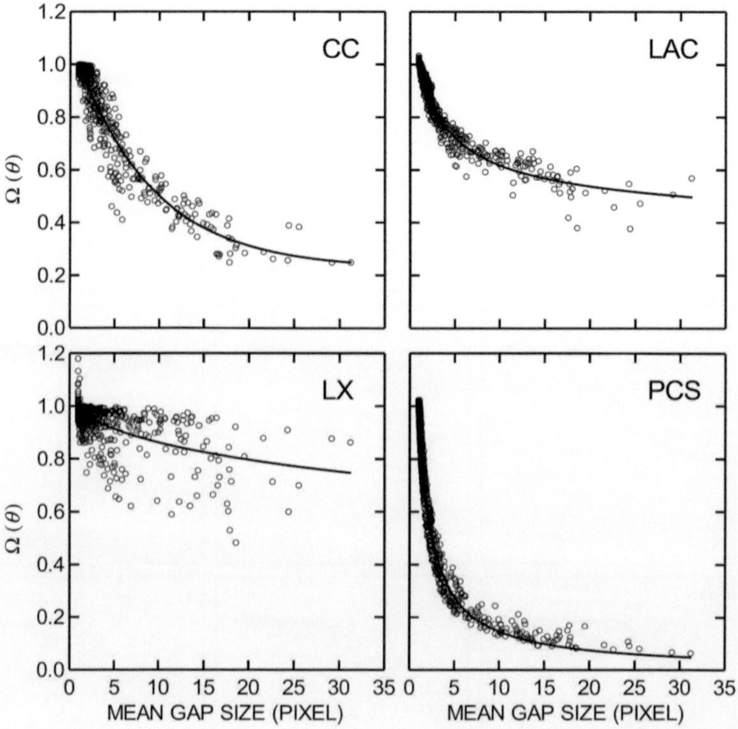

Fig. 6.6 Nonlinear relationship between $\Omega(\theta)$ and mean gap size: CC, $r = 0.96$; LAC, $r = 0.98$; LX, $r = 0.61$; PCS, $r = 0.99$. These data were derived from the simulated fisheye photoset. The weak relationship between $\Omega_{LX}(\theta)$ and mean gap size is due to the long segment-lengths that were used to estimate $P_i(\theta)$

was more sensitive than CC to small increases in gap size, especially at larger view angles where the average gap size was typically very small (Fig. 6.5). We believe that this higher sensitivity was a consequence of the logarithmic transformations of $\Lambda(r)$ and r (Fig. 6.8), which caused the normalized $\overline{\Lambda_{Obs}(r)}$ curve to shift rapidly (logarithmically) away from the random curve $\Lambda_{CSR}(r)$ when the average gap size was small, and more slowly when the average gap size was large. CC estimates of $\Omega(\theta)$, in contrast, were often lower than LAC at small values of θ for two reasons. First, LAC explicitly measures spatial non-randomness across a range of spatial scales and, therefore, was sensitive to the coarse-scale (i) random arrangement of foliage clumps within a single simulated canopy volume V, and (ii) uniform spatial distribution of foliage clumps that was created by the side-by-side placement of multiple copies of V. Second, any single large gap that was removed using CC accounted for a relatively large portion of the accumulated canopy gap fraction when the instrument FOV was limited to a small, overhead region of the canopy.

Gap-size distributions vary markedly with view angle and direction when forests are highly structured in both vertical and horizontal dimensions. In this study, we

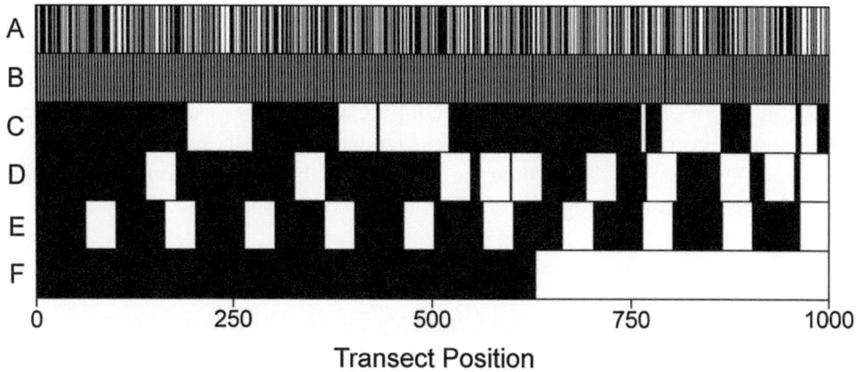

Fig. 6.7 Simulated binary sets displaying six unique scale-dependent patterns of foliage (*black*) and gap (*white*) dispersion: (*A*) random, (*B*) uniform, (*C*) randomly distributed gaps of random width, (*D*) randomly distributed gaps of equal width, (*E*) uniformly distributed gaps of equal width, and (*F*) single gap. Each set has the same number of occupied sites ($a = 630$), total length ($n = 1000$), and density p ($a/n = 0.63$)

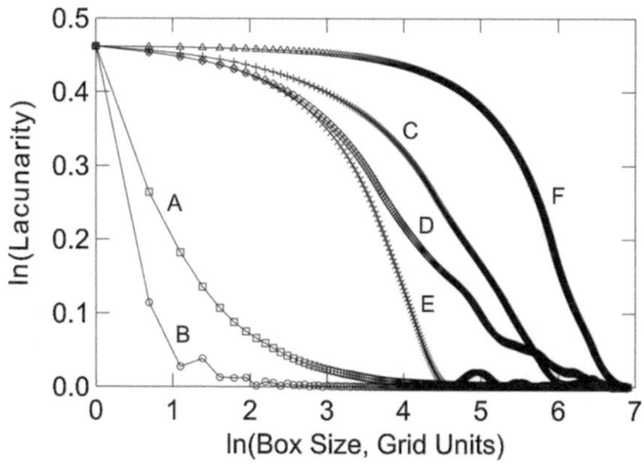

Fig. 6.8 Natural logarithm of $\Lambda(r)$ plotted against the natural logarithm of box size r. Lacunarity *curves A–F* correspond directly to binary sets *A–F* in Fig. 6.7. Lacunarity curves that lie above or fall below the $\Lambda(r)$ that is expected under CSR (*curve A*) indicate the kind, magnitude, and scale of non-randomness. For example, *curve B* falls substantially below CSR and thus indicates spatial uniformity across all measurement scales. *Curves E, D, C,* and *F*, on the other hand, all lie above CSR and display, in ascending order, increasing quantities and scales of spatial aggregation or clumping

found that all four clumping indices revealed a very different angular pattern of response to spatial heterogeneity. LX, for instance, was the least responsive to angular changes in gap-size distribution, while PCS exhibited extreme sensitivity across all zenith view angles. CC and LAC were situated between LX and PCS in terms of responsiveness. Therefore, we expect that certain clumping indices may perform better in one forest type compared to another, especially when paired with the optimal range of view angles over which L_t is integrated. For example, PCS performed best in the presence of severe spatial clumping and when L_t was integrated across a narrow range of view angles at 55°–60°. LX provided satisfactory results only under low levels of spatial clumping and when L_t was integrated across the widest range of possible view angles. Understanding these inherent methodological differences may help guide the selection of the most appropriate clumping index and integration strategy for a given forest condition.

6.4.4 Sensitivity of $\Omega(\theta)$ to Changes in Image Resolution

Image resolution (pixel size) can have a significant effect on photograph-based estimates of $\Omega(\theta)$, because it determines the 'minimum' gap size that can be measured, and also influences the number and proportion of pure (canopy or sky) and mixed (canopy and sky) image pixels (Frazer et al. 2001; Walter et al. 2003). Pixel mixing specifically interferes with image classification (i.e., the correct separation of pixels into sky and canopy classes), and the delineation and measurement of gap size, while pixel size alone arbitrarily defines the lower size limit of the measured gap-size distribution. We would expect that higher image resolutions will facilitate the capture of smaller canopy gaps, alleviate some occurrences of local saturation, and improve the reliability of gap size and gap fraction measurements. Because all four clumping methods depend directly upon the canopy gap-size distribution or its gap fraction, we also expect that each clumping index will exhibit some inherent sensitivity to changes in image resolution (Walter et al. 2003; Frazer 2007).

LAC and PCS methods were specifically designed to measure the spatial relationship (i.e., random, aggregated, or uniform) between black (foliage) and white (gap) pixels within a binary set. One important consideration to remember when using these two techniques is that the number of pixels comprising any discrete canopy gap or foliage clump will arbitrarily rise with an increase in image resolution. Thus, sequences of black and white pixels will appear more spatially clumped at higher image resolutions, i.e., when more pixels of the same type are found together. PCS is particularly sensitive to changes in image resolution, because it is based solely on the average run-lengths of black and white pixels (Walter et al. 2003). LAC, on the other hand, is derived from the mean and variance of black- and white-pixel counts measured at several spatial scales and, therefore, less sensitive to shifts in image resolution.

CC will also be affected by image resolution, because pixel size arbitrarily defines the minimum gap size λ of the measured gap-size distribution $F_m(\lambda, \theta)$ (Eq. 6.7). For example, coarser (lower) image resolutions will truncate $F_m(\lambda, \theta)$ at larger minimum gap sizes λ and, thus, have the potential to reduce the magnitude of the accumulated gap fraction $F_m(0, \theta)$. Forest canopies characterized by many small gaps will be particularly sensitive to this problem. Since image resolution has little effect on the removed portion (i.e., the gap fraction associated with the largest canopy gaps) of the 'compacted' gap fraction $F_{mr}(0, \theta)$, estimates of CC from low-resolution images could be smaller than those that are retrieved from high-resolution images. LX does not directly rely upon measurements of gap size and, therefore, is relatively insensitive to changes in image resolution. Nevertheless, we recommend capturing fisheye photographs at the highest image resolution possible to minimize local saturation problems, and to implement an image re-sampling experiment to better understand the potential effects of image resolution on $\Omega(\theta)$ and L_t (Pekin and Macfarlane 2009).

6.5 Conclusion

Our results demonstrate the validity of the Poisson model when the spatial arrangement of foliage satisfies the required assumption of complete spatial randomness. At LAI > 6, canopy gaps become rare to nonexistent in random canopies (the global saturation point), and gap fraction analysis can no longer be used to obtain accurate estimates of LAI. Therefore, an LAI of 5 or 6 has been widely regarded as the practical upper limit for optical measurements of LAI (Gower et al. 1999). We found that the Poisson model increasingly underestimated LAI when the simulated pattern of leaves departed from a random distribution toward progressively more clumped spatial arrangements. Inclusion of $\Omega(\theta)$ in the Markov model reduced the underestimation of LAI only when CC, LAC or LX methods were invoked. The effect of PCS on L_t was more idiosyncratic, less predictable, and strongly dependent upon canopy structure and the range of view angles over which L_t was integrated. Therefore, we would recommend further experimentation with PCS in stands of known LAI and with different image resolutions prior to its operational use. LAC was the only technique that explicitly measured the integrated effect of spatial non-randomness across a broad range of scales. Further research is required to determine whether or not the estimation of $\Omega(\theta)$ improves when LAC is restricted to a more limited range of spatial scales. CC produced results similar to those of LAC, but was computationally far simpler to estimate. The LX approach was the least difficult to implement; however, segment-length and azimuthal position are critical input parameters that will influence the magnitude of $\Omega_{LX}(\theta)$ in heterogeneous forest canopies. As van Gardingen et al. (1999) have suggested, 'optimal' segment-length can be found when the measurement scale has maximized the total number of segment-lengths i, while still minimizing the number of occurrences where $P_i(\theta) = 0$. We made no attempt to optimize segment length in this study, nor did we investigate alternative values of n in the estimation of $\Omega_{LX}(\theta)$. Our findings, therefore, may not reflect the true potential of LX to improve LAI estimation in clumped canopies.

We suggest two main reasons why photograph-based estimates of L_t have continued to underestimate LAI. First, local saturation will always occur at one or more spatial scales and locations within a fisheye photograph, particularly when forest canopies are clumped and support a large LAI. Once saturation occurs, optical methods ignore the LAI that is contained within discrete foliage clumps and, instead, measure the 'imaginary' surface area of the clumps themselves. Second, foliage clumping is intrinsically a 3D geometrical phenomenon with strong scale dependencies, i.e., foliage clumps occupy a discrete volume of canopy space, have an associated leaf density, and their size depends entirely upon the scale of observation. It is unlikely then that photo-based measurements acquired in a dimensionally reduced space (i.e., 1D sequences of black and white pixels that are extracted from a 2D binary image) will ever adequately describe the complex spatial heterogeneity that occurs within the original three dimensions of a forest canopy.

We believe that the local saturation problem can be alleviated to some extent (i) with higher spatial and spectral image resolutions, (ii) by more intensive, view angle-specific sampling using longer focal length (non-fisheye lenses) photographic lenses, and (iii) through the application of non-binary, local thresholding algorithms. Little can be done, however, to compensate for the reduced spatial dimensionality (1D and 2D) of optical data, and we see this as the biggest obstacle and challenge for reliable LAI estimation, particularly in tall, spatially complex, high biomass forests. We also recognize that the scale and pattern of foliage clumping displayed in our simulated fisheye photographs were not entirely realistic representations of the leaf distributions and canopy structure found in real forests. Nevertheless, these simulated canopies have provided a convenient framework and baseline to compare the behavior and performance of foliage clumping indices. Further research is required to determine how well these photograph-based clumping indices operate in a broad range of real forest canopies.

Acknowledgements We thank Drs. Jean-François Côté and Kamel Soudani for production of the simulated fisheye photosets. We also gratefully acknowledge the time, effort and expertise provided by two reviewers. Other editorial improvements were contributed by Bill Parsons.

Appendix: Lacunarity as a Scale-Dependent Measure of Spatial Non-randomness in Forest Canopies

Gliding-Box Algorithm[3]

The gliding-box algorithm requires five main steps to compute the lacunarity of a binary set composed of 1s (occupied/foliage) and 0s (unoccupied/gap) (Allain and Cloitre 1991; Plotnick et al. 1996). First, a 'gliding' or moving window of box size r (in grid cell units) is shifted one position i at a time along the entire length n of the set. At each grid cell position i along the set, the sum or box mass s of grid cell values

[3]See Plotnick et al. (1996) for a detailed description and working examples of the gliding-box algorithm.

x_j contained within the box of size r is computed until all positions i along the set have been sampled. The result is a frequency distribution of box mass $n(s, r)$ sampled at r. Second, each box mass s of the frequency distribution $n(s, r)$ sampled at r is divided by the total number of sampled boxes of size r to generate a probability distribution $Q(s, r)$. Third, the first and second statistical moments of the probability distribution $Q(s, r)$ are computed (Cheng 1997; Dale 1999) as:

$$Z_Q^{(1)}(r) = \frac{1}{(n+1-r)} \sum_{i=1}^{n+1-r} \sum_{j=1}^{i+1-r} x_j \tag{6.11}$$

$$Z_Q^{(2)}(r) = \frac{1}{(n+1-r)} \sum_{i=1}^{n+1-r} \left(\sum_{j=1}^{i+r-1} x_j \right)^2 \tag{6.12}$$

Fourth, lacunarity at r is estimated by dividing the second moment of the probability distribution $Q(s, r)$ by the square of the first moment (Plotnick et al. 1996):

$$\Lambda(r) = \frac{Z_Q^{(2)}(r)}{\left[Z_Q^{(1)}\right]^2} = 1 + \frac{\sigma_Q^2(r)}{\left[Z_Q^{(1)}\right]^2} \tag{6.13}$$

where $\sigma_Q^2(r)$ and $Z_Q^{(1)}$ are, respectively, the sample variance and mean of the probability function $Q(s, r)$. Last, repeating the lacunarity calculations at box sizes $1 \le r \le n$ reveals a distinct pattern of variation in the logarithm of $\Lambda(r)$ when it is plotted against the logarithm of r.

Scale-Dependency and Interpretation of Λ(r)

Three general patterns of spatial dispersion are recognized in binary sets composed of occupied (black = 1) and unoccupied (white = 0) sites, viz., random, uniform, and aggregated (Dale 1999). Random patterns arise when occupied sites occur randomly and independently of one another along the length of a set (e.g., set A, Fig. 6.7). Uniform and aggregated patterns, in contrast, are neither random nor independent and exhibit characteristics of spatial dependence or interaction. For example, uniform patterns form a 'regular' spacing of occupied sites, where the presence of one occupied site reduces the probability that another will be found in close proximity to it (e.g., set B, Fig. 6.7). Aggregated patterns appear 'clumped,' and the presence of an occupied site increases the probability that another will be found close by (e.g., sets C–F, Fig. 6.7). Spatial pattern, however, is strongly scale-dependent and those patterns that appear random, uniform, or aggregated at one measurement scale may be very different at finer or at coarser scales (Plotnick et al. 1996; Dale 2000).

Scatterplots of the logarithm of $\Lambda(r)$ against the logarithm of r reveal the scale-dependent spatial structure, heterogeneity, and non-randomness that is associated with each of the binary sets described below (Fig. 6.8). The magnitude of $\Lambda(r)$ is a measure of the spatial heterogeneity that is found in the set at scale r (Plotnick et al. 1996). For example, values of $\Lambda(r)$ that are equal or close to 1 indicate translational invariance or spatial homogeneity at r, while increasingly larger values of $\Lambda(r)$ denote increasing spatial heterogeneity. Lacunarity curves A to F show an overall decline in $\Lambda(r)$ with increasing r; however, the rate and pattern of this decline differs according to the set's scale-dependent spatial structure or 'gappiness'. All six curves share the same X and Y intercepts, because each set has the same number of occupied sites (a = 630), total length (n = 1000), and pattern density ($p = a/n = 0.63$). The magnitude of $\Lambda(r)$ at r = 1 will vary inversely with p, since $\Lambda(1)$ is equal to $1/p$ (Plotnick et al. 1996).

Lacunarity curves that were generated for sets A and B approach translational invariance ($\Lambda(r) = 1$) more rapidly than curves representing the coarser-grained sets C–F (Fig. 6.8). The steepest decline in $\Lambda(r)$ with increasing r occurs when discrete occupied sites are regularly spaced (set B). When occupied sites are dispersed randomly and independently of one another, as is the case in set A, the decline in $\Lambda(r)$ occurs at a rate predicted by the Binomial Theorem (Dale 2000):

$$\Lambda_{CSR}(r) = 1 + 1/rp - 1/r \tag{6.14}$$

where $\Lambda_{CSR}(r)$ is the expected lacunarity statistic at box size r under the condition of complete spatial randomness (CSR), and p is the pattern density of the binary set. Lacunarity curves that lie above or which fall below $\Lambda(r)$ expected under CSR (set A) indicate the type, magnitude, and scale of non-randomness. For example, curves E, D, C, and F all lie above CSR and display, in ascending order, increasing quantities and scales of spatial aggregation (clumping). Curve E exhibits the least amount of coarse-scale aggregation, because gaps are relatively small, homogeneous in size, and uniformly dispersed across the set. Set F is composed of one large gap, and displays the greatest amount of aggregation at all spatial scales. Curve B falls substantially below CSR and, therefore, indicates the presence of spatial uniformity at all measurement scales.

Derivation of $\Omega_{LAC}(\theta)$

From the brief description and examples of lacunarity analysis that were outlined in the previous sections, it is evident that the integrated cross-scale differences between observed estimates of $\Lambda(r)$ and those expected under CSR provide both the conceptual and analytical bases for the derivation of $\Omega(\theta)$. Three important analytical steps were necessary to develop this index. First, binary sets extracted from fisheye photographs have no logical start or end due to the circular nature of the sample transect. Nevertheless, there are implications to and rules for selecting a starting point: (i) sets must always start at a gap edge to avoid the fragmentation of discrete gaps or clumps; (ii) $\Lambda(r)$ can be sensitive to the set's starting point, because at coarse measurement scales the gliding-box algorithm samples pixels (grid cells) in the middle of the set more often than ones near the start or end (Cheng 1997). It is, therefore, imperative to resample the set using randomly or regularly placed starting points and to compute the mean observed lacunarity statistic for all resampled sets:

$$\overline{\Lambda_{Obs}(r)} = \frac{1}{m}\sum_{t=1}^{m}\Lambda_t(r) \tag{6.15}$$

where $\Lambda_t(r)$ is the lacunarity statistic estimated at r for each resampled transect t, and m is the total number of resampled transects. Short, spatially heterogeneous sets require more resampling than long, homogeneous sets; however, we found that resampling the binary set four to eight times using regularly spaced starting points was usually sufficient to stabilize $\overline{\Lambda_{Obs}(r)}$.

Second, the magnitude of $\Lambda(r)$ is a function of p and its spatial variability or dispersion across the binary set. Lacunarity curves generated for sets with identical lengths n and different densities p are therefore not directly comparable unless the effect of density p on $\Lambda(r)$ is first eliminated. Normalization to a common Y intercept can be accomplished by dividing each value of $\Lambda(r)$ by $\Lambda(r = 1)$. In doing so, the effect of the first statistical moment $Z_Q^{(1)}$ on all values of $\Lambda(r)$ is removed, which allows direct comparison of the scale-dependent variation in lacunarity for sets with identical lengths n and different densities p (Plotnick et al. 1996; Feagin, 2003).

Last, the total integrated area found between the normalized $\overline{\Lambda_{Obs}(r)}$ and $\Lambda_{CSR}(r)$ curves across all measurement scales r gives a quantitative estimate of a set's departure from CSR, where areas below and above $\Lambda_{CSR}(r)$ indicate the amount of spatial uniformity and clumping, respectively. Also knowing the area beneath a hypothetical curve of 'maximal clumping' allows us to derive an estimate of $\Omega(\theta)$:

$$\Omega_{LAC}(\theta) = \left[\frac{A_{Max}(\theta) - A_{Obs}(\theta)}{A_{Max}(\theta) - A_{CSR}(\theta)}\right] \tag{6.16}$$

where $A_{Obs}(\theta)$, $A_{CSR}(\theta)$, and $A_{Max}(\theta)$ are respectively the integrated areas under the normalized $\overline{\Lambda_{Obs}(r)}$ $\Lambda_{CSR}(r)$, and 'maximally clumped' $\Lambda_{Max}(r)$ lacunarity curves computed for a binary set extracted at θ (Fig. 6.9). Values of $\Omega_{LAC}(\theta)$ range from greater than unity for uniform sets, unity at CSR, and less than unity for clumped sets. Previous work by Plotnick et al. (1996) indicates that a binary set defining $\Lambda_{Max}(r)$ will have two important properties: (i) all occupied sites (black pixels) will be completely separate from all unoccupied sites (white pixels), and (ii) p will approach zero. To understand this conceptually, imagine that all shoots or leaves

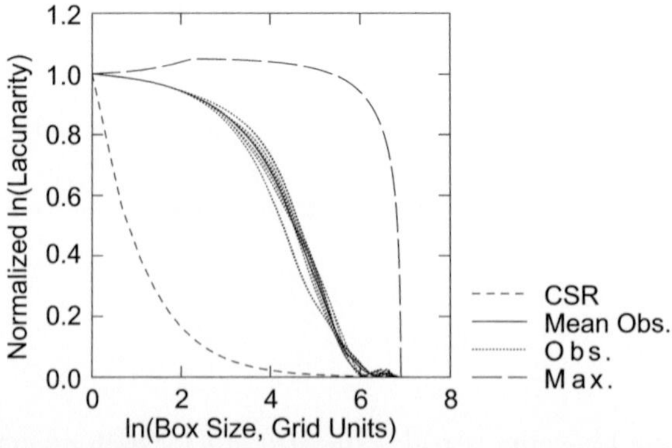

Fig. 6.9 Normalized lacunarity curves identifying the position of the mean 'observed' curve (set *C* in Fig. 6.7) relative to CSR and maximal clumping. We divide the integrated area between maximal clumping and the mean 'observed' curve by the integrated area between maximal clumping and CSR to estimate $\Omega_{LAC}(\theta)$. The range of variation in the 14 'observed' curves is due to the resampling of transect starting and end points. The estimated mean and standard deviation of $\Omega_{LAC}(\theta)$ for set *C* are 0.36 ± 0.01 ($0.34 \leq \Omega_{LAC}(\theta) \leq 0.40$) using all 14 possible transect start and end points

visible in a fisheye photograph at θ were rearranged so that they appear at only one point in the hemispherical object region, with a single foliage unit obstructing the view of all other units that are stacked behind it. Although this kind and degree of clumping would never occur in natural forests, $\Lambda_{Max}(r)$ does represent a reasonable endpoint for spatial clumping from a purely geometrical perspective. We found that $\Lambda_{Max}(r)$ reached its maximum value when p was equal to 0.01.

References

Allain C, Cloitre M (1991) Characterizing the lacunarity of random and deterministic fractal sets. Phys Rev A44:3552–3558

Baldocchi DD, Hutchison BA, Matt DR, McMillen RT (1985) Canopy radiative transfer models for spherical and known leaf inclination angle distributions: a test in an oak-hickory forest. J Appl Ecol 22:539–555

Barclay HJ, Trofymow JA (2000) Relationship of readings from the LI-COR canopy analyzer to total one-sided leaf area index and stand structure in immature douglas-fir. For Ecol Manage 132:121–126

Black TA, Chen J-M, Lee X, Sagar RM (1991) Characteristics of shortwave and longwave irradiances under a douglas-fir forest stand. Can J For Res 21:1020–1028

Blumenfeld R, Mandelbrot BB (1997) Lévy dusts, Mittag-Leffler statistics, mass fractal lacunarity, and perceived dimension. Phys Rev E 56:112–118

Bréda NJJ (2003) Ground-based measurements of leaf area index: a review of methods, instruments and current controversies. J Exp Bot 54:2403–2417

Castro F, Fetcher N (1999) The effect of leaf clustering in the interception of light in vegetal canopies: theoretical considerations. Ecol Model 116:125–134

Cescatti A (1998) Effects of needle clumping in shoots and crowns on the radiative regime of a Norway spruce canopy. Ann For Sci 55:89–102
Chason JW, Baldocchi DD, Huston MA (1991) A comparison of direct and indirect methods for estimating forest canopy leaf area. Agric For Meteorol 57:107–128
Chen JM, Black TA (1991) Measuring leaf area index of plant canopies with branch architecture. Agric For Meteorol 57:1–12
Chen JM, Black TA (1992) Defining leaf area index for non-flat leaves. Plant Cell Environ 15:421–429
Chen JM, Cihlar J (1995) Plant canopy gap-size theory for improving optical measurements of leaf-area index. Appl Opt 34:6211–6222
Chen JM, Rich PM, Gower ST, Norman JM, Plummer S (1997) Leaf area index of boreal forests: theory, techniques, and measurements. J Geophys Res 102(D24):29429–29443
Cheng Q (1997) Multifractal modeling and lacunarity analysis. Math Geol 29:919–932
Cooper CF (1961) Pattern in ponderosa pine forests. Ecology 42:493–499
Dale MRT (1999) Spatial pattern analysis in plant ecology. Cambridge University Press, Cambridge
Dale MRT (2000) Lacunarity analysis of spatial pattern: a comparison. Landscape Ecol 15: 467–478
Deblonde G, Penner M, Royer (1994) Measuring leaf area index with the LI-COR LAI-2000 in pine stands. Ecology 75:1507–1511
Fassnacht KS, Gower ST, Norman JM, M cMurtrie RE (1994) A comparison of optical and direct methods for estimating foliage surface area index in forests. Agric For Meteorol 71:183–207
Feagin RA (2003) Relationship of second-order lacunarity, Hurst exponent, Brownian motion, and pattern organization. Phys A 328:315–321
Fournier RA, Landry R, August NM, Fedosejevs G, Gauthier RP (1996) Modelling light obstruction in three conifer forests using hemispherical photography and fine tree architecture. Agric For Meteorol 82:47–72
Fournier RA, Mailly D, Walter J-MN, Soudani K (2003) Indirect measurement of forest canopy structure from in situ optical sensors. In: Wulder MA, Franklin SE (eds) Remote sensing of forest environments: concepts and case studies. Kluwer, Dordrecht, pp 77–113
Fournier RA, Rich PM, Landry R (1997) Hierarchical characterization of canopy architecture for boreal forest. J Geophys Res 102(D24):29445–29454
Frazer GW (2007) Fine-scale, multi-dimensional spatial patterns of forest canopy structure derived from remotely sensed and simulated datasets. Dissertation, University of Victoria, BC. 251 pp
Frazer GW, Fournier RA, Trofymow JA, Hall RJ (2001) A comparison of digital and film fisheye photography for analysis of forest canopy structure and gap light transmission. Agric For Meteorol 109:249–263
Frazer GW, Trofymow JA, Lertzman KP (1997) A method for estimating canopy openness, effective leaf area index, and photosynthetically active photon flux density using hemispherical photography and computerized image analysis techniques. Natural Resources Canada, Canadian Forest Service, Pacific Forestry Centre, Victoria, BC. Information Report BC-X-373, 73 pp
Frazer GW, Wulder MA, Niemann KO (2005) Simulation and quantification of the fine-scale spatial pattern and heterogeneity of forest canopy structure: a lacunarity-based method designed for analysis of continuous canopy heights. For Ecol Manage 214:65–90
Gefen Y, Meir Y, Mandelbrot BB, Aharony A (1983) Geometric implementation of hypercubic lattices with noninteger dimensionality by use of low lacunarity fractal lattices. Phys Rev Lett 50:145–148
Gonsamo A, Pellikka P (2009) The computation of foliage clumping index using hemispherical photography. Agric For Meteorol 149:1781–1787
Gower ST, Kucharik JK, Norman JM (1999) Direct and indirect estimation of leaf area index, f_{APAR}, and net primary production of terrestrial ecosystems. Remote Sens Environ 70:29–51
Gower ST, Norman JM (1991) Rapid estimation of leaf area index in conifer and broad-leaf plantations. Ecology 72:1896–1900

Hwang Y, Ryu Y, Kimm H, Jiang C, Lang M, Macfarlane C (2016) Correction for light scattering combined with sub-pixel classification improves estimation of gap fraction from digital cover photography. Agric For Meteorol 222:32–44

Jonckheere I, Fleck S, Nackaerts K, Muys B, Coppin P, Weiss M, Baret F (2004) Methods for leaf area index determination. Part I: Theories, sensors and hemispherical photography. Agric For Meteorol 121:19–35

Jonckheere I, Nackaerts K, Muys B, Coppin P (2005) Assessment of automatic gap fraction estimation of forests from digital hemispherical photography. Agric For Meteorol 132:96–114

Jonckheere I, Nackaerts K, Muys B, van Aardt J, Coppin P (2006) A fractal dimension-based modelling approach for studying the effect of leaf distribution on LAI retrieval in forest canopies. Ecol Model 197:179–195

Kucharik CJ, Norman JM, Gower ST (1999) Characterization of radiation regimes in nonrandom forest canopies: theory, measurements, and a simplified modeling approach. Tree Physiol 19:695–706

Kucharik CJ, Norman JM, Murdock LM, Gower ST (1997) Characterizing canopy non-randomness with a multiband vegetation imager (MVI). J Geophys Res 102 (D24):29455–29473

Lang ARG (1991) Application of some of Cauchy's theorems to estimation of surface area of leaves, needles and branches of plants, and light transmittance. Agric For Meteorol 55:191–212

Lang ARG, Xiang Y (1986) Estimation of leaf area index from transmission of direct sunlight in discontinuous canopies. Agric For Meteorol 35:229–243

Law BE, Van Tuyl S, Cescatti A, Baldocchi DD (2001) Estimation of leaf area index in open-canopy ponderosa pine forests of different successional stages and management regimes in Oregon. Agric For Meteorol 108:1–14

Leblanc SG, Chen JM (2001) A practical scheme for correcting multiple scattering effects on optical LAI measurements. Agric For Meteorol 110:125–139

Leblanc SG, Chen JM, Fernandes R, Deering DW, Conley A (2005) Methodology comparison for canopy structure parameters extraction from digital hemispherical photography in boreal forests. Agric For Meteorol 129:187–207

Leblanc SG, Fournier RA (2014) Hemispherical photography simulations with an architectural model to assess retrieval of leaf area index. Agric For Meteorol 194:64–76

Leblanc SG, Fournier RA (2017) Measurement of forest structure with hemispherical photography. In: Fournier RA, Hall RJ (eds) Hemispherical photography in forest science: theory, methods, applications. Springer, New York

Macfarlane C (2011) Classification method of mixed pixels does not affect canopy metrics from digital images of forest overstory. Agric For Meteorol 151:833–840

Macfarlane C, Hoffman M, Eamus D, Kerp N, Higginson S, McMertrie R, Adams M (2007a) Estimation of leaf area index in eucalypt forest using digital photography. Agric For Meteorol 143:176–188

Macfarlane C, Grigg A, Evangelista E (2007b) Estimating forest leaf area using cover and fullframe fisheye photography: thinking inside the circle. Agric For Meteorol 146:1–12

Macfarlane C, Ryu Y, Ogden GN, Sonnentag O (2014) Digital canopy photography: exposed and in the raw. Agric For Meteorol 197:244–253

Mandelbrot BB (1983) The fractal geometry of nature. W.H. Freeman and Co., New York

Miller JB (1967) A formula for average foliage density. Aust J Bot 15:141–144

Neumann HH, den Hartog G, Shaw RH (1989) Leaf area measurements based on hemispheric photographs and leaf-litter collection in a deciduous forest during autumn leaf-fall. Agric For Meteorol 45:325–345

Nilson T (1971) A theoretical analysis of the frequency of gaps in plant stands. Agric For Meteorol 8:25–38

Nilson T (1999) Inversion of gap frequency data in forest stands. Agric For Meteorol 98–99:437–448

Nilson T, Kuusk A (2004) Improved algorithm for estimating indices from gap fraction data. Agric For Meteorol 124:157–169

Oker-Blom P, Kaufmann MR, Ryan MG (1991) Performance of a canopy light interception model for conifer shoots, trees and stands. Tree Physiol 9:227–243
Parker GG (1995) Structure and microclimate of forest canopies. In: Lowman MD, Nadkarni NM (eds) forest canopies. Academic Press, San Diego, pp 73–106
Pekin BK, Macfarlane C (2009) Measurement of crown cover and leaf area index using digital cover photography and its application to remote sensing. Remote Sensing 1:1298–1320
Pielou EC (1962) Runs of one species with respect to another in transects through plant populations. Biometrics 18:579–593
Planchais I, Pontailler J-Y (1999) Validity of leaf areas and angles estimated in a beech forest from analysis of gap frequencies, using hemispherical photographs and a plant canopy analyzer. Ann For Sci 56:1–10
Plotnick RE, Gardner RH, Hargrove WW, Prestegaard K, Perlmutter M (1996) Lacunarity analysis: a general technique for the analysis of spatial patterns. Phys Rev E 53:5461–5468
Ryu Y, Sonnentag O, Nilson T, Vergas R, Kobayashi H, Wenk R, Baldocchi D (2010) How to quantify tree leaf area index in an open savanna ecosystem: a multi-instrument and multi-model approach. Agric For Meteorol 150:63–76
Smith NJ (1993) Estimating leaf area index and light extinction coefficients in stands of douglas-fir (*Pseudotsuga menziesii*). Can J For Res 23:317–321
Smith NJ, Chen JM, Black TA (1993) Effects of clumping on estimates of stand leaf area index using the LI-COR LAI-2000. Can J For Res 23:1940–1943
Smolander H, Stenberg P, Linder S (1994) Dependence of light interception efficiency of Scots pine shoots on structural parameters. Tree Physiol 14:971–980
Stenberg P (1996) Correcting LAI-2000 estimates for the clumping of needles in shoots of conifers. Agric For Meteorol 79:1–8
Stenberg P (2006) A note on the *G*-function for needle leaf canopies. Agric For Meteorol 136:76–79
van Gardingen PR, Jackson GE, Hernandez-Daumas S, Russell G, Sharp L (1999) Leaf area index estimates obtained for clumped canopies using hemispherical photography. Agric For Meteorol 94:243–257
Walter J-MN, Fournier RA, Soudani K, Meyer E (2003) Integrating clumping effects in forest canopy structure: an assessment through hemispherical photographs. Can J Remote Sens 29:388–410
Warren Wilson J (1960) Inclined point quadrats. New Phytol 59:1–8
Weiss M, Baret F, Smith GJ, Jonckheere I, Coppin P (2004) Review of methods for in situ leaf area index (LAI) determination: Part II. Estimation of LAI, errors and sampling. Agric For Meteorol 121:37–53
Welles JM, Cohen S (1996) Canopy structure measurement by gap fraction analysis using commercial instrumentation. J Exp Bot 47:1335–1342
Welles JM, Norman JM (1991) Instrument for indirect measurement of canopy architecture. Agron J 83:818–825
Whitford KR, Colquhoun IJ, Lang ARG, Harper BM (1995) Measuring leaf area index in a sparse eucalypt forest: a comparison of estimates from direct measurement, hemispherical photography, sunlight transmittance and allometric regression. Agric For Meteorol 74:237–249

Chapter 7
Comparison of Software Tools for Analysis of Hemispherical Photographs

Ronald J. Hall, Jean-François Côté, Daniel Mailly and Richard A. Fournier

Abstract A question of particular interest is which software program should be used for processing of hemispherical photographs (HP) of forest canopies? While a few studies have attempted to address this question, varying approaches to comparing different HP software have not resulted in conclusive evidence as to which software is best. Our objective was to select from available software and to conduct a functional analysis of software features and capabilities, followed by a quantitative analysis of photographs representing a range of forest canopies at four sites, to provide an improved basis for software selection. Four criteria encompassing availability, maintenance, documentation and active use were used to select among available HP programs for assessment. Six programs were selected (Gap Light Analyzer (GLA), HemiView, CAN-EYE, CIMES, WinSCANOPY, SOLARCALC), but one (WinSCANOPY) was not made available by the manufacturer for further assessment. The programs were used to estimate three structural-canopy and two light-regime attributes. There was a high degree of agreement among some of the attributes evaluated, and statistically significant differences among others. Differences were attributed, in part, to variations in the algorithm selected and in its implementation in the software for a given attribute.

R.J. Hall (✉)
Canadian Forest Service, Northern Forestry Centre, Natural Resources Canada, 5320—122 Street, Edmonton, AB T6H 3S5, Canada
e-mail: Ron.Hall@Canada.ca

J.-F. Côté
Canadian Wood Fibre Centre, Natural Resources Canada, 1055 Du P.E.P.S. Street, P.O. Box 10380, Québec, QC G1V 4C7, Canada
e-mail: Jean-Francois.Cote@Canada.ca

D. Mailly
Decizio, 929 ave Sir Adolphe-Routhier, Quebec City, QC G1S 3P4, Canada
e-mail: daniel.mailly.1@ulaval.ca

R.A. Fournier
Département de géomatique appliquée, Centre d'Applications et de Recherches en Télédétection (CARTEL), Université de Sherbrooke, 2500, boul. de l'Université, Sherbrooke, QC J1K 2R1, Canada
e-mail: Richard.Fournier@USherbrooke.ca

R.A. Fournier and R.J. Hall (eds.), *Hemispherical Photography in Forest Science: Theory, Methods, Applications*, Managing Forest Ecosystems 28,
DOI 10.1007/978-94-024-1098-3_7

Identifying which software to select is a function of many factors governing interoperability, availability, software functionality, user experience and intended application. The decision depends on knowledge of these factors, of the algorithms used by the programs, and of the potential for error in image acquisition and processing.

Keywords Canopy structure · Gap fraction · Hemispherical photography software · Leaf area index (LAI) · Light regime

7.1 Introduction

Hemispherical (fisheye) photography of forest canopies is the process of obtaining photographs from the ground looking upward through an extremely wide–angle lens. These photographs result in a permanent record of the spectral and spatial attributes of the forest canopy. Attributes characterizing the penetration of direct and diffuse light can be estimated from these photographs (Rich 1990; Frazer et al. 2001). While hemispherical photographs (HP) were previously analyzed using a manual process (Anderson 1964), they could be assessed more objectively and consistently when software programs became available. Examples of early software for processing HP include SOLARCALC (Chazdon and Field 1987), CANOPY (Rich 1989), SYLVA (Becker et al. 1989), and Hemiphot/Winphot (ter Steege 1996). As digital technologies and computing capabilities have evolved, so have the advancements in digital HP acquisition hardware, leading to its increasing use (Englund et al. 2000; Frazer et al. 2001; Hale and Edwards 2002; Leblanc et al. 2005). There has been a corresponding interest in computer software programs to process HP (Frazer et al. 1997; Bréda 2003; Promis et al. 2011), for at least three reasons. First, preparing HP in analog or digital format (image format, linkage to camera parameters, image enhancement and classification for analysis of such parameters as the estimation of canopy structure) is cumbersome without analytical capability. Second, the computations for extracting estimates of canopy architecture and light regime from HP are complex. Third, image processing software allows users to quickly and consistently process large numbers of photographs.

Given the array of software that is available for processing HP, choosing a program has become a primary question. Within the past 10 years, relatively few studies have focused on assessment and comparisons of HP programs (Bréda 2003; Jarčuška et al. 2010; Promis et al. 2011). In a review of the literature, Bréda (2003) briefly mentioned two free software programs for computing leaf area index (LAI), GLA and Winphot, and followed with a more detailed tabular comparison of two commercial programs, HemiView and WinSCANOPY, but the author provided no analysis. Jarčuška et al. (2010) compared GLA and WinSCANOPY for calculations of canopy openness, LAI, relative total and diffuse radiation and relative direct transmittance. They found differences in the threshold for classifying the image between the software programs. Promis et al. (2011) compared four programs (HemiView, GLA, hemiIMAGE, and Winphot) for estimating canopy structure (e.g., canopy openness and effective plant area index) and solar radiation (e.g., direct, diffuse and global solar radiation) for three different broadleaf forest ecosystems that were located in Chile, Germany, and Venezuela. While canopy openness and effective plant area index were highly correlated among the HP programs, there was no correlation or the relationship was weak for effective LAI (LAI_{eff}) and canopy gaps between HemiView and Winphot (Promis et al. 2011). Estimates of solar radiation transmittance were similar among the HP programs that were evaluated. The evaluation also included comments about software functionality, technical support, computer system requirements, and costs. It was obvious from these studies that choosing a particular HP program is complicated because of multiple factors that are related to interoperability, software functionality and intended applications; also, knowledge about the functional differences among different programs is not readily available.

Addressing the differences among HP programs entails familiarity with the features of a given program and an assessment of its empirical performance in processing HP. Hence, the first objective of this chapter is to summarize the features of current software for processing HP. The second objective is to identify differences in software functionality (i.e., functionality analysis) and compare the estimation of several structural canopy and light regime variables in four test sites that represent coniferous and deciduous forest vegetation characterized by open or closed canopies and by random or clumped foliage distributions (i.e., quantitative analysis).

To determine which HP program should be considered in such an assessment, four criteria were defined as a basis for selection:

1. Is the program commercially available or readily available in the public domain?
2. Is the program being maintained, updated or supported?
3. Are program documentation and user instructions available?
4. Is the program in active use, as supported by scientific references, within the most recent 10 years?

Based on these criteria, six HP programs (Gap Light Analyzer [GLA], HemiView, CAN–EYE, CIMES, WinSCANOPY, SOLARCALC) were identified for comparison of features and functionality, with four being used to process HP for the quantitative analysis (Table 7.1). Of the six programs, four are freely available for download (GLA, CAN-EYE, CIMES, SOLARCALC), and two are from commercial vendors (HemiView, WinSCANOPY). The functional analysis consisted of comparing image registration and display, calibration tools, image classification and output results. The quantitative analysis involved selecting a set of

Table 7.1 Software packages for analyzing hemispherical photographs

Software	Version evaluated	Availability	Source and References[a]
Gap Light Analyzer (GLA)	2.0	Freeware	Source: http://www.ecostudies.org/gla/ or https://www.sfu.ca/rem/forestry/downloads/gap-light-analyzer.html Frazer et al. (1999), Nobis and Hunziker (2005), Jarčuška (2008)
HemiView	2.0	Commercial	Source: Delta-T Devices Ltd., Cambridge, UK http://www.delta-t.co.uk/ (Rich et al. (1999), Hale and Edwards (2002), Bréda (2003), Valladares and Guzmán (2006)
CAN-EYE	5.0[b]	Freeware	Source: https://www6.paca.inra.fr/can-eye Baret and Weiss (2004), Demarez et al. (2008)
CIMES	2009	Freeware	Source: http://jmnw.free.fr/ Walter (2009), Gonsamo and Pellikka (2009), Gonsamo et al. (2011)
WinSCANOPY[c]	2012a	Commercial	Source: Regent Instruments Inc., Quebec, Canada Available in 3 versions: Basic, Regular, Pro DSLR http://www.regent.qc.ca/ Bréda (2003), North et al. (2004), Macfarlane et al. (2007)
SOLARCALC[d]	7.0	Freeware	Source: https://sites.google.com/site/solarcalcseven/downloads Mailly et al. (2013)

[a]References include samples of papers that illustrate use and application
[b]Version 6.4.7 now available
[c]Software was not made available for loan or purchase for this study from the vendor. Version 2016a is now available from the vendor
[d]Software was not released until recently

forest structure and light regime variables that were common to the four programs that were analyzed, i.e., GLA, HemiView, CAN-EYE, and CIMES. SOLARCALC was not included in the comparison because the most recent version of this program was not available until after the analyses for this study were completed. Four sets of HP were processed with each HP software program, representing a range of conditions that are typical of boreal forest canopies and a Mediterranean forest canopy with these characteristics: (a) open-canopy, clumped conifer forest; (b) closed-canopy, clumped conifer forest; (c) open-canopy, randomly distributed foliage conifer forest; and (d) closed-canopy, randomly distributed foliage deciduous forest.

7.2 Tools to Process Hemispherical Photographs

Processing HP consists primarily of three basic steps in the following sequence: image acquisition, digitization and editing, and analysis (Rich 1990). The acquisition of HP has been summarized in Chap. 4 by Fournier et al. (2017). Image digitization and editing involves the preparation of images for analysis. Digital cameras are now in prevalent use; previous traditional film-based systems relied upon a digital conversion process to translate the analog photograph to a digital format (Frazer et al. 2001; Hale and Edwards 2002; Macfarlane 2011). Image editing involves decomposing color images into their red, green and blue components, followed by optional image enhancement to reduce the presence of mixed pixels, thereby facilitating subsequent classification to discriminate between sky and foliage components (Walter 2009). Most software provides little or no capability for altering the gray-level intensity of an image to be processed, although GLA provides functions to alter image brightness, contrast, hue and saturation (Frazer et al. 1999). As a result, any image editing and enhancements are often undertaken outside of the HP software environment.

For the six HP programs that are identified in Table 7.1, we compiled a list of their main features (Table 7.2).

7.2.1 Software Features

GLA (v2.0) and HemiView (v2.1) run on a Microsoft Windows platform with a graphical user interface (GUI). CAN-EYE (v5.0) was tested under Windows XP but can run under any platform where a runtime version of MATLAB Compiler can be installed (Baret and Weiss 2004). CIMES (v2009) is a suite of programs written in C and C++ running under Windows, Linux/Unix, and OS X operating systems; the programs are continuously revised and updated (Walter 2009; Gonsamo et al. 2011). SOLARCALC (v7.0) is a newer implementation of a previous version of the program (Ackerly et al. 1995) running under Windows, Linux/Unix, and OS X operating systems (Mailly et al. 2013).

Software providers usually provide no access or very restricted access to the source code to prevent users from performing modifications within the application that could result in program malfunctions and incorrect results. Furthermore, proprietary software packages provide the user with a license specifying that the manufacturer retains all rights. In implementing and validating the different components of the software package, the user must therefore trust the developer(s). Non-commercial software packages may differ in these respects because they operate under free or open-source licensing agreements. Under such licenses, software is freely available for inspection of its code, for modification, and for distribution.

Table 7.2 Compilation of features for selected software programs for processing hemispherical photographs

Name/characteristics			Gap Light Analyzer	HemiView	CAN-EYE	CIMES	WinSCANOPY[a]	SOLARCALC
General	Current version	Feature	2.0	2.1	6.1	2009	2012a PRO	7.0
	Software version tested		2.0	2.0	6.1	2007	Software not available to evaluate. Version 2016a now available	Software not tested (just released)
	Source code access		Upon request: VB (GUI), C++ (Engine)	No	MATLAB source files	Yes: C and C++ language	No	No
	Computer requirement		Windows 95 or later; Windows NT 4.0 or later	Windows NT v.4.0, Windows 95, Windows 98, or later	All platforms where a MATLAB compiler runtime (free) can be installed	Windows (32 and 64 bits), Linux/Unix, OSX	Windows	Windows, Linux/Unix, OSX
	Batch processing		No	No	No. *Yes, except classification process. can concurrently process several images	Yes: command line programs	Yes: Interactive or batch. No limit on the number of images. Verification of analysis can be performed over random or all images	No
Image registration	Input format		Most common graphics formats (e.g. BMP, JPG, TIFF, PCX, GIF, etc.)	BMP, JPG, TIFF, PCX, TARGA and photo CD	JPG, TIFF	BMP with B-channel the grey mode	BMP, JPG, TIFF	Most common graphics formats supported by Quicktime (e.g. JPG, TIFF, BMP, PCX, GIF, etc.)

(continued)

Table 7.2 (continued)

Name/characteristics			Gap Light Analyzer	HemiView	CAN-EYE	CIMES	WinSCANOPY[a]	SOLARCALC
	Image resolution		Not specified	Not specified	Any size	Any size	Not specified	Not specified
	Geographic orientation	*North position*	Yes	Yes	No	Yes	Yes: automatically extract the geographic location and orientation when available from image files (e.g., camera is connected to a GPS)	Yes
		Magnetic declination	Yes	Yes	No	Yes, if you compute gap fraction	Yes	No
	Circular extent	*Zoom*	Yes	Yes	Automatic	No	Yes	No
		Mouse device	Yes	Yes	Automatic	No	Yes	Yes
		Manual registration	Yes	No	Yes	Yes: knowing 3 points and choose to extract background	Yes, automatic extraction when using Regent's Instruments	Yes
Configuration settings	Sky sector calibration	*Zenith resolution*	2°–22.5°	0.5°–30°	2.5°, 5°, 10°	2.5°–22.5° for gap fraction, 1° for gap size	Any, equal and unequal view angle specification	2°–22.5°
		Azimuth resolution	4°–22.5°	2°–120°	5°, 10°, 20°	2.5°–360°	Any	4°–45°
		Sub sample factor	No	No	Yes, 1-2-3	No	Yes	Yes
		Circle of Interest	No	No	Yes	No	No	No
		fCOVER max zenith angle	No	No	10°, 15°, 20°	90° nadir view	No	No

(continued)

Table 7.2 (continued)

Name/characteristics			Gap Light Analyzer	HemiView	CAN-EYE	CIMES	WinSCANOPY[a]	SOLARCALC
	Site calibration	*Latitude*	Yes	Yes	Yes	Yes	Yes: automatic extraction of GPS data available	Yes
		Longitude	Included but not implemented	Yes	No	Yes	Yes: automatic extraction of GPS data available	Yes
		Altitude	Yes, optional	Yes	No	Yes	Yes: automatic extraction of GPS data available	No
		Inclination	Yes, slope and aspect	No	No	Yes, slope and aspect	Yes, slope and aspect	No
		Topographic mask	Yes	No	No	No	Yes	No
	Lens calibration	*Lens projection*	Lambert (Equal-Area), Orthographic, Polar, Stereographic, and user-defined	Linear 180, Coolpix 800, Sigma 8 mm	Projection law derived from a polynomial (1° = polar, up to degree 3)	Polar equidistant, orthographic	Calibration coefficients for: 2009 version: Linear, FC-E8 Nikon Coolpix (950, 990, 995, 4500), FC-E9 Nikon Coolpix (5000, 5400, 8400, 8700, 8800), Nikkor 8 mm, Fisheye DSLR 4.9 mm, Sigma EX 8 mm, Canon EOS300D, EOS20D, Kodak DCS Pro 14 N User can create lens (.cal) file	Polar

(continued)

Table 7.2 (continued)

Name/characteristics			Gap Light Analyzer	HemiView	CAN-EYE	CIMES	WinSCANOPY[a]	SolarCalc
		User-defined projection	Binary search and linear interpolation between user-defined calibration points	Polynomial (4th order)	No	No	Polynomial (9th order)	No
	Field of view	Not adjustable (180°)	Adjustable	Adjustable: Horizon Pixel	Adjustable	Adjustable	Adjustable	
	Solar model	*Solar transmission*	Units, solar constant, cloudiness index (Kt), beam fraction, spectral fraction, clear-sky transmission coefficient	Units, transmittivity, diffuse proportion, external solar flux, intercepting surface	No	units, solar constant, external solar flux, atmospheric transmissivity, beam fraction, spectral fraction	version 2006a: solar constant, Rad to PAR conversion factor, diffuse radiation fraction of direct radiation, Atmospheric transmissivity, Sun size (pixel, real size, real size + penumbra)	Solar constant, cloudiness Index (Kt), beam fraction, spectral fraction, clear-Sky Transmission Coefficient
		Diffuse distribution	Standard or Uniform Overcast Sky, Clear-Sky Transmission Coefficient. Or user-defined data per month	Standard or Uniform Overcast Sky	No	Standard or Uniform Overcast Sky, Clear-sky, Scattered radiation, Diffuse proportion	Standard or Uniform Overcast Sky	Standard (b = 2 or b = 1.23) or Uniform Overcast Sky

(continued)

Table 7.2 (continued)

Name/characteristics			Gap Light Analyzer	HemiView	CAN-EYE	CIMES	WinSCANOPY[a]	SOLARCALC
		User-defined model	Monthly or growing season Kt, beam fraction, and spectral fraction; PAR or total shortwave; absolute or relative units	Not available	No	No	User defined radiation data	No
		Reflectance calibration	No	No	No	No	Reflectance calibration can be done using multi-steps targets to get reflectance measurements data and image	No
	Suntrack/Daytrack		User-defined solar time step (in minutes) integrated daily for each day of user-defined growing season	Day of year	Date of acquisition. *Default: Date of acquisition from the photo. Possible to define it interactively	Day or Year	Time Zone, Growing Season,	User-defined solar time step (in minutes) integrated daily for each day of user-defined growing season
	Variables in output		Yes: output by sky-region or for entire hemispherical view; utilities for gap fraction, sunfleck, extraterrestrial radiation, and solar tracks in graphical and tabular form	Yes: selection of available outputs	Yes: HTML report, MATLAB outputs, ASCII outputs, binarised images (binary files)	Summary or detailed reports, Each variable set linked to a specific program, Ascii outputs for spreadsheets	Yes: selection of available outputs	Yes: selection of available outputs

(continued)

Table 7.2 (continued)

Name/characteristics			Gap Light Analyzer	HemiView	CAN-EYE	CIMES	WinSCANOPY[a]	SolarCalc
Image classification	Thresholding		Single Grey Value	Single Grey Value	Color classes	Single Grey Value	Global single value; Adaptive [adapt for lighting variations in function of the location in the image]; Hemispherical [light variations of hemispherical lenses]; Solar [sky's light variations: sun, etc.]; Color based [sky conditions variations: dark blue]	Single Grey Value, Automatic thresholding methods (3): isodata, minimum error, Otsu's algorithm
	Improving classification	*Image editing tool*	Yes: brightness, contrast, hue, saturation, histogram, gamma correction, sharpen, emboss, RGB plane splitting, paint	No	Yes, gamma correction	Sharpening tool	Yes: editing and sharpening tools, histogram, panoramic view transform	No
		Region of interests	Yes: inside or outside	No	Yes	No	No	No

(continued)

Table 7.2 (continued)

Name/characteristics			Gap Light Analyzer	HemiView	CAN-EYE	CIMES	WinSCANOPY[a]	SolarCalc
		View pixel classification	Yes	Yes	Yes	No	Yes	Yes
		Interactive pixels classification modifications	Yes, paint function	No	Yes	No	Yes	No
		Mask	No	Yes, ignored color values	Yes, selected regions	Yes, selected zenith annuli	Yes	No
Output parameters	Output format		Text/ASCII	Excel 5.0 spreadsheet	MATLAB (*.mat), ASCII. *Excel	Text/ASCII	Text/ASCII, Excel spreadsheet with XLScanopy option	Text/ASCII
	Geometric indices	*Gap fraction*	Yes	Yes	Yes	Yes	Yes	Yes
		Mean contact frequency	No	Yes	No	Yes, variance only of contact number	No	No
		Gap size	No	No	No	Yes	Yes	No
		Spherical statistics	No	No	No	Yes: mean, variance, moment of inertia, direction of maximum gap, direction of mean	No	No

(continued)

Table 7.2 (continued)

Name/characteristics			Gap Light Analyzer	HemiView	CAN-EYE	CIMES	WinSCANOPY[a]	SOLARCALC
	Canopy indices	*LAI*	Yes: LAI4 integrated over 4 rings; LAI5 integrated over 5 rings (same method as Li-Cor LAI-2000 PCA; see Welles and Norman 1991)	Yes: effective LAI, best fit to the measured gap fraction values (theoretical ellipsoidal distribution)	Yes, LAI at 57.5°, LAIeff assuming an ellipsoidal distribution of the leaf inclination using look-up-table techniques (Knyazikhin et al. 1998; Weiss et al. 2000), and true LAI clumped corrected	Yes: Lang's graphical and iterative method, Miller's integral (adjusted from Leblanc et al. 2005), Campbell's ellipsoidal distribution model; LAI at any range of zenith angles; corrections for slope and clumping	6 index where most of them are available in linear and the log average (Lang and Xiang 1986): (1) LAI at 57.5° (Bonhomme and Chartier 1972), (2) Miller (1967), (3) Welles and Norman (1991), (4) Spherical leaf distribution, (5) Ellipsoidal leaf distribution (Campbell 1986), (6) 2D projected area (Lindsey and Bassuk 1992; Peper and McPherson 1998)	LAI5 integrated over 5 rings (same method as Li-Cor LAI-2000 PCA; see, Welles and Norman 1991); LAI based on Campbell's ellipsoidal distribution model
		Leaf inclination angle	No	No	Yes: effective and true averaged	Yes: average	Yes, LAD, mean leaf angle (MLA) and Leaf projection coefficient in function of view zenith angle	No

(continued)

Table 7.2 (continued)

Name/characteristics			Gap Light Analyzer	HemiView	CAN-EYE	CIMES	WinSCANOPY[a]	SolarCalc
		ELADP or 'shape parameter'	No	Yes: best fit to the measured gap fraction values (theoretical ellipsoidal distribution)	No	Yes	No	No
		Openness	Yes	Yes	No	Yes, in addition with 'closure'	Yes	Yes
		Clumping	No	No	Yes, Lang and Yueqin (1986): log gap fraction method	Yes: Chen and Cihlar (1996), Lang and Xiang (1986), Leblanc et al. (2005) combined gap size and log, Pielou (1962) coefficient of segregation	Yes: Lang and Xiang (1986), Chen and Cihlar (1996)	No
		Fraction of the soil covered by vegetation	No	No	Yes	Yes in addition with 'fraction soil', derived from nadir view	No	No
	Solar radiation indices	*Diffuse radiation above canopy*	Yes, based on gap light index (Canham 1988); estimated for each sky region	Yes, amount of incoming radiation that scatters from each sky direction	No	Yes, amount of incoming radiation that scatters from each sky direction	Yes, Theoretical models (user modifiable)	Yes

(continued)

Table 7.2 (continued)

Name/characteristics			Gap Light Analyzer	HemiView	CAN-EYE	CIMES	WinSCANOPY[a]	SolarCalc
		Direct radiation above canopy	Yes, based on gap light index (Canham 1988); estimated for each sky region	Yes, depends upon the relative position of the sun as well as upon atmospheric conditions	No	Yes, amount of direct radiation for each relative position of the sun and atmospheric conditions	Yes, Theoretical models (user modifiable)	Yes
		Diffuse radiation below canopy	Yes, above-canopy diffuse times gap fraction per sky region	Yes, Gap fraction of a sunmap sector *the radiation originating from that sky sector, and correcting for angle of incidence	No	Yes, fraction of diffuse radiation for UOC, SOC and clear sky radiation originating from each sky sector	Yes, Theoretical models (user modifiable)	Yes
		Direct radiation below canopy	Yes, above-canopy direct times gap fraction along solar track per sky region	Yes, Gap fraction of a skymap sector *the radiation originating from that sky sector, and correcting for angle of incidence	No	Yes, Gap fraction of a skymap sector *the radiation originating from that sky sector, and correcting for angle of incidence	Yes, Theoretical models (user modifiable)	Yes

(continued)

Table 7.2 (continued)

Name/characteristics			Gap Light Analyzer	HemiView	CAN-EYE	CIMES	WinSCANOPY[a]	SolarCalc
		Sunfleck	Yes, size-frequency distribution and daily duration	Yes, duration and radiation value	No	Yes, distribution and daily duration	Yes, distribution and daily duration	Yes, distribution and daily duration
		Site factor	Yes, indirect, direct, global	Yes, indirect, direct, global	No	Yes, indirect, direct, global	Yes, indirect, direct, total	Yes, indirect, direct, global
		Interception/Absorption of PAR (Photosynthetically Active Radiation)	No	No	Yes: instantaneous, integrated and white sky fAPAR assuming black leaves (fAPAR = fIPAR)	Yes: PPFD in the PAR, intercepted and transmitted by canopies, under clear-sky, overcast conditions	No	No

[a]Further information available on WinSCANOPY website: http://www.regent.qc.ca/assets/winscanopy_about.html

Hence, source files are available for the three open-source programs that were being evaluated. Access to the source code for GLA, which is implemented in Microsoft Visual Basic for the GUI and in C++ for the main engine, is available upon request. The CAN-EYE package was developed under the MATLAB environment, and the source files can be accessed to rebuild the package using MATLAB version 6.4 or higher (Baret and Weiss 2004). The developers of CIMES also provide access to the source code that is implemented in language C and C++, which allows any user to recompile the source for its specific platform. In these cases, additional tests can be performed to validate each algorithm or to enhance the existing suite of algorithms according to specific application needs. Processing several images concurrently or through a batch script is possible with CAN-EYE and CIMES owing, in part, to access to the source files, which allows the use of individual programs or functions in the software packages.

7.2.2 *Image Registration and Software Parameterization*

The input image formats vary among the different software programs, but many applications can be used to convert from one image format to another with no appreciable loss in data quality. Once the image is loaded, registration is performed by selecting a circular extent and, optionally, setting the geographic orientation. In most cases, the circular extent can be selected by using a mouse to orient the image on the display monitor; automatic selection is possible with CAN-EYE and WinSCANOPY. By default, CIMES assumes that the circular extent is positioned at the image's center, with a diameter equal to the image dimension (minimum between height and width), using a Nikon Coolpix and lens fisheye adapter. All programs that are listed in Table 7.2, with the exception of HemiView, allow the user to select the circular extent of the HP by manually entering its coordinates and diameter in pixel units. Geographic orientation is not available with version 6.1 of CAN-EYE.

Software functions allow the user to specify or derive the parameters (such as sky sector, site, lens, solar model, and suntrack/daytrack) that will be used in the calculation of different variables. Variable settings to adjust the sky sector angular resolution (zenith, azimuth) are standard in all programs, as well as for a covered area in CAN-EYE. Site location attributes include its geographic coordinates and inclination. In GLA, only latitude and elevation are currently used, although longitude is an optional input parameter that can be defined (Frazer et al. 1999). HemiView and SOLARCALC do not implement site inclination. In CAN-EYE, only latitude can be set. It is also possible to build a topographic mask to take into account surrounding terrain obstruction (e.g., in a mountainous area) with GLA and WinSCANOPY.

Each program provides functionality specific to the camera lens to define the projection and field of view to use in image processing. HemiView and WinSCANOPY provide pre-defined settings for a suite of specific lens models, for example, while standard or user-defined projections are used when a lens model is not specified.

Solar models are available in all programs except CAN-EYE and are used to calculate radiation indices (diffuse, direct, and total [global] radiation). Parameters controlling solar transmission and diffuse distribution can be adjusted. A particularly useful feature is the ability to define a solar model using collected measurement data, which is available in GLA and WinSCANOPY. The track of the sun across the sky, for example, can be obtained by specifying its geographic location and date of HP acquisition. This information can then be used to calculate the visibility of the solar disk and to associate solar radiation values with the course of a time series (e.g., day, growing season).

7.2.3 Image Classification

Classification involves segmenting an image by selecting an optimal threshold brightness value that discriminates the foliage from canopy openings that represent the sky (Jonckheere et al. 2004). While it is essential to ensure the correct exposure of HP for accurate extraction of forest canopy variables (Zhang et al. 2005), determining the optimal threshold has been a challenging exercise (Jonckheere et al. 2005; Nobis and Hunziker 2005). Often a manual, subjective process, threshold determination has proven to be a source of inconsistency and error (Englund et al. 2000). When imaging forest canopies, the technique often entails a compromise between overestimating large areas of visible sky (large canopy openings) near the zenith and underestimating small areas of visible sky (small canopy openings) near the horizon (Fournier et al. 1997; Leblanc et al. 2005).

Single-value thresholding is used in GLA, HemiView, CIMES, WinSCANOPY, and SOLARCALC (Table 7.2). SOLARCALC also provides three methods to perform automatic thresholding using the image RGB information or preferably the blue channel, since absorption of leafy material is maximal and scattering of sky tends to be lowest in this color channel (Jonckheere et al. 2005). CAN-EYE uses a classification algorithm that involves an interactive selection of sky components that can comprise multiple classes that are each assigned a non-vegetation label (Baret and Weiss 2004). The colors associated with each class depend on the illumination conditions and on the objects themselves. CAN-EYE also permits use of photographs taken from above the canopy looking downward, allowing separation between the vegetation and soil background. When operating the software, it is generally more efficient to select a single, relatively uniform class such as the sky. There is less potential for gray-value variation when the smallest feature is selected for classification. For sparse canopies, the least-represented feature in the image, such as green vegetation, should be selected, whereas under dense canopies the sky occupies the smallest amount of the image. It is also possible to select two or more classes. If pixels are not allocated to one class or the other, they are considered "mixed" and can be processed later. At the beginning of this classification process, a default color allocation is proposed that is refined interactively by the user. Once class colors are assigned, HPs are transformed into classified images. WinSCANOPY offers several methods to classify the pixels, but their availability depends upon which product version is purchased; the Pro DSLR version provides access to all of them. The global threshold pixel classification method uses the light intensity (gray level) to determine whether the pixels belong to sky or canopy. Adaptive thresholding considers the non-linear decline in light sensitivity (due to limitations of the optical lens and camera charge-coupled device [CCD] array) and the anisotropy that is associated with sky brightness at any time. Hemispherical threshold pixel classification uses light intensity and takes into account the light variations of hemispherical lenses, which are brighter at the zenith and darker at the horizon. Solar threshold pixel classification uses light intensity and takes into account the sky's light variations (due to sun or other factors). The attenuation compensation can be linear or similar to the standard overcast sky (SOC) diffuse radiation distribution. Another approach is the color-based pixel classification that uses true color information (i.e., Red-Green-Blue, and hue, intensity and saturation) to determine whether pixels belong to sky or canopy. This algorithm is more tolerant of variations in sky conditions. Once the classification is performed, it can be verified and modified interactively. Depending upon software selection, there are multiple approaches to discriminating foliage from sky in a hemispherical image.

Image pre-processing can improve the classification. For example, utility tools can be used to mask parts of the images that are undesirable, or to brighten or darken the images that might provide better visual discrimination between the vegetation elements and the background. Only GLA, CAN-EYE and WinSCANOPY offer such tools explicitly. However, it is relatively simple to find external programs that can pre-process images (e.g., Microsoft Photo Editor, GIMP, Adobe Photoshop, etc.), and there are third-party, automated solutions for image thresholding (e.g., Sidelook, http://www.appleco.ch/).

7.2.4 Program Functionality and Output

After image thresholding and classification are completed, canopy structure attributes and light regime variables are computed using specific algorithms with user-defined input parameters. These variables can be presented in several output formats that generally include ASCII text, although HemiView does not appear to offer this option. Instead, HemiView provides outputs in Microsoft Excel spreadsheet format, from which ASCII text can be generated if desired. CAN-EYE can also generate output in a MATLAB file format. WinSCANOPY has an optional extension (XLScanopy) to create Microsoft Excel spreadsheet output. In fact, the Microsoft Excel format is likely the preferred format for data summarization, reporting and analysis. In addition to standard output files, some programs have alternative outputs. GLA offers a utility menu where the user can select a particular parameter that can be plotted or output directly (e.g., gap fraction, sunflecks)

(Frazer et al. 1999). CAN-EYE can generate an HTML report containing main figures (plots) and tables, as well as a summary of global derived parameters.

There are differences in how various HP programs function, particularly in how extensively a user may provide input to the different models and algorithms, and calculate canopy structure or light regime variables (Tables 7.2 and 7.3). In most cases, the methods or algorithms used to estimate the HP variables are described to some degree in the software documentation. Most of the variables extracted from HP follow a basic routine (e.g., gap fraction) and mathematical procedure to translate the spatial distribution of gaps into measures of canopy structure (e.g., LAI). The development and validation of these procedures has been the subject of multiple studies and review (Chap. 3, Leblanc and Fournier 2017). Knowing what method is used in a given program, and which assumptions (if any) the method is based on, is essential to correctly interpret the results of HP analysis. This is particularly true in the calculation of LAI, in which different assumptions about canopy structure can generate different values for the same variables.

HP was initially developed and is best used for estimating gap-light transmission because of the geometry of the fisheye projection. To estimate LAI and clumping requires knowledge of canopy height, volume and density; these are more difficult to reliably measure from a two-dimensional photograph.

Gap fraction distribution is offered by all software and constitutes one of the core variables from which other parameters are derived. Output indices of interest can be grouped into those associated with forest canopy structure and those associated with solar radiation.

LAI is a common canopy structural variable extracted from HP, and all programs provide one or more algorithms from which to compute this variable. The LAI extracted from HP is referred to as the effective LAI (LAI_{eff}) because the contribution from clumping has not been incorporated: $LAI_{eff} = \Omega$ LAI, where Ω is the clumping index (sometimes referred to as the aggregation or dispersion parameter) (Equation 3, Chen et al. 2005; Chap. 6, Frazer et al. 2017). The clumping index depends on the spatial distribution of the element within the canopy at the plant/tree level (e.g., foliage, shoot, trunk, branch) but also at the canopy level (e.g., the relative position of the plants). The presence of wood material can complicate the visualization of foliage hidden behind stems, branches or trunk, contributing to errors in LAI estimation (Chen et al. 1997). To account for these factors, Chen (1996) formulated an equation that calculates LAI as a function of the woody-to-total plant area ratio (α), LAI_{eff}, needle-to-shoot area ratio (γ_E) and foliage element clumping index (Ω_E) where: $L = (1 - \alpha)\, LAI_{eff}\, \gamma_E/\Omega_E$ In addition to calculation of LAI_{eff}, CAN-EYE, CIMES and WinSCANOPY will estimate clumping factors that can be used to calculate LAI (Table 7.2).

In regard to solar radiation indices, diffuse/direct radiation indices above/below canopy can be computed by all software programs except CAN-EYE. The diffuse radiation above canopy refers to the amount of incoming radiation that scatters from each sky direction. The direct radiation above canopy is the amount of direct radiation for each relative position of the sun and for each atmospheric condition. The diffuse radiation below canopy is the fraction of diffuse radiation and clear sky radiation originating from each sky sector, corrected by incidence angle. The direct radiation below canopy is computed following the gap fraction of a skymap sector multiplied by the radiation originating from that sky sector, corrected by incidence angle. Diffuse and direct radiation calculations depend on the parameterization of the solar models available in the software. A sunfleck may be defined as the brief period for which a point on the forest floor experiences direct sunlight. Over the course of a day, a site may experience numerous sunflecks of variable length and intensity depending on the distribution and size of canopy gaps. Sunfleck frequency distribution, therefore, describes the frequency of sunflecks of a specific duration that a site could experience between the beginning and end dates of the growing season. Sunfleck duration describes the number of minutes of potential direct sun that a site could experience over the course of a day for each day of the growing season. Site factors are indices of the proportion of radiation reaching a given location. HemiView, CIMES, WinSCANOPY and SOLARCALC can estimate the indirect, direct and total site factors corresponding to the proportion of diffuse, direct and total solar radiation reaching a given location, relative to a location with no sky obstructions (Anderson 1964). CAN-EYE and CIMES can estimate the intercepted (absorbed) radiation in the PAR waveband. CAN-EYE calculates the instantaneous integrated white sky fraction of absorbed PAR (FAPAR, Andrieu and Sinoquet 1993) assuming black leaves. CIMES can calculate the photosynthetic photon flux density (PPFD), intercepted and transmitted by canopies, in the PAR waveband (McCree 1981) under a clear-sky (cloudless) model; the calculation of PPFD is not typically found in other software.

7.3 Comparative Analysis of Selected HP Programs

A comparative analysis was undertaken to determine whether empirical estimates generated from a common set of HPs would vary when processed by different software programs. Of the five software programs identified for the comparative analysis, four were acquired or made available for this study: GLA, HemiView, CAN-EYE and CIMES (Table 7.1). We attempted to identify a set of canopy structure and light regime variables that would be common to all four programs. The variables selected included three canopy structural variables (canopy openness, LAI_{eff} and clumping) and two light regime variables (fraction of transmitted diffuse radiation and fraction of direct radiation). The algorithms used to compute a given canopy structure or light regime variable are summarized in Table 7.3, and these descriptions provide perspectives on differences found from the statistical analysis.

Table 7.3 Forest canopy structure and light regime variables selected for analysis of hemispherical photographs

Software	Variables selected for analysis				
	Canopy structure			Light regime	
	Effective LAI	Openness	Clumping	Fraction of diffuse radiation	Fraction of direct radiation
1. GLA	Integrated over the zenith angles 0°–60° equivalent to LAI 4th ring (Stenberg et al. 1994)	Area fraction of the sky hemisphere that is unobstructed by vegetation (Frazer et al. 1997)	Not available	Computed as ratio: Diffuse solar radiation transmitted by the canopy and topographic mask (if one has been defined) OVER Diffuse radiation incident on a horizontal or tilted surface located at the earth surface without any effect of topography or forest canopy	Computed as ratio: Direct solar radiation transmitted by the canopy and topographic mask (if one has been defined) AND Direct (beam) radiation incident on a horizontal or tilted surface located at the earth surface without any effect of topography or forest canopy
2. HemiView	Based on the ellipsoidal leaf angle distribution (Campbell 1986)	Overall proportion of the sky hemisphere that is visible	Not available	Computed as ratio: Diffuse radiation above the canopy, corrected for intercepting surface orientation OVER Diffuse radiation below the canopy, corrected for intercepting surface orientation	Computed as ratio: Direct radiation above the canopy, corrected for intercepting surface orientation OVER Direct radiation below the canopy, corrected for intercepting surface orientation

(continued)

Table 7.3 (continued)

Software	Variables selected for analysis				
	Canopy structure			Light regime	
	Effective LAI	Openness	Clumping	Fraction of diffuse radiation	Fraction of direct radiation
3. Can-Eye	Effective LAI from gap fraction at 57.5°		Lang and McMurtie (1992), computed here as the ratio: True LAI/Effective LAI	Not available	Not available
4. CIMES	Effective LAI based on Miller's integral (Miller 1967), computed at zenith view angle of 57° (calculated from 55° to 60°) (CLMPMLSH)	Similar to (Frazer et al. 1997), computed from cumulated gap fraction (OPENNESS)	Computes clumping from gap size distributions over a range of zenith angles using (Chen and Cihlar 1995; Leblanc et al. 2005; Lang and Xiang 1986) (CLMPMLSH)	Cumulated fractional diffuse sky light computed at 87.5° and extracted from the OPENNESS program	Not available

7.3.1 Test Site Description

The HPs that are used for the comparison of software outputs were previously acquired for other studies from four sites which were selected to represent four specific canopy types (Table 7.4). Ten HPs were analyzed from each study site. The first site was from the Boreal Ecosystem–Atmosphere Study (BOREAS) in the Southern Study Area–Old Black Spruce (BOREAS SSA-OBS, in and around Prince Albert National Park and Narrow Hills Provincial Forest, Saskatchewan, Canada) and exemplified a highly clumped open-canopy conifer forest (Fournier et al. 1997; Sellers et al. 1997; Fig. 7.1a). This site was characterized by a patchy distribution of the highest-density stands, with the shortest and smallest-diameter trees among the four sites. The second site was from the ECOLEAP project (Extended Concentration to Link Ecophysiology And forest Productivity) in Forêt Montmorency (Québec, Canada), representing a clumped and closed-canopy conifer forest (Bernier et al. 2002; Fig. 7.1b). This site was dominated by *Abies balsamea* (L.) Miller, with a small proportion of *Betula papyrifera* Marsh., and had the second highest stem density among the four sites. The third and fourth sites have been described as containing randomly distributed foliage distributions, although all natural forest canopies are inherently clumped to varying degrees (Chen et al. 2005).

Table 7.4 Description of field sites for hemispherical photographs

Site characteristics	Boreas SSA-OBS[a]	ECOLEAP Forêt Montmorency	Le Bray[b]	ECOLEAP Duchesnay
Forest cover type	Coniferous	Coniferous	Coniferous	Deciduous
Canopy type	Open and clumped	Closed and clumped	Open and random	Closed and random
Location	Saskatchewan, Canada	Québec, Canada	Bordeaux, France	Québec, Canada
Dominant species	*Picea mariana* (Mill.) BSP	*Abies balsamea* (L.) *(91%), Betula papyrifera* Marsh. *(9%)*	*Pinus Pinaster* Ait.	*Acer saccharum* Marsh. *(65%), Betula alleghaniensis* Britt. *(20%), Fagus grandifolia* Ehrh. *(14%)*
Longitude	105°07′12″ W	71° 06′ 00″W	00° 46′ 00′ ′W	71° 40′ 00″W
Latitude	53° 59′ 06″N	47° 19′ 00″N	44° 42′ 00′ ′N	46° 56′ 36″N
Altitude (m)	629	800	60	280
Area (m^2)	2400	3080	42,920	3660
Basal area (m^2/ha)	30.0	44.5	34.2	26.5
Stems/ha density	2537	1961	621	577
Mean stem diameter (cm)	10.5	17.0	26.0	26.5
Average tree height (m)	11.0	17.0	18.0	20.0
Slope (degrees)	<3°	15°–25°, north west exposure	0°	15°, south exposure
Reference	Sellers et al. (1997); BOREAS CD-ROM 1, TE23: 94-00-00_multiplesites_msm.csv	Bernier et al. (2001, 2002)	Porté et al. (2000), Valentini (2003)	Bernier et al. (2002), Raulier et al. (2002)

[a]http://daac.ornl.gov/BOREAS/bhs/Sites/SSA-OBS.html
[b]https://fluxnet.ornl.gov/site/466

The third site was from the EUROFLUX project (Le Bray, Bordeaux, France), and it was considered a randomly distributed, closed-canopy coniferous forest consisting of a single species, *Pinus pinaster* Ait. (Valentini 2003; Fig. 7.1c). The fourth site was also from the ECOLEAP project (Duchesnay, Québec, Canada)

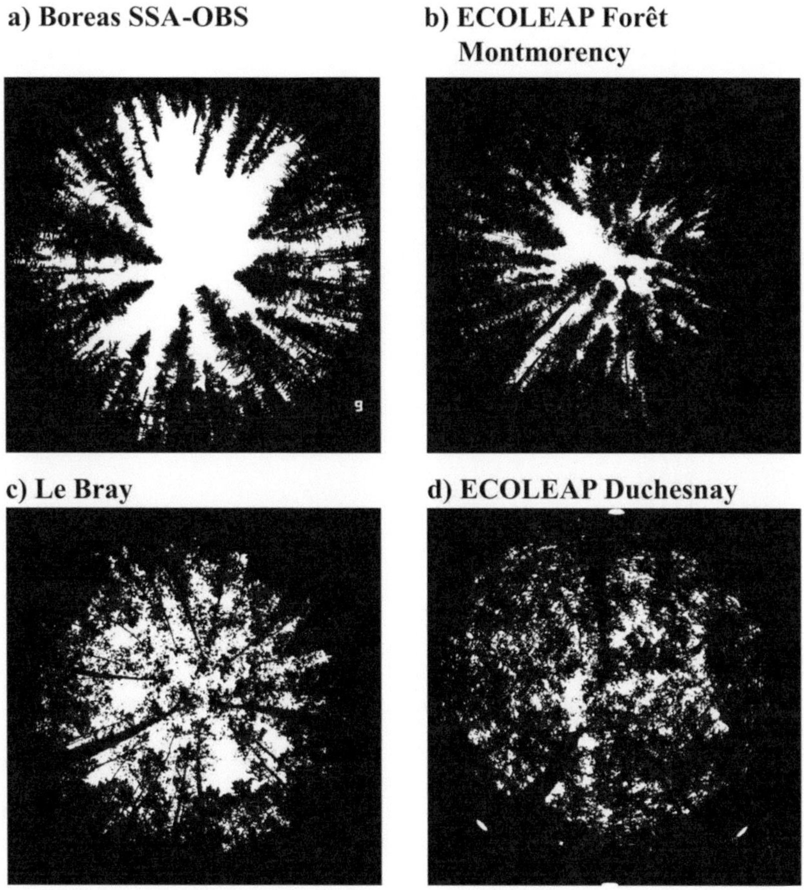

Fig. 7.1 Examples of classified hemispherical photographs from each study site

and it was described as a randomly distributed, closed-canopy deciduous forest consisting of three species (Table 7.4), which were the largest trees among the four sites (Bernier et al. 2002; Fig. 7.1d).

7.3.2 Hemispherical Photo Classification and Processing

In deriving a threshold, the images were classified to separate canopy and sky elements using a consistent and objective approach based on the image histogram for each photograph. Generally, two visible peaks can be observed, corresponding to a dark and to a bright region. Between these extremes, a “gray” zone indicates pixels containing both canopy and sky. The threshold was defined as the center of this “gray” zone (Fig. 7.2). HemiView was used to extract pixel values over the

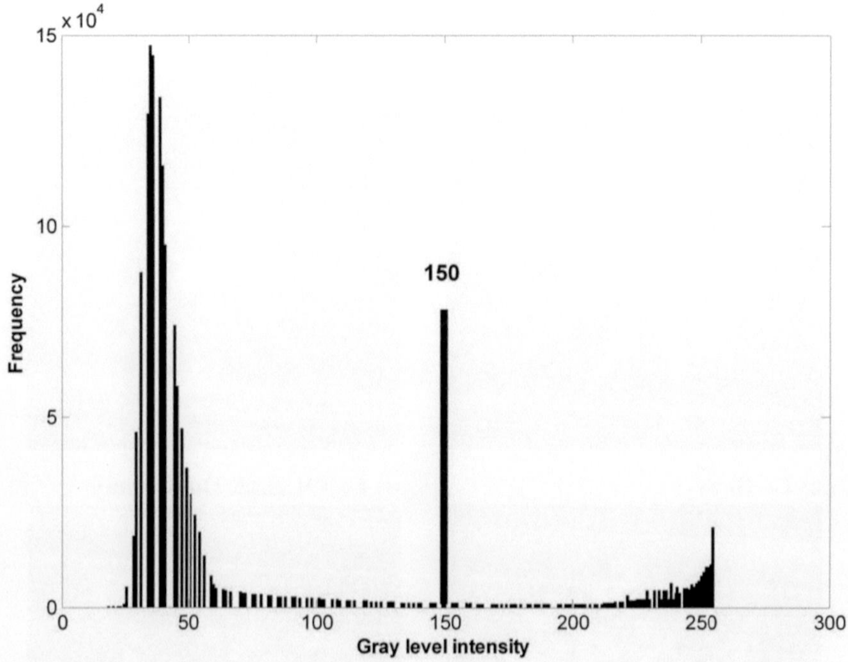

Fig. 7.2 Example of hemispherical photo histogram and determination of threshold between canopy and sky

circular extent of the HP image because of the software's simplicity in adjusting the circle over the images. Adobe PhotoShop's histogram tool was subsequently used to determine the threshold value. GLA was used to apply the threshold value and to save the working image into a binary BMP format, which was either read directly by the four HP software programs or converted to a standard image format such as JPG or TIFF for programs (e.g., CAN-EYE) that require this input image format.

Once circular extents were extracted and a single threshold was applied, the classified HPs were imported into the HP software for analysis. First, the area covered by a zenith angle interval of 5° (of a possible 90°) and an azimuth angle interval of 20° (of a possible 360°) was defined, resulting in 18 zenith regions (or rings) and 18 azimuth regions. North was set in the "up" direction without correction for magnetic declination, and geographic locations were provided as summarized in Table 7.4. The slope and aspect for the second site (Forêt Montmorency) and the fourth site (Duchesnay) were considered only for GLA and CIMES, which, for instance, had an effect on the calculation of LAI_{eff}. The lens model that was employed was the default polar lens without distortion or the equivalent (e.g., HemiView uses a Linear 180 model). The solar or light models were used with their default parameter values and universal overcast sky (UOC) for the diffuse radiation sky brightness model. Estimation of light transmission of radiation regime was applied to a single day (July 3). A possible limitation involves the short time period over which the HPs were acquired at each site. Differences that were observed for direct gap-light transmission may have been smaller had results been obtained over an entire growing season and, thus, differences among the software programs may average out over longer periods.

All image inputs into the software were binary, containing only black-and-white pixels corresponding to forest canopy or sky. Every program used its own classification method: threshold value of 128 of 256 gray values for GLA and HemiView, and automatic selection of two classes (sky or vegetation) with CAN-EYE. In CIMES, applying a threshold was unnecessary because one could specify that the image being imported was previously classified. Table 7.5 presents a summary of the specifications for processing the 10 photographs used to represent each of the four sites.

7.3.3 *Statistical Analysis*

Within each of the four stand types and four HP software programs, the descriptive statistics for openness, LAI_{eff}, clumping, fraction of diffuse radiation, and fraction of direct radiation were computed to ascertain whether there were differences in these parameters that were computed from the 10 HPs used as the sample for each study area. This was followed by one-way analysis of variance (ANOVA) to determine whether these parameter estimates were statistically different as a result of the HP program employed. If a statistical difference was found, Bonferroni multiple means comparison tests were conducted to determine differences among the HP software. A paired *t*-test was employed when there were only two parameter estimates among the four programs (for example, clumping, true LAI, and fraction of direct radiation). To test the effect of the clumping algorithm, the clumping implementation that was based on Chen and Cihlar (1995) and Lang and Xiang (1986) in CIMES was compared with the CAN-EYE implementation based on Lang and McMurtie (1992). All statistical tests were performed at the 5% probability level.

Table 7.5 Summary hemispherical photo image processing parameters

Threshold	GLA: 128/256 grey values HemiView: 128/256 gray values CAN-EYE: Automatic classification of vegetation and sky pixels (2 classes) CIMES: Binary (no need for specifying a threshold)
Azimuth résolution	20° = 18 sectors
Zenith resolution	5° = 18 rings
North position	Up direction: 0° without magnetic declination
Slope and aspect	Adjusted from the site characteristic (Table 7.4)
Lens model	Polar (without distortion) or equivalent
Solar/Light model	Default, UOC (Universal Overcast Sky)
Day of radiation regime computation	July 3rd

7.4 Results

7.4.1 Canopy Openness

Before results could be interpreted, we determined whether the angular distribution of gap fractions from each HP program was comparable. Because the same number of rings and annuli that were used for each HP and binary HPs (already segmented) were submitted to each program, we expected the angular distributions that were computed to be very similar or identical, as shown from the derived canopy openness values. A cursory analysis verified that this assumption was correct.

Canopy openness was computed by all programs except CAN-EYE; this parameter was clearly and similarly defined in GLA, HemiView, and CIMES (Table 7.3). As a result, it was not surprising that the average value among the 10 HP samples from each study site were either similar or identical (Table 7.6a). In addition, the relative magnitude of canopy openness could be interpreted relative to the characteristics of the forest that it represented. For example, the value for openness from the open and clumped Boreas SSA-OBS site was larger, suggesting a larger proportion of canopy gaps compared with the closed and clumped ECOLEAP Forêt Montmorency site. One-way ANOVA showed no statistical differences in the estimates of canopy openness among GLA, HemiView and CIMES across all of the study sites (Boreas SSA-OBS, $p = 0.99$; ECOLEAP Forêt Montmorency, $p = 0.52$; Le Bray, $p = 0.99$; ECOLEAP Duchesnay, $p = 0.91$).

7.4.2 Effective LAI

For each of the study sites, there were small differences in the calculation of LAI_{eff} among the four HP programs (Table 7.6b). Across three of the study sites, CIMES generated the smallest estimate of LAI_{eff}, but the range was within those computed from the other HP programs. Regardless of stand type, the estimates of LAI_{eff} were comparable among the four HP programs. Three observations regarding the interpretability of the computed LAI_{eff} values can be made from Table 7.6b. First, the largest LAI_{eff} was observed for the ECOLEAP Forêt Montmorency site that was characterized by closed and clumped canopy, and which appears reasonable when compared with an open and clumped canopy conifer forest such as the Boreas SSA-OBS site, which would be expected to have smaller LAI_{eff} values. Second, a randomly distributed foliage distribution for a conifer stand is expected to be less variable than that of a deciduous stand, and this supposition was confirmed by observations from the Le Bray site compared with the ECOLEAP Duchesnay site. Third, among the conifer stands, the most variable LAI_{eff} was observed at the Boreas SSA-OBS site, with an open and clumped foliage distribution, which would logically be more variable than a closed-density stand such as ECOLEAP Forêt Montmorency and Le Bray. These observations suggest that computations of LAI_{eff}

Table 7.6 Descriptive statistics of canopy and light regime parameters derived by study area and hemispherical photo software

Study site	Mean (N = 10)	Standard deviation	Minimum	Maximum	Coefficient of variation
(a) *Openness*					
Boreas SSA-OBS: Open canopy, clumped conifer					
1. GLA	0.29	0.07	0.20	0.41	0.24
2. HemiView	0.30	0.07	0.20	0.41	0.24
3. CIMES	0.30	0.07	0.20	0.41	0.24
ECOLEAP Forêt Montmorency: Closed canopy, clumped conifer					
1. GLA	0.10	0.01	0.08	0.11	0.10
2. HemiView	0.10	0.01	0.09	0.11	0.08
3. CIMES	0.11	0.01	0.09	0.12	0.10
Le Bray: Open canopy, randomly-distributed foliage conifer					
1. GLA	0.22	0.02	0.19	0.25	0.09
2. HemiView	0.22	0.02	0.19	0.25	0.09
3. CIMES	0.22	0.02	0.19	0.25	0.09
ECOLEAP Duchesnay: Closed canopy, randomly-distributed foliage deciduous					
1. GLA	0.13	0.03	0.09	0.18	0.23
2. HemiView	0.13	0.03	0.09	0.18	0.23
3. CIMES	0.13	0.03	0.10	0.19	0.22
(b) LAI_{eff}					
Boreas SSA-OBS: Open canopy, clumped conifer					
1. GLA	1.29	0.37	0.76	1.87	0.29
2. HemiView	1.27	0.35	0.82	1.93	0.27
3. CAN-EYE	1.38	0.33	0.93	1.93	0.24
4. CIMES	1.16	0.24	0.87	1.53	0.20
ECOLEAP Forêt Montmorency: Closed canopy, clumped conifer					
1. GLA	2.68	0.16	2.39	2.94	0.06
2. HemiView	2.91	0.19	2.57	3.16	0.06
3. CAN-EYE	2.68	0.32	2.27	3.29	0.12
4. CIMES	2.59	0.08	2.45	2.73	0.03
Le Bray: Open canopy, randomly-distributed foliage conifer					
1. GLA	1.44	0.11	1.31	1.65	0.08
2. HemiView	1.53	0.10	1.39	1.70	0.07
3. CAN-EYE	1.78	0.19	1.49	2.07	0.11
4. CIMES	1.60	0.12	1.44	1.80	0.08
ECOLEAP Duchesnay: Closed canopy, randomly-distributed foliage deciduous					
1. GLA	2.29	0.32	1.71	2.71	0.14
2. HemiView	2.44	0.64	1.72	3.96	0.26
3. CAN-EYE	2.31	0.45	1.42	2.91	0.19
4. CIMES	1.98	0.37	1.49	2.58	0.19

(continued)

Table 7.6 (continued)

Study site	Mean (N = 10)	Standard deviation	Minimum	Maximum	Coefficient of variation
(c) *Clumping*					
Boreas SSA-OBS: Open canopy, clumped conifer					
1. CAN-EYE	0.53	0.04	0.47	0.58	0.07
2. CIMES (Chen and Cihlar 1995)	0.58	0.05	0.50	0.66	0.09
3. CIMES (Lang and Xiang 1986)	0.48	0.07	0.40	0.61	0.15
ECOLEAP Forêt Montmorency: Closed canopy, clumped conifer					
1. CAN-EYE	0.63	0.02	0.60	0.65	0.03
2. CIMES (Chen and Cihlar 1995)	0.73	0.03	0.69	0.77	0.04
3. CIMES (Lang and Xiang 1986)	0.51	0.03	0.45	0.56	0.06
Le Bray: Open canopy, randomly-distributed foliage conifer					
1. CAN-EYE	0.67	0.02	0.65	0.70	0.02
2. CIMES (Chen and Cihlar 1995)	0.67	0.03	0.61	0.72	0.05
3. CIMES (Lang and Xiang 1986)	0.51	0.04	0.46	0.56	0.07
ECOLEAP Duchesnay: Closed canopy, randomly-distributed foliage deciduous					
1. CAN-EYE	0.66	0.02	0.61	0.69	0.03
2. CIMES (Chen and Cihlar 1995)	0.74	0.03	0.67	0.77	0.04
3. CIMES (Lang and Xiang 1986)	0.65	0.04	0.55	0.70	0.07
(d) *Fraction of diffuse radiation*					
Boreas SSA-OBS: Open canopy, clumped conifer					
1. GLA	0.41	0.11	0.28	0.58	0.26
2. HemiView	0.41	0.11	0.28	0.58	0.26
3. CIMES	0.41	0.11	0.28	0.58	0.26
ECOLEAP Forêt Montmorency: Closed canopy, clumped conifer					
1. GLA	0.14	0.02	0.12	0.17	0.12
2. HemiView	0.16	0.02	0.13	0.19	0.13
3. CIMES	0.17	0.02	0.14	0.20	0.12
Le Bray: Open canopy, randomly-distributed foliage conifer					
1. GLA	0.34	0.03	0.29	0.37	0.08
2. HemiView	0.34	0.03	0.29	0.37	0.08
3. CIMES	0.34	0.03	0.29	0.37	0.07
ECOLEAP Duchesnay: Closed canopy, randomly-distributed foliage deciduous					
1. GLA	0.17	0.05	0.07	0.26	0.31
2. HemiView	0.15	0.06	0.09	0.27	0.38

(continued)

Table 7.6 (continued)

Study site	Mean (N = 10)	Standard deviation	Minimum	Maximum	Coefficient of variation
3. CIMES	0.19	0.04	0.14	0.27	0.21
(e) *Fraction of direct radiation*					
Boreas SSA-OBS: Open canopy, clumped, conifer					
1. GLA	0.39	0.14	0.23	0.63	0.37
2. HemiView	0.30	0.12	0.18	0.51	0.41
ECOLEAP Forêt Montmorency: Closed canopy, clumped conifer					
1. GLA	0.17	0.08	0.03	0.32	0.48
2. HemiView	0.14	0.03	0.09	0.21	0.25
Le Bray: Open canopy, randomly-distributed foliage conifer					
1. GLA	0.43	0.07	0.35	0.56	0.16
2. HemiView	0.34	0.03	0.30	0.39	0.09
ECOLEAP Duchesnay: Closed canopy, randomly-distributed foliage deciduous					
1. GLA	0.19	0.11	0.01	0.41	0.59
2. HemiView	0.19	0.05	0.10	0.25	0.24

were sensitive to the composition and structural differences that were exemplified by the four study sites.

There were no statistical differences in estimates of LAI_{eff} among the software programs for Boreas SSA-OBS (One-way ANOVA: $p = 0.51$) or for ECOLEAP Duchesnay ($p = 0.16$); however, statistical differences were observed for ECOLEAP Forêt Montmorency ($p = 0.02$) and Le Bray ($p = 0.02$). Upon further assessment of ECOLEAP Forêt Montmorency, HemiView and CIMES were significantly different ($p = 0.01$). For Le Bray, multiple means comparisons identified several statistical differences between CAN-EYE and GLA (Bonferroni test: $p < 0.001$), CAN-EYE and HemiView ($p = 0.001$), and CAN-EYE and CIMES ($p = 0.03$), which were likely attributable to the relatively larger and more variable estimate of LAI_{eff} from CAN-EYE (mean = 1.78, standard deviation = 0.19; Table 7.6b).

7.4.3 Clumping

Only CAN-EYE and CIMES offered routines to estimate clumping (Table 7.3). Values were within 10% of each other and identical for one site (Le Bray) when clumping in CIMES was based on Chen and Cihlar (1995), and as subsequently modified by Leblanc et al. (2005) (Table 7.6c). Clumping values were smallest and most variable (largest coefficient of variation) when based on Lang and Xiang (1986). Clumping measures were statistically different between CAN-EYE and CIMES (Chen and Cihlar 1995) at all study sites (One-way ANOVA: $p = 0.01$)

except for Le Bray, for which the differences were not statistically significant ($p = 0.52$). However, clumping measures varied only slightly between CAN-EYE and CIMES (Lang and Xiang 1986) when clumping was calculated as the ratio of LAI_{eff}:LAI at an angle of 57.5° for CAN-EYE. In this comparison, differences between programs were statistically significant (Bonferroni test: $p = 0.01$) for all sites, except for ECOLEAP Duchesnay ($p = 0.51$). Although Le Bray and the ECOLEAP Duchesnay sites were described as having random foliage distributions, an average clumping index was computed for these sites, thereby supporting the notion that most natural forest stands are inherently clumped (Table 7.6c). Given the trends across the various sites, the estimates of clumping are generally considered to differ between CAN-EYE and CIMES. With an estimate of clumping and woody fraction, LAI can be computed from LAI_{eff}. As a result of statistical differences in either LAI_{eff} or clumping, the estimation of true LAI from CAN-EYE and from CIMES (Chen and Cihlar 1995) differed for all of the study sites (One-way ANOVA: $p < 0.01$). This result is attributable to the implementation of different algorithms used to determine LAI_{eff} and clumping in the different programs.

7.4.4 *Fraction of Transmitted Diffuse Radiation*

GLA, HemiView and CIMES provide the fraction of diffuse radiation (Table 7.3). Within each study site, values from all three programs were either identical or similar, and were equally variable (Table 7.6d). The magnitude of the fraction appeared to be highly sensitive to the characteristics of the forested areas when the values were compared across the four study sites. For each site, estimates did not differ significantly between HP programs (Boreas SSA-OBS, $p = 1.00$; Le Bray, $p = 0.99$; ECOLEAP Duchesnay, $p = 0.36$), except for ECOLEAP Forêt Montmorency, where estimates between GLA and CIMES differed significantly ($p = 0.02$). The fraction values were larger from CIMES than from GLA, and when combined with the small standard deviations, would have resulted in statistical distributions in which there would be insufficient overlap to accept the null hypothesis of no statistical difference. While reasons for this result were not clear, differences in transmitted diffuse radiation were likely attributable to the implementation and resolution of the UOC model. Any inference that is made from the statistical result may need to be tempered by the empirical reality that the fraction of diffuse radiation at the ECOLEAP Forêt Montmorency site was the smallest among sites, with the smallest standard deviations (Table 7.6d). This site was characterized by a homogeneous conifer canopy with closed density and clumped foliage distribution. As a result, significant differences among programs were more likely to be found for this site. Based on the results that were obtained, and for practical application purposes, there appears to be no difference in the estimates of the transmitted diffuse radiation fraction among GLA, HemiView and CIMES across the four study sites. These results were, in part, consistent with a prior study by

Promis et al. (2011), which compared GLA, HemiView, hemIMAGE, and Winphot.

7.4.5 *Fraction of Direct Radiation*

GLA and HemiView provided the fraction of direct radiation (Table 7.3). In contrast to the fraction of transmitted diffuse radiation (Table 7.6e), average values for direct radiation were generally smaller and less variable from HemiView than from GLA across all study sites, with the exception of ECOLEAP Duchesnay, where they were identical (Table 7.6e). The fraction of direct radiation for the two ECOLEAP sites was similar (Table 7.6e), and the relatively larger standard deviations from GLA may have led to non-significant differences for the paired *t*-tests that were computed for the Forêt Montmorency (p = 0.08) and Duchesnay (p = 0.84) sites. In contrast, the relatively large differences in average fraction of direct radiation values led to statistical differences that were determined from the paired *t*-tests between the two programs for the BOREAS SSA-OBS ($p \leq 0.001$) and Le Bray (p = 0.001) sites. Some solar radiation models compute clear-sky estimates, while others, like the one used in GLA, compensate for seasonal patterns of cloud cover with a cloudiness index. The selection of the solar radiation model, therefore, may account for the differences that were observed between software programs.

7.5 Discussion and Conclusions

Based on selection criteria that were oriented towards documentation, availability, support and recent use, five software programs for processing HP were identified (GLA, HemiView, CAN-EYE, CIMES, WinSCANOPY; Table 7.1), of which four were available for analysis. The four software programs that were evaluated (GLA, HemiView, CAN-EYE, CIMES) all run under Microsoft Windows, but CAN-EYE functions within a MATLAB environment. The import image formats varied among these programs, but tended to include the more common graphic file formats (Table 7.2). Recent digital cameras generate propriety formats that may not be supported by certain HP software programs, necessitating graphic file format conversion (Promis et al. 2011). While format conversion is not necessarily a problem, some care and attention is needed to minimize losses in quality. Fournier et al. (2017) describe the digitization process and recommend storing photographs from digital cameras in their native format to retain the original information in its most compact format, and then converting to more conventional formats (e.g., TIFF, JPEG, BMP) as needed.

Documentation, ease of use, technical support, and cost are attributes of concern for users of HP software. Detailed documentation in user manuals was available for GLA (Frazer et al. 1999), HemiView (Rich et al. 1999), CAN-EYE

(Weiss and Baret 2010), and CIMES (Walter 2009; Gonsamo et al. 2011). CIMES documentation was particularly notable, as it included a theory section and an extensive list of references. GLA and HemiView were the most user-friendly, with graphical user interfaces. CAN-EYE requires a commercial program (MATLAB), which is available from MathWorks (http://www.mathworks.com/products/matlab/), to run a set of scripts. This entails creating a work directory where the image files reside and providing parameters on image and site characteristics in a configuration window. CIMES consists of a package of self-contained programs in a toolbox with command-line capability that is conducive to creating user scripts and to batch processing of large numbers of photographs (Walter 2009; Gonsamo et al. 2011). Its design is the most flexible, allowing it to be used in several operating systems including Microsoft Windows, MAC OS X, Unix and Linux (Walter 2009; Gonsamo et al. 2011). In most cases, CIMES offers a short version of the programs, providing a compact output in order to process multiple HPs in batch mode and to display the results in a spreadsheet-like format. Technical support from GLA, CAN-EYE and CIMES is often from the developers themselves, as the programs are open-source. This has been cited as a positive aspect; customers have direct contact with the developers and receive almost immediate response to queries (Promis et al. 2011). As a commercial program, HemiView offers technical support via the company's website, which also provides information for direct contact with the vendor (Table 7.1). Of these four programs, CIMES and CAN-EYE apparently are among the most active in terms of software development and updates; no future development plans are indicated for GLA and HemiView.

The degree to which attributes of canopy structure or solar radiation can be estimated is influenced greatly by the threshold that is used to discriminate between sky and canopy elements. Jarčuška et al. (2010) reported differences in output results arising from the approach to binary image pixel classification. Human interpretation in determining an optimal threshold has been one of the main problems and sources of error which is encountered during the analysis of an HP (Rich 1990; Nobis and Hunziker 2005) because it is subjective (Jonckheere et al. 2004). This problem becomes magnified when large numbers of photographs must be processed. This issue has led to increasing interest in automated thresholding algorithms (Jonckheere et al. 2005; Nobis and Hunziker 2005). For example, a freely available software program, which is called Hemisfer (http://www.wsl.ch/dienstleistungen/produkte/software/hemisfer/index_EN?redir=1), was designed to estimate LAI from several algorithms (Schleppi et al. 2007). Its implementation includes automatic threshold determination, based on Nobis and Hunziker (2005). An approach that is based on a two-value threshold has also been proposed to achieve a closer alignment with the camera's gamma function (Macfarlane et al. 2007). Of the software that was evaluated in this study, GLA, HemiView and CIMES all employ a single-value threshold, and CAN-EYE can generate multiple color classes from which a binary classification can be generated (Table 7.2). In this study, the varying effects on image quality resulting from the thresholding capabilities within the software were not evaluated. Fixing the threshold at the same value for all images that were being analyzed facilitated assessment of outputs, but it provided no insights into the effects of software specific thresholding capabilities on the accuracy of canopy structure or light regime variable outputs.

LAI is a forest structure parameter that is frequently measured from in situ direct and indirect optical and remote sensors (Fassnacht et al. 1994; Gower et al. 1999; Zheng and Moskal 2009). It is a highly sought-after parameter because of its relevance as an ecophysiological measure of the photosynthetic surface within the canopy (Chen et al. 1997) and its governing role in ecosystem process-based models of forest productivity (Liu et al. 1999). While LAI_{eff} values were similar for two sites, of particular interest was why these values differed between programs for the ECOLEAP Forêt Montmorency and Le Bray sites. A consistent result from these two sites was the statistical difference in values between HemiView and CIMES. The primary difference was that HemiView computes LAI_{eff} based on the ellipsoidal leaf angle distribution that is described by Campbell (1986), whereas CIMES (using the program "CLMPML") computes LAI_{eff} based on Miller's theorem (Miller 1967). Would these two values be similar if LAI_{eff} was based on the same algorithm? The CIMES program"LAICMP" computes LAI_{eff} using the Campbell (1986) ellipsoidal leaf angle distribution (Walter 2009), which should then be equivalent to that used by HemiView. Recomputing LAI_{eff}, however, resulted in no differences in statistical interpretation, especially since computed LAI_{eff} values were very similar to those reported in Table 7.6b.

While the value of HP for optical measurements of leaf area has been well demonstrated, foliage clumping has caused errors in LAI determination. Moreover, a completely random foliage distribution is not a likely occurrence in natural forest canopies (Fournier et al. 2003; Jonckheere et al. 2004; Chen et al. 2005; Frazer et al. 2017). If accurate estimation of LAI is important, the ability to estimate clumping with HP software becomes an important feature. While three of the five software programs offer the ability to estimate foliage clumping (Table 7.2), estimated values were not equivalent between two of the programs (CAN-EYE, CIMES) that were being evaluated (Table 7.6c). There are several algorithms for estimation of foliage clumping index, and knowledge of their application to the types of forest canopies to be measured is necessary to ensure that results are appropriate for the forest type being evaluated. Lang and Xiang (1986), Chen and Cihlar (1995), Leblanc et al. (2005), Gonsamo and Pellikka (2009), Walter (2009) and Frazer et al. (2017) provide information specific to foliage clumping, including algorithms and approaches to its measurement from HP.

The spatial organization of the vegetation elements greatly influences the radiation regime above and within the canopy. At any given level in the canopy, the incoming direct and diffuse solar radiation may vary considerably from the effects of canopy structure on sunfleck distribution (Parker et al. 2002). Interspecies competition for space as well as regeneration and replacement dynamics between individuals contribute to the horizontal and vertical organization of canopy elements, subsequently influencing the understory radiation regime. These factors, and the biophysical responses of forest canopy structure, are correlated with plant growth, survival and reproduction. It is therefore not surprising that all of the software that was evaluated allows the calculation of one or more variables that are related to the light regime within the forest canopy, as this was what HP was designed for. Diffuse and direct radiation (above and below canopy), sunflecks and site factors are currently found in GLA, HemiView, CIMES, and SOLARCALC.

Empirical results demonstrated a high degree of agreement for some of the parameters that were evaluated, such as openness and fraction of diffuse radiation, whereas differences were observed for other output parameters, including LAI_{eff}, clumping, and fraction of direct radiation. Rich (1990) identified the three stages at which errors may arise: image acquisition, digitization, and analysis. The results of an HP analysis is highly dependent upon the software program that is employed and the algorithms underlying the estimation of a particular attribute (Tables 7.2 and 7.3).

Computer processing of HP is essential for estimating canopy structure and quantifying forest light environments. While several software tools have been developed for this purpose, there is relatively little information available regarding their functionality and comparability. This chapter undertook a software description and empirical comparison exercise to provide insights into four HP software programs that were available to this study. Value judgements concerning which software is considered best for a given function and application were avoided. The choice of software depends upon many factors governing interoperability, availability, software functionality, user experience, and cost.

A primary question that remains is which software should be selected for processing HP. The algorithms that were employed do not appear to be a primary factor because all programs that were being evaluated in this study used algorithms that are well documented and which have been reported in the literature (Tables 7.2 and 7.5). Free access to GLA, CAN-EYE, CIMES, and SOLARCALC have contributed greatly to their wide and frequent use in the HP application community. While the selection of software remains with the users and depends upon their intended application, knowledge of algorithms that are employed by software programs and of the potential for errors during image acquisition and processing are essential to generate consistent estimates of canopy structure and solar radiation.

Acknowledgements The authors gratefully acknowledge the assistance of Drs. Gordon Frazer (Gap Light Analyzer), Marie Weiss and Frédéric Baret (CAN-EYE), and Jean-Michel Walter (CIMES) for their reviews and contributions toward the hemispherical photo software features that are summarized in Table 7.2. Technical peer review by Gordon Frazer, Jean-Michel Walter, and Kamel Soudani resulted in an improved version of this chapter and is greatly appreciated.

References

Ackerly DD, Bazzaz FA (1995) Seedling crown orientation and interception of diffuse radiation in tropical forest gaps. Ecology 76:1134–1146

Anderson MC (1964) Studies of the woodland light climate I. The photographic computation of light condition. J Ecol 52:27–41

Andrieu B, Sinoquet H (1993) Evaluation of structure description requirements for predicting gap fraction of vegetation canopies. Agric For Meteorol 65:207–227

Baret F, Weiss M (2004) CAN-EYE: processing digital photographs for canopy structure characterization. INRA-EMMAH UMR 114, Avignon, France

Becker P, Erhart DW, Smith AP (1989) Analysis of forest light environments part I. computerized estimation of solar radiation from hemispherical canopy photographs. Agric For Meteorol 44:217–232

Bernier PY, Fournier RA, Ung C-H, Robitaille G, Larocque GR, Lavigne MB, Boutin R, Raulier F, Paré D, Beaubien J, Delisle C (2002) Linking ecophysiology and forest productivity: an overview of the ECOLEAP project. Forest Chronicle 75:417–421

Bernier PY, Raulier F, Stenberg P, Ung C-H (2001) Importance of needle age and shoot structure on canopy net photosynthesis of balsam fir (*Abies balsamea*): a spatially inexplicit modeling analysis. Tree Physiol 21:815–830

Bonhomme R, Chartier P (1972) The interpretation and automatic measurement of hemispherical photographs to obtain sunlit foliage area and gap frequency. Israel J Agric Res 22:53–61

Bréda NJJ (2003) Ground-based measurements of leaf area index: a review of methods, instruments and current controversies. J Exp Bot 54:2403–2417

Campbell GS (1986) Extinction coefficients for radiation in plant canopies calculated using an ellipsoidal inclination angle distribution. Agric For Meteorol 36:317–321

Canham CD (1988) An index for understory light levels in and around canopy gaps. Ecology 69:1634–1638

Chazdon RL, Field CB (1987) Photographic estimation of photosynthetically active radiation: evaluation of a computerized technique. Oecologia 73:525–532

Chen JM (1996) Optically-based methods for measuring seasonal variation of leaf area index in boreal conifer stands. Agric For Meteorol 80:135–163

Chen JM, Cihlar J (1995) Plant canopy gap-size analysis theory for improving optical measurements of leaf-area index. Appl Opt 34:6211–6222

Chen JM, Cihlar J (1996) Retrieving leaf area index of boreal conifer forests using landsat TM images. Remote Sens Environ 55:153–162

Chen JM, Menges CH, Leblanc SG (2005) Global mapping of foliage clumping index using multi-angular satellite data. Remote Sens Environ 97:447–457

Chen JM, Rich PM, Gower ST, Norman JM, Plummer S (1997) Leaf area index of boreal forests: theory, techniques, and measurements. J Geophys Res 102(D24):29429–29443

Demarez V, Duthoit S, Baret F, Weiss M, Dedieu G (2008) Estimation of leaf area and clumping indexes of crops with hemispherical photographs. Agric For Meteorol 148:644–655

Englund SR, O'Brien JJ, Clark DB (2000) Evaluation of digital and film hemispherical photography and spherical densiometry for measuring forest light environments. Can J For Res 30:1999–2005

Fassnacht KS, Gower ST, Norman JM, M cMurtrie RE (1994) A comparison of optical and direct methods for estimating foliage surface area index in forests. Agric For Meteorol 71:183–207

Fournier RA, Mailly D, Walter J-MN, Jonckheere IGC (2017) Acquiring hemispherical photographs in forest environments: from planning to archiving photographs. In: Fournier RA, Hall RJ (eds) Hemispherical photography in forest science: theory, methods, applications. Springer, Berlin

Fournier RA, Mailly D, Walter J-MN, Soudani K (2003) Indirect measurement of forest canopy structure from *in situ* optical sensors. In: Wulder MA, Franklin SE (eds) Methods and applications for remote sensing of forests: concepts and case studies. Kluwer, Dordrecht, pp 77–113

Fournier RA, Rich PM, Landry R (1997) Hierarchical characterization of canopy architecture for boreal forest. J Geophys Res 102(D24):29445–29454

Frazer GW, Canham CD, Lertzman KP (1999) Gap Light Analyzer (GLA), Version 2.0: Imaging software to extract canopy structure and gap light transmission indices from true-colour fisheye photographs, users manual and program documentation. Simon Fraser University, Burnaby, British Columbia, and the Institute of Ecosystem Studies, Millbrook, New York, 52 pp

Frazer GW, Fournier RA, Leblanc SG, Walter J-MN (2017) View angle-dependent clumping indices for indirect LAI estimation. In: Fournier RA, Hall RJ (eds) Hemispherical photography in forest science: Theory, methods, applications. Springer, Berlin

Frazer GW, Fournier RA, Trofymow JA, Hall RJ (2001) A comparison of digital and film fisheye photography for analysis of forest canopy structure and gap light transmission. Agric For Meteorol 109:249–263

Frazer GW, Trofymow JA, Lertzman KP (1997) A method for estimating canopy openness, effective leaf area index, and photosynthetically active photon flux density using hemispherical photography and computerized image analysis techniques. Natural Resources Canada, Canadian Forest Service, Pacific Forestry Centre, Victoria, BC. Information Report BC-X-373, 73 pp

Gonsamo A, Pellikka P (2009) The computation of foliage clumping index using hemispherical photography. Agric For Meteorol 149:1781–1787

Gonsamo A, Walter J-MN, Pellikka P (2011) CIMES: A package of programs for determining canopy geometry and solar radiation regimes through hemispherical photographs. Comput Electron Agric 79:207–215

Gower ST, Kucharik JK, Norman JM (1999) Direct and indirect estimation of leaf area index, f_{APAR}, and net primary production of terrestrial ecosystems. Remote Sens Environ 70:29–51

Hale SE, Edwards C (2002) Comparison of film and digital hemispherical photography across a wide range of canopy densities. Agric For Meteorol 112:51–56

Jarčuška B (2008) Methodological overview to hemispherical photography, demonstrated on an example of the software GLA. Folia Oecologica 35:66–69

Jarčuška B, Kucbel S, Jaloviar P (2010) Comparison of output results from two programmes for hemispherical image analysis: Gap Light Analyser and WinScanopy. J For Sci 56:147–153

Jonckheere I, Fleck S, Nackaerts K, Muys B, Coppin P, Weiss M, Baret F (2004) Methods for leaf area index determination. part I: theories, sensors and hemispherical photography. Agric For Meteorol 121:19–35

Jonckheere I, Nackaerts K, Muys B, Coppin P (2005) Assessment of automatic gap fraction estimation of forests from digital hemispherical photography. Agric For Meteorol 132:96–114

Knyazikhin Y, Martonchik JV, Myneni RB, Diner DJ, Running SW (1998) Synergistic algorithm for estimating vegetation canopy leaf area index and fraction of absorbed photosynthetically active radiation from MODIS and MISR data. J Geophys Res 103(D24):32257–32275

Lang ARG, McMurtrie RE (1992) Total leaf areas of single trees of *Eucalyptus grandis* estimated from transmittances of the sun's beam. Agric For Meteorol 58:79–92

Lang ARG, Xiang YQ (1986) Estimation of leaf area index from transmission of direct sunlight in discontinuous canopies. Agric For Meteorol 37:229–243

Leblanc SG, Chen JM, Fernandes R, Deering DW, Conley A (2005) Methodology comparison for canopy structure parameters extraction from digital hemispherical photography in boreal forests. Agric For Meteorol 129:187–207

Leblanc SG, Fournier RA (2017) Measurement of forest structure with hemispherical photography. In: Fournier RA, Hall RJ (eds) Hemispherical photography in forest science: Theory, methods, applications. Springer, Berlin

Lindsey PA, Bassuk NL (1992) A nondestructive image analysis technique for estimating whole-tree leaf area. HortTechnology 2:66–72

Liu J, Chen JM, Cihlar J, Chen W (1999) Net primary productivity distribution in the BOREAS study region from a process model driven by satellite and surface data. Journal of Geophysical Research 104(D22):27735–27754

Macfarlane C (2011) Classification method of mixed pixels does not affect canopy metrics from digital images of forest overstorey. Agric For Meteorol 151:833–840

Macfarlane C, Grigg A, Evangelista C (2007) Estimating forest leaf area using cover and fullframe fisheye photography: thinking inside the circle. Agric For Meteorol 146:1–12

Mailly D, Turbis S, Chazdon RL (2013) SOLARCALC 7.0: An enhanced version of a program for the analysis of hemispherical canopy photographs. Comput Electron Agric 97:15–20

McCree KJ (1981) Photosynthetically active radiation. In: Lange OL, Nobel PS, Osmond CB, Ziegler H (eds) Encyclopedia of plant physiology, vol 12A. Springer, Berlin, pp 41–55

Miller JB (1967) A formula for average foliage density. Aust J Bot 15:141–144

Nobis M, Hunziker U (2005) Automatic thresholding for hemispherical canopy-photographs based on edge detection. Agric For Meteorol 128:243–250

North M, Chen J, Oakley B, Song B, Rudnicki M, Gray A, Innes J (2004) Forest stand structure and pattern of old-growth western hemlock/Douglas-fir and mixed-conifer forests. For Sci 50:299–311

Parker GG, Davis MM, Chapotin SM (2002) Canopy light transmittance in Douglas-fir–western hemlock stands. Tree Physiol 22:147–157

Peper PJ, McPherson EG (1998) Comparison of five methods for estimating leaf area index of open-grown deciduous trees. J Arboric 24:98–111

Pielou EC (1962) Runs of one species with respect to another in transects through plant populations. Biometrics 18:579–593

Porté A, Bosc A, Champion I, Loustau D (2000) Estimating the foliage area of maritime pine (*Pinus pinaster* Ait.) branches and crowns with application to modelling the foliage area distribution in the crown. Ann For Sci 57:73–86

Promis A, Gärtner S, Butler-Manning D, Durán-Rangel C, Reif A, Cruz G, Hernández L (2011) Comparison of four different programs for the analysis of hemispherical photographs using parameters of canopy structure and solar radiation transmittance. Waldökologie Landschaftsforschung und Naturschutz 11:19–33

Raulier F, Bernier PY, Ung C-H' Boutin R (2002) Structural differences and functional similarities between two sugar maple (*Acer saccharum*) stands. Tree Physiol 22:1147–1156

Rich PM (1989) A manual for analysis of hemispherical canopy photography. Los Alamos National Laboratory Report LA-11733-M, Los Alamos, New Mexico. http://www.creeksidescience.com/files/rich_et_al_1989_canopy_manual.pdf

Rich PM (1990) Characterizing plant canopies with hemispherical photographs. In: Goel NS, Norman JM (eds) Instrumentation for studying vegetation canopies for remote sensing in optical and thermal infrared regions. Remote Sens. Rev. 5:13–29

Rich PM, Wood J, Vieglais DA, Burek K, Webb N (1999) HemiView user manual. Delta-T Devices Ltd., Cambridge, UK, 79 pp

Schleppi P, Conedera M, Sedivy I, Thimonier A (2007) Correcting non-linearity and slope effects in the estimation of the leaf area index of forests from hemispherical photographs. Agric For Meteorol 144:236–242

Sellers PJ, Hall FG, Kelly RD, Black A, Baldocchi D, Berry J, Ryan M, Ranson KJ, Crill PM, Lettenmaier DP, Margolis H, Cihlar J, Newcomer J, Fitzgerald D, Jarvis PG, Gower ST, Halliwell D, Williams D, Goodison B, Wickland DE, Guertin FE (1997) BOREAS in 1997: Experiment overview, scientific results, and future directions. J Geophys Res 102 (D24):28731–28769

Stenberg P, Linder S, Smolander H, Flower-Ellis J (1994) Performance of the LAI-2000 plant canopy analyzer in estimating leaf area index of some Scots pine stands. Tree Physiol 14:981–995

ter Steege H (1996) WINPHOT 5: a programme to analyze vegetation indices, light, and light quality from hemispherical photographs. Tropenbos-Guyana Programme and Utrecht University, Utrecht, 52 pp

Valentini R (ed) (2003) Fluxes of carbon, water and energy of European forests. Springer, Berlin, 273 pp

Valladares F, Guzmán B (2006) Canopy structure and spatial heterogeneity of understory light in an abandoned Holm oak woodland. Ann For Sci 63:749–761

Walter J-MN (2009) CIMES-FISHEYE. Hemispherical photography of forest canopies: A package of programs for the assessment of canopy geometry and solar radiation regimes through hemispherical photographs [online]. Université de Strasbourg, Strasbourg, France. Available from http://equinoxe.u-strasbg.fr/cimes

Weiss M, Baret F (2010) CAN-EYE V6.1 User manual. INRA-UMR, Avignon, France [online]. Available from http://www.paca.inra.fr/can_eye

Weiss M, Baret F, Myneni RB, Pragnère A, Knyazikhin Y (2000) Investigation of a model inversion technique to estimate canopy biophysical variables from spectral and directional reflectance data. Agronomie 20:3–722

Welles JM, Norman JM (1991) Instrument for indirect measurement of canopy architecture. Agron J 83:818–825

Zhang Y, Chen JM, Miller JR (2005) Determining digital hemispherical photograph exposure for leaf area index estimation. Agric For Meteorol 133:166–181

Zheng G, Moskal LM (2009) Retrieving leaf area index (LAI) using remote sensing: theories, methods and sensors. Sensors 9:2719–2745

Chapter 8
Hemispherical Photography in Support of Forest Inventory and Silviculture

Daniel Mailly

Abstract Hemispherical photography (HP) can be used in support of silviculture and forest inventory. At the tree level, HP can be used to estimate attributes such as tree architecture, crown dimensions, and volume of standing trees. At the stand level, there are relationships between indices extracted from HP and stand attributes such as density, basal area, canopy cover, and gap size. As an aid to silvicultural activities, the technique can provide valuable information for monitoring the effects of thinning and patch cuttings on regeneration processes and stand development. The HP technique is versatile, scalable, and the results compare well with other methods that are used to estimate canopy cover and light transmittance in forests. This chapter provides examples of applications of HP at both tree and forest stand levels. The strengths and weaknesses of the technique will be examined, and a comparison will be made with other techniques available in support of silviculture and forest inventory.

Keywords Canopy cover · Fisheye · Light transmittance · Regeneration · Stand-level attributes

8.1 Introduction

The hemispherical photography (HP) literature abounds with studies in forests that compare various techniques for characterizing light transmittance (Salminen et al. 1983; Comeau et al. 1998; Gendron et al. 1998; Ferment et al. 2001) and LAI, the leaf area index (van Gardingen et al. 1999; Jonckheere et al. 2004; Seidel et al. 2011). The HP technique generally fares well in comparison with other approaches such as light sensors and optical leaf area index meters. In addition, HP has several important strengths and is widely accepted in the forest ecology and forestry

D. Mailly (✉)
Decizio, 929 ave Sir Adolphe-Routhier, Québec, QC G1S 3P4, Canada
e-mail: daniel.mailly.1@ulaval.ca

R.A. Fournier and R.J. Hall (eds.), *Hemispherical Photography in Forest Science: Theory, Methods, Applications*, Managing Forest Ecosystems 28,
DOI 10.1007/978-94-024-1098-3_8

literature (Rich 1990; Mitchell and Whitmore 1993). HP has also been used in other areas such as tree architecture and fruit tree cropping (e.g., Lakso 1976), forest meteorology (e.g., Blennow 1998), and stream hydrology and riparian ecology (e.g., Teti 2001). The application of the technique to forest inventory and silvicultural activities has received less attention.

For most forest jurisdictions around the world, forest inventory is a huge undertaking involving intensive field measurements on a routine basis. Inventory programs serve several purposes such as the preparation of silvicultural prescriptions and data collection to supply models that are used in allowable cut calculations. While advances have recently been made in remote sensing applications for forest inventory (e.g., Gougeon and Leckie 2003; Maltamo et al. 2006; van Leeuwen et al. 2011), forest inventory activities mainly rely upon aerial photo-interpretation and field sampling to collect tree and stand measurements. In forest inventories, variables such as the canopy cover and light transmittance are sometimes acquired; they are usually measured using conventional equipment and techniques, such as spherical densiometers, sensors, moosehorn crown-closure estimators, and the line-intercept method (Garrison 1949; McIntosh et al. 2012). The use of HP for such purposes is a relatively new but promising approach; accurate results can be obtained if the proper equipment is used and the technique applied correctly. However, its use may remain limited because hemispherical photographs must be taken under uniform overcast sky conditions or when the sun is low on the horizon (Rich 1990).

The overall goal of this chapter is to review the use of HP for forest inventory surveys, silvicultural treatments, and forest management in general. This goal will be achieved in the following way: (i) providing selected examples of its application, (ii) providing an overview of the advantages and areas of improvement concerning HP, and (iii) comparing HP to other methods in support of forest inventory and silviculture.

8.2 HP in Support of Forest Inventory

In forestry-based economies around the world, inventory programs are a costly operation and a large undertaking. These programs require extensive planning and logistics, and they must follow strict procedures for collecting data that represent a broad range of attributes at the stand level, such as ecosystem type and canopy cover. More specific attributes that are often measured at the tree level include crown dimensions, tree diameter, and tree height. These attributes form the basis of various statistics or tables; for example, height-diameter curves are used in tariff (yield) tables to estimate tree (merchantable) volume. Stand-level attributes such as basal area (BA) and canopy cover are measured or estimated in most inventory programs. With the increasing emphasis that has been placed on ecosystem-based management over the past two decades, sampling programs have been expanded to collect more ecologically related measures such as light transmittance, canopy cover, and leaf area index. While these measures are relatively well documented in the research literature, they are absent from many forest inventory reference manuals or reports (e.g., Avery and Burkhart 2002; Curtis and Marshall 2005). Examples of such applications will be presented in the following sections.

8.2.1 Tree Level

The projected area and diameter of tree crowns are often used as surrogate measures for crown variables such as photosynthetic potential and foliage area (Jahnke and Lawrence 1965; Brown et al. 2000b). Precise measurement of such attributes is important for forest simulation modeling; this is particularly true for broadleaf tree species that may have irregular crown shapes. Although crown size can be estimated using predefined crown shapes and simple measurements that are made in the field, the latter are prone to errors and may be imprecise. As a consequence, there have been a few attempts to use other techniques such as vertical photography (e.g., Decourt 1956; Wilkinson 1995) or HP (e.g., Ducrey and Bartoli 1974; Brown et al. 2000b) to estimate crown dimensions.

Ducrey and Bartoli (1974) used HP to develop mathematical formulas to estimate crown dimensions of standing trees. Their method also included a correction factor for cases where trees are located on a sloped surface. The authors used a conventional Nikon camera (Nikon Corporation, Tokyo, Japan) that was equipped with a 7.5 mm Nikkor lens with a 180° field of view (FOV). They also provided mathematical equations to estimate the tree height and diameter. They were particularly interested in estimating the crown diameter of trees with asymmetrical crown shapes, and they subsequently recommended that several pictures be taken at various view angles to increase the precision of their method. Their technique, which is mostly adapted to deciduous forests, requires that fisheye photographs be taken during the leafless season and that the distance between the camera's optical axis and each tree is known. To avoid the problem of taking the photographs during the leafless season, the near-infrared (NIR) hemispherical imagery technique that was developed by Chapman (2007) may be more appropriate. This technique involves the use of a modified digital camera that is capable of capturing NIR images through the removal of an infrared blocking filter and the addition of an inexpensive, screw-on, cold mirror filter (i.e. a specialized mirror that reflects the entire visible light spectrum while very efficiently transmitting infrared wavelengths) to block out visible light. By using a simple clear filter of the same thickness as the cold mirror filter, the camera can then be used as a standard digital camera. With this modification in place, it is much easier to differentiate between the bole and branch elements in hemispherical images using a single image threshold (Chapman 2007).

Ducrey and Bartoli (1974) expanded their technique to permit calculation of the volume of individual trees. For this purpose, they used a method similar to that used with a Barr and Stroud dendrometer (Avery and Burkhart 2002, p. 151). The volume is calculated from bole diameters that are measured at heights of 0.5, 1.3 m and, subsequently, in 2–5 m increments. The observer must be positioned at a distance of about one-third to one-half of total tree height from the tree being measured. Under such conditions, the apex of the tree in the photograph is located at angular heights between 60° and 70°.

HP can be used not only to estimate the crown dimensions of standing trees, but also to document interspecific variation in the openness of individual crowns of tree species. Crown openness is defined as the fraction of sky that on average, can be seen through the crown of an individual tree (Astrup and Larson 2006). The procedure for assessing species-specific crown openness requires that a number of hemispherical photographs be taken of individual trees with their entire crown. The outlines of tree crowns that are projected on the photograph are then submitted to image analysis to estimate individual crown openness (i.e., the percent of sky that is visible through the crown; Canham et al. 1999; Beaudet et al. 2002). These values of canopy openness are particularly useful in model simulations of forest light.

Brown et al. (2000b) used vertical HP, together with calibration scales, to obtain measurements of crown diameter and projected area. Their procedure is based, to a large extent, on a hemispherical analysis program (WINPHOT, ter Steege 1997), which is used as a standard against which simpler modifications of the photographic technique could be compared. Their image analysis method was relatively accurate, particularly for the comparatively shallow crowns of tall trees. Application of their method is slower than conventional techniques for ground-based measurements, and it requires additional equipment. However, it has the advantage of allowing the measurement of changes in crown size and foliage projected area over time.

Some HP applications to tree-level measurement techniques that are mentioned in this section are theoretically valid but have not been systematically used in forest inventory programs. For instance, the method that was developed by Ducrey and Bartoli (1974) apparently has not been used on a routine basis. The same applies to Chapman's (2007) near-infrared (NIR) hemispherical imagery technique, which is fairly new. Species-specific crown openness that is determined directly from HP has been used more frequently than the techniques cited above (Beaudet et al. 2002; Astrup and Larson 2006). Although the determination of crown openness using HP is of general interest for understanding light transmission through forest canopies, its application has generally been limited to the calibration of light simulation models (Stadt and Lieffers 2000; Beaudet et al. 2002). Nevertheless, these promising methods have practical applications in forest inventory when tree-level-specific and detailed data are required for tree growth assessments or for the development of simulation models.

8.2.2 *Stand Level*

Stand level variables such as density and basal area are important in forestry because of their close relationships with stand volume. Stand density refers to the number of trees per sample plot, whereas the plot basal area is the sum of the area in square meters of the cross section at breast height (1.3 m) of all trees in a plot.

Foresters have developed various instruments to measure or estimate stand variables such as density and basal area (e.g., prism, relascope; Avery and Burkhart 2002). However, the increasing application of HP in forest ecology and forestry over the last few years has generated an interest in using the technique in the field and to relate, for instance, light transmittance to stand basal area.

8.2.2.1 Light Transmittance

Because light transmittance is a crucial environmental factor governing many processes in forest understories, several researchers have tried to relate light transmittance to stand basal area using HP. For instance, Comeau (2000) found a negative exponential relationship between light transmittance and basal area ($r^2 = 0.88$) in aspen stands that were younger than 40-years-old. Hale (2003) successfully used HP to relate transmittance to stand basal area but found that the technique proved to be less useful in dense canopies, where it discriminated very poorly at transmittance values below 0.10. These studies show that relationships can be established between conventional forest inventory measurements, such as stand basal area, and light transmittance that is estimated using HP. These relationships must be established for each forest type, as shown in Fig. 8.1.

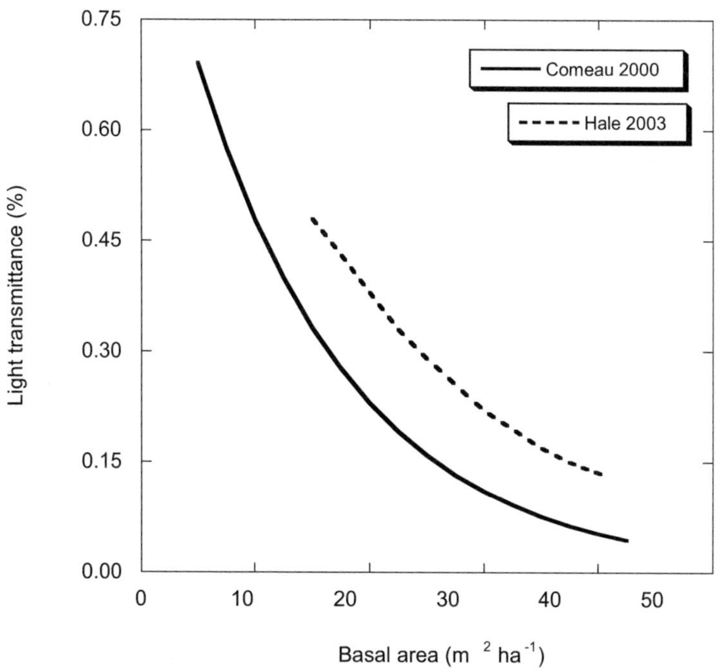

Fig. 8.1 Light transmittance (%) estimated from HP versus stand basal area for stands of trembling aspen (Comeau 2000) and Sitka spruce (Hale 2003)

When the objective is to measure understory light levels using HP (or other indirect means), it is recommended that these measurements be calibrated against quantum sensors (Comeau et al. 1998). Quantum sensors should be installed at selected measurement points and left in operation throughout a long period (i.e., ranging between seven days and the entire growing season).

8.2.2.2 Canopy Structure and Stand-Level Attributes

The concept of canopy cover has long been used in forest ecology (e.g., Robinson 1947; Lemmon 1956; Evans and Coombe 1959; Paletto and Tosi 2009) and to assist in management decisions. Fiala et al. (2006) defines vertical canopy cover as the vertical projection of plant foliage onto a horizontal surface with a canopy view angle equal to zero. In other words, vertical canopy cover refers to the area that is projected onto the ground by the canopy area that is positioned directly overhead. For example, the line-intercept method (Canfield 1941) provides accurate estimates of vertical canopy cover. In contrast, angular canopy cover (Fiala et al. 2006) refers to canopy cover that is measured with a view angle by using, for example, a wide angle lens (e.g., 30 mm lens with a 154° field of view, Lacinová and Zboja 1985) or a fish-eye lens (Nikkor 8 mm lens with a 180° FOV, Fournier et al. 1996). Riaño et al. (2004) used HP to estimate the vertically projected canopy area per unit ground area (which they called ground cover). It is calculated assuming that the canopy has an ellipsoidal distribution such that

$$\mathrm{GC} = 1 - \exp(-K(x, 0) \cdot \mathrm{LAI}) \tag{8.1}$$

where *GC* is the ground cover, $K(x, 0)$ is the extinction coefficient for a zenith angle of zero, and x is the ellipsoidal leaf angle distribution (Riaño et al. 2004). In their study, they found that there was a highly significant correlation between the airborne laser scanner (LIDAR) and HP estimates of canopy properties, but the relationship was affected by the type of forest and the LIDAR radius size. According to the study, *GC* was better estimated using a small radius size (2.5 m, equivalent to one-fourth of the canopy height), while LAI was better estimated using a larger radius size (7.5–12.5 m, equivalent to the entire canopy height).

In the mixed-conifer forests of the Sierra Nevada (California, USA), North et al. (2004) used HP to obtain estimates of angular canopy cover. They found that plots exhibited either open-gap or closed-canopy conditions; one group of plots had ≥60% canopy cover, whereas canopy cover in the second group ranged between 5 and 40%. They had hypothesized that a minimum canopy cover was necessary for tree establishment in these forests. They also suggested that specific experimentation is needed to determine if there is a canopy cover threshold that substantially affects tree regeneration. This can be achieved by taking fisheye photographs at regular grid point locations within a study site and analyzing the images to determine the exact relationships between canopy cover values and regeneration success.

Ducrey and Bartoli (1974) developed, but did not test, a technique to calculate stand basal area directly from HP. The technique is based on the Bitterlich relascope method (Bitterlich 1948; Grosenbaugh 1952) for estimating stand basal area. With this method, the probability of inclusion of a given tree in the data is proportional to the basal area of the tree rather than to the area of the plot sampled. The method is based on the relationship between the distance to a given tree and its diameter; these data are easily converted to basal area in $m^2\ ha^{-1}$ (Grosenbaugh 1952). In order to use the method, a photograph must be taken at 1.3 m height (breast height); each tree at the periphery of the image the diameter of which in the photograph is greater than a length determined by the value of the scan angle (viewing angle γ, in radians, of the Bitterlich sampling method) must be counted (total number of trees = N). The radius of the hemispherical image must also be recorded. According to the following formula (Pardé and Bouchon 1988), basal area G (in m^2 on a per hectare basis) is calculated as

$$G = 2500 \cdot \gamma^2 \cdot N \tag{8.2}$$

It is recommended that 20–30 trees be counted to minimize errors (i.e., that the scan angle γ be adapted to the estimated basal area of the stand) (Ducrey and Bartoli 1974; Table 8.1). The length of the segment to be displaced at the periphery of the hemispherical photograph is obtained by multiplying the scan angle γ that has been chosen by the radius of the photograph. The method developed by Ducrey and Bartoli (1974), although more tedious than using a relationship between stand basal area and light transmittance (Sect. 8.2.2.1), has the advantage of permitting direct estimation of stand basal area; however, the photographs must be taken during the leafless season or the separation between the bole and branch elements in the hemispherical images must be done using NIR imagery (Chapman 2007).

As is the case for light transmittance, some researchers have tried to relate canopy cover to stand basal area using HP. In a study that was conducted in Norway spruce (*Picea abies* L.) and Scots pine (*Pinus sylvestris* L.) stands, Kuusipalo (1985) found a good relationship (r^2 = 0.63) between total basal area of the stand and canopy cover that was computed from HP. Incorporating the relative

Table 8.1 Values for scan angles* (γ, radians) of the Bitterlich horizon tour as a function of basal area**

Basal area ($m^2\ ha^{-1}$)										
Number of trees per plot	5	10	15	20	25	30	35	40	45	50
20	0.0100	0.0141	0.0173	0.0200	0.0224	0.0245	0.0265	0.0283	0.0300	0.0316
25	0.0089	0.0126	0.0155	0.0179	0.0200	0.0219	0.0237	0.0253	0.0268	0.0283
30	0.0082	0.0115	0.0141	0.0163	0.0183	0.0200	0.0216	0.0231	0.0245	0.0258

*Calculated from Eq. 8.2

**Adapted and expanded from Ducrey and Bartoli (1974)

proportion of spruce that was in the stand increased the fit of the relationship to $r^2 = 0.75$.

As can be seen from the studies that have been presented above, the literature includes many examples of HP being used to estimate parameters of canopy structure and other stand-level attributes. These studies show that HP can be used effectively to estimate or monitor attributes such as canopy cover or basal area, provided that a suitable sampling strategy is used and the techniques are well mastered. The application of this approach also requires that the user is familiar with HP equipment and various theoretical aspects of forest mensuration calculations. With these elements in place, the user benefits from a powerful tool for obtaining valuable information, such as canopy cover and basal area values, which is important in forestry.

8.3 HP in Support of Silviculture

Silviculture has been defined as the art and science of applying various treatments to forest stands in order to maintain and enhance their utility for any purpose (Smith 1986). Part of the role of foresters and forest managers is to manipulate forest cover by tree harvesting while also providing optimal conditions for tree regeneration and growth. Knowledge of the degree, or percent, of canopy cover is important when designing treatments that would allow sufficient light for a regeneration response while also maintaining a permanent forest cover without excessive competition for site resources (Delong et al. 2005; Lhotka and Loewenstein 2008). As is the case with forest inventory, the application of HP methods to silviculture has been beneficial over the last few years, and some notable applications are worth mentioning.

8.3.1 Regeneration

HP provides a convenient means of quantifying the correlation between seedling growth and understory light conditions or canopy structure in a forest. The technique is thought to be more precise and less subjective than ocular estimates (Chan et al. 1986). HP techniques can also incorporate the effects of slope and distant objects on the horizon, which can reduce the hemispherical percent visible sky below 100%, even for open-grown trees, as shown in the study by Chan and Walstad (1987). In this particular study, basal diameter of Douglas-fir (*Pseudotsuga menziesii* (Mirbel) Franco) seedlings was related to percentage of sky visible. Forty-three to 64% of the variation during the first year of growth, and 48–74% for year two in basal diameter could be attributed to the percentage of sky visible. In another study, Brandeis et al. (2001) compared the performance of HP estimates of percent canopy cover (PCC), total basal area (TBA), and visual estimates of canopy cover (VCC) as predictors of understory conifer growth. They found that of these three stand density descriptors, VCC had the strongest relationship with under-planted seedling growth. They concluded that one possible reason for this result was the fact that with VCC, the observer's eye better integrates information about the seedling's operational environment than does either absolute basal area or estimates that are based on HP. However, they also noted that the differences between VCC and PCC may be attributed to the fact that VCC measurements were taken over each seedling, while PCC measurements were made only at the centers of seedling groups.

Turner (1990) used HP to relate tree seedling growth and survival to levels of available light in a Malaysian rain forest. Results from this study indicated that seedling height growth increased significantly with the availability of PAR (photosynthetically active radiation), but the relationship was weak. Three reasons were hypothesized to explain the poor relationship between PAR and seedling growth: (i) the inaccuracy of the technique at low light levels, (ii) the importance of other factors in explaining seedling growth, and (iii) the stochasticity that is inherent to processes of seedling growth and survival. Using a similar approach, Mailly and Kimmins (1997) found that light transmittance $\geq 20\%$ was necessary for seedling survival, and at least 40% light transmittance was necessary for optimum growth of Douglas-fir that was underplanted within a mature Douglas-fir-Western hemlock stand in coastal British Columbia. In the Mediterranean region, Puerta-Piñero et al. (2007) found that irradiance, as estimated using HP, significantly influenced holm oak (*Quercus ilex* L.) seedling survival and growth. Their results indicated that seedling survival decreased with irradiances greater than 50%, whereas better growth results were obtained at intermediate irradiance exposures of around 50%.

The examples that have been described above demonstrate the versatility of HP for studies on the growth and survival of seedlings under a forest canopy. Indeed, indices that are derived from HP such as canopy cover and percent light transmittance can explain a large proportion of the variation that is observed in the field. Thresholds below or above which seedling growth or survival are affected can be obtained relatively quickly in the field by taking a series of photographs under a gradient of understory conditions. It should be noted, however, that growth and survival of seedlings in the understory is related to a number of factors other than light and canopy structure. As a consequence, any attempt to establish relationships between these factors will be limited by the fact that other factors influence tree regeneration.

8.3.2 Thinning and Partial Cutting

Foresters have long been interested in manipulating the forest canopy to create growing space and light environments that are suitable for seedling establishment and growth (Jennings et al. 1999). They have devised and used various silvicultural methods such as thinning (i.e., partial cutting or spacing operations that are made in immature stands to accelerate the growth of the remaining trees). In a study conducted in Sitka spruce (*Picea sitchensis* (Bong.) Carr) stands for instance, Hale (2001) used HP to relate gap fraction to basal area for a range of stands in northern Britain. A good relationship was found between gap fraction and basal area; it was concluded that a basal area lower than 30 m^2 ha^{-1} was necessary to produce a gap fraction that was greater than 0.1, which allows sufficient light penetration to encourage natural regeneration.

Battaglia et al. (2002) took hemispherical photographs in longleaf pine (*Pinus palustris* Miller) forests that had been subjected to various silvicultural treatments to establish a relationship between light transmittance and spatial structure. The harvest treatments retained similar residual basal area, but the spatial patterns of the residual trees differed. Their results demonstrated that an aggregated canopy structure of longleaf pine forest facilitates longleaf pine seedling regeneration. Zhu et al. (2003) found that patch-pattern thinning is potentially a viable silvicultural measure for the management of coastal black pine (*Pinus thunbergii* Parl.) forests in Japan. Using HP, they had shown that canopy openness values around 30% were required for seedling survival in canopy gaps. To encourage further development of seedlings into saplings, canopy openness values above 40% were necessary in canopy gaps. Delong et al. (2005) used HP to relate the survival and growth response of seedlings in root disease-infected (*Armillaria ostoyae* [Romagnesi] and *Inontus tomentosus* [Fr.Fr] S. Teng.) partial cuts in the Interior Cedar Hemlock zone of southeastern British Columbia. They found that growth rates increased as the level of canopy openness increased; they concluded that a variety of partial retention systems could be used to regenerate mixed species forests in this part of Canada.

HP can also be used to establish relationships between the size of patch cuts (or canopy gap size) and the light environment under the canopy. A hemispherical analysis program such as WINPHOT (ter Steege 1997) can measure physical gap size, especially for regular gaps, although problems in defining the gap edge render the method less useful in irregularly shaped gaps. For example, Martins and Rodrigues (2002) successfully used WINPHOT to measure gap size while studying gap-phase regeneration in a semi-deciduous mesophytic forest in southeastern Brazil (São Paulo State). For gap sizes ranging from 20 to 468 m^2, they found that species richness increased significantly with gap size ($r^2 = 0.93$). Expanding on the approach initiated by ter Steege (1997), Hu and Zhu (2008) introduced a new method called the geometric calculation (GeoCalc) of gap coordinates, from which the effects of gap size, canopy height, and topography on canopy gap light regimes could be estimated. The program uses gap coordinates for any point specified within a canopy gap and takes into account the tri-dimensional profile of the gap, including the slope and aspect.

Various studies have demonstrated that the potential of HP is not limited to traditional silvicultural operations. Overall, the studies that were previously cited show that HP can be successfully used to estimate canopy cover and light transmittance after implementing a diverse range of silvicultural treatments. HP is likely to have a promising future, given the increasing popularity of alternative silvicultural treatments that require precise canopy measurements in the field. In particular, HP can be used to account for the interaction between light transmittance and canopy structure; this cannot be done using conventional devices for measuring light transmittance or LAI. HP has proven to be useful for determining values or thresholds of light transmittance that are required for the survival or establishment of tree regeneration under patch cuts.

8.4 Implementing HP in Forest Inventory and Silviculture

Although HP is increasingly used in support of forest inventory and silviculture, its acceptance as an inventory tool has not yet reached the same level of popularity as other methods or instruments. The technical and scientific basis of HP is not new and dates back around 55 years (Evans and Coombe 1959). Nevertheless, the popularity of HP is growing fast, as evidenced by the number of research and field projects citing its use. Furthermore, the potential for extending the use of HP to a broader range of measures is much greater than for more conventional instruments. The adoption of HP techniques by potential users requires, however, knowledge of their strengths and weaknesses.

8.4.1 Advantages

HP as a technique for estimating light transmittance or canopy cover has several advantages. For instance, HP provides a permanent record of the canopy or overstory and, thus, it permits comparisons of various stages over the course of its development. The computerized analysis of fisheye images provides estimates of several canopy parameters, including the following: percent angular canopy openness, minutes of direct light, length and duration of sunflecks, and leaf area index (LAI). It has also been shown that the technique is significantly faster than directly collecting the data using light sensors, especially if numerous sites must be measured (Roxburgh and Kelly 1995). The equipment for acquiring and analyzing HP is now relatively affordable. Finally, digital cameras and faster computers have greatly reduced the time that is required to treat images and obtain results.

8.4.2 Areas of Improvement

HP has recently become increasingly popular, particularly because of the aforementioned advantages that have been cited, such as the arrival of new and affordable digital equipment and powerful image analysis tools (Fournier et al. 2017). Nevertheless, several factors must be taken into account when using this technique to estimate light transmittance or canopy cover in forestry applications that are typical of inventory programs. For instance, the main disadvantage of the method is the requirement for suitable environmental conditions for taking the photographs, which is not always the case when field crews are at work. HP can be taken during overcast days to obtain a uniform background. Alternatively, taking the pictures at dusk/dawn for the period before the sun corolla is visible may be suitable. Rainy, bright sunny, or partly cloudy conditions that do not pose major problems for conventional forest inventory operations are not suitable conditions for collecting hemispherical photographs. Unsuitable lighting conditions influence image quality by affecting the contrast between plant/foliage elements and the sky. No easy solution exists to resolve this problem.

Another problem that is often mentioned in the literature is the difficulty in applying a threshold to HP to separate pixels that are associated either with plant/foliage elements or with the sky. Manual thresholding of HP is somewhat difficult because it is, to a certain extent, an arbitrary and subjective operation (Nobis and Hunziker 2005). One way to overcome this difficulty is to apply separate threshold values to sensitive regions within the image when using advanced hemispherical image analysis programs. With the advent of high resolution digital cameras, however, the problem of manual thresholding appears to be less critical since the frequency of mixed pixels is reduced in comparison to the aggregations of pixels that are produced by cameras with lower resolution (Blennow 1995). Moreover, recent research efforts into the development of automatic thresholding algorithms (Ishida 2004; Jonckheere et al. 2004, 2017; Nobis and Hunziker 2005; Mailly et al. 2013) demonstrate that it is now possible to utilize an objective and reproducible method which is applicable to large sets of images. It should also be noted that the arrival of digital photographic equipment has helped overcome some of the limitations of traditional (film) HP (Fournier et al. 2017). This new technology appears to be a much more effective means of estimating the light environment. Still, recent studies suggest that attention must be given to image size and the type of camera that is being used (Inoue et al. 2004), as well as to optimum exposure (Zhang et al. 2005). Consequently, image thresholding should not be a problem, if due diligence in image acquisition under favorable conditions is maintained.

Another problem with HP that is reported in forestry is its relative imprecision under low light conditions, as evidenced in studies that were conducted by Roxburgh and Kelly (1995), Machado and Reich (1999), Frazer et al. (2001) and Hale (2003). It appears that HP performs poorly at transmittance values under 10%, and that values of indirect light under 5% are rarely measured (Hale 2003). For instance, Frazer et al. (2001) observed that consumer-grade digital camera equipment had difficulty discriminating the fine-scale architecture of canopies, especially under closed canopies when openness falls below 10%. They attributed these problems to the presence of image aberrations that interfere with quantitative measurements. Obtaining similar results, Roxburgh and Kelly (1995) had hypothesized that at low light levels (<3%), there is at least one leaf blocking almost every area of sky. In contrast, light sensors have the ability to record some light that is transmitted or reflected through the canopy. Therefore, it appears that HP is very sensitive to error that is attributable to photographic quality such as contrast and threshold analysis (Machado and Reich 1999).

HP, like any other forest inventory measuring technique, has its strengths and weaknesses that must be balanced against other methods prior to its adoption in support of forest inventory and silviculture. The reader should bear in mind that the technique is evolving rapidly, that the equipment is constantly improving and becoming more affordable, and that standardized procedures for data acquisition (Fournier et al. 2017) or for image processing (Jonckheere et al. 2017) do exist. However, there are other ground-based methods for estimating light transmittance or canopy cover for forest inventory and silvicultural purposes, which to some extent provide better alternatives to the use of HP. Some of these methods share the fisheye-view approach for estimating understory light environments or other canopy indices in forests (Sect. 8.5), while others are based on other approaches (see Appendix).

8.5 Other Optical Ground-Based Methods Using the Fisheye Approach

In addition to HP, several other instruments can be used for estimating canopy cover and understory light levels. These include the Plant Canopy Analyzer, Digital Plant Canopy Imager, and the convex spherical densiometer. The reader is referred to review papers (e.g., Lieffers et al. 1999; Ferment et al. 2001; Bréda 2003; Teti and Pike 2005; Fiala et al. 2006; Seidel et al. 2011) and to the Appendix for a comparison and a list of the advantages and disadvantages of the various instruments and methods. Table 8.2 summarizes the main differences between the instruments and methods for measuring canopy cover or light transmittance through forest canopies.

As its name implies, HP uses the fisheye-view approach to estimate understory light environments or other canopy indices in forests. It shares this feature with other instruments such as the Plant Canopy Analyzer (LAI-2000 PCA, now LAI-2200, LI-COR Biosciences, Lincoln, NE, USA), the Digital Plant Canopy Imager (CI-110, CID Bio-Science, Camas, WA, USA), and the Horizontoscope (Institut für Tageslichttechnik, Stuttgart, Germany). Among these instruments, the LAI-2000 is probably the most widely used, especially for leaf area index and diffuse non-interceptance (DIFN) assessments (Welles and Norman 1991). The instrument measures the diffuse sky radiation on five concentric annuli in the ranges 0°–12°, 15°–28°, 31°–43°, 45°–58°, and 61°–74° from zenith (Ferment et al. 2001). It collects the information at 480 nm, which is a light wavelength that provides good contrast between the foliage and sky. The LAI-2000 has been used in several studies comparing its performance with other instruments, such as HP and handheld light sensors (e.g., Comeau et al. 1998; Gendron et al. 1998; Ferment et al. 2001;

Table 8.2 Comparison of selected instruments and methods for measuring canopy cover or light transmittance through forest canopies[a]

Instrument	Field of view	Advantages	Disadvantages	Price range[b]	Source of information
Convex spherical densiometer (CSD)	Conical, about 110° wide centred on zenith, if original procedure is followed	Widely used as a general index of canopy. Compact instrument	Not designed to measure shade. If adapted to estimate ACD, optical quality limits accuracy	*	Lemmon (1956, 1957)
Angular canopy density (ACD)	Sun's path from 1000 to 1400	Can be used to estimate ACD. Provides a clear image	Awkward to carry and use in small streams. Reflection can be dangerous to eyes	*	Brazier and Brown (1973)
Spherical angular canopy density (SACD)	Sun's path from 1000 to 1400	Can be used to estimate ACD and other parameters. Folds flat for easy transport. Provides high quality image	Does not account for shade before 1000 or after 1400	*	Teti (2001)
Moosehorn	Narrow angle around the zenith	Effective for estimating vertical canopy cover	Loses resolution when canopy openness exceeds 30 %	*	Robinson (1947), Garrison (1949)
Vertical tube	~45°	Reliable method for estimating crown projection	Systematic sampling may introduce bias when used in a regularly spaced plantation. Measurements must be performed on days with little or no wind	*	Johansson (1985)
Crown illumination index	~45°	Allows simple visual assessments of tree crown illumination	Poor level of repeatability between observers (subjectivity). Analyses based on this index are constrained to ranking statistics. Loses resolution when canopy openness exceeds 30%	*	Dawkins (1958), Clark and Clark (1992)
Crown illumination ellipse	~45°	Improves both within- and between-observer repeatability when compared to the crown illumination index method	Loses resolution when canopy openness exceeds 30%	*	Brown et al. (2000a)

(continued)

Table 8.2 (continued)

Instrument	Field of view	Advantages	Disadvantages	Price range[b]	Source of information
Canopy-scope	46° (when using a 20 cm cord)	Lower within- and between-observer errors than similar methods. Correlates very well with canopy openness measured with hemispherical photographs. Inexpensive, small and robust	Loses resolution when canopy openness exceeds 30%	*	Brown et al. (2000a)
Multiple light sensors		Can provide precise measurement of below- and above-canopy light	Ideally, two units are needed for below- and above-canopy measurements, one attached to a data logger	***[c] **[d] ****[e]	Accupar[c]; Line Quantum Sensor[d]; Sunscan[e]
Plant Canopy Analyzer (LAI PCA 2200)	150° (adjustable with view caps)	Can provide objective measure of cosine-weighted canopy density in the field of view	Ideally, two units are needed for below- and above-canopy measurements. Measurements should be made preferably in diffuse radiation or under a clear sky at sunset or sunrise	****	Welles and Norman (1991)[f]
Digital Plant Canopy Imager (CI-110)	150°	Takes and processes coloured hemispherical images in real-time in the field. The lens is mounted on an auto-leveling device	No image can be captured at ground level (minimum is 10 cm high). Maximum resolution is 768 × 494 pixels	****	Digital Plant Canopy Imager[g]; Bréda (2003)
Horizontoscope or Solar Pathfinder	Sun's path from sunrise to sunset	Can be used to estimate effective shade if used in combination with solar data	Unmirrored plastic dome provides poor canopy image under some lighting conditions	*	Schütz and Brang (1995)[h], Swenson and Beilfuss (2001)[i]

(continued)

Table 8.2 (continued)

Instrument	Field of view	Advantages	Disadvantages	Price range[b]	Source of information
Hemispherical photography	180°	Can be used to calculate any canopy density parameter. Provides permanent record. More accurate than ocular methods	Relatively high capital cost. Most difficult to use in the field. Skill and time required to analyze images. Measurements should be made preferably in diffuse radiation or under a clear sky at sunset or sunrise	**	For a review of equipment, see Fournier et al. (2017)
Terrestrial 3D laser scanner (TLS)	Omni-directional	Can provide lots of information at high resolution. Provides a permanent record	Very high capital cost. More or less still in development stage. Data analysis more tedious	****	Seidel et al. (2011, 2012), Seidel and Ammer (2014), Stanley (2013)[j]

*below $500, this and all subsequent price quotes are in 2014 US dollars
**between $500 and $2000
***between $2000 and $5000
****over $5000
[a]Adapted and expanded from Bréda (2003), Teti and Pike (2005), and Fiala et al. (2006)
[b]Price range
[c]AccuPAR, Decagon Devices, Pullman, WA, USA (http://www.decagon.com/products/canopy-atmosphere/light/lp-80-ceptometer-leaf-area-index-lai-fpar/), price ~$3500 US
[d]Line Quantum Sensor, LI-COR Biosciences, Lincoln, NE, USA (http://www.licor.com/env/products/light/quantum_sensors/index.html), price ~$2200 US
[e]Sunscan, Delta-T Devices Ltd, Cambridge, UK (http://www.delta-t.co.uk/product-display.asp?id=944&div=Plant%20Science&cat=Canopy%20Analysis), ~$6800 US for standard kit (SS1-STD3)
[f]Plant Canopy Analyzer LAI-2200, LI-COR Biosciences, Lincoln, NE, USA (http://www.licor.com/env/products/leaf_area/LAI-2200/price ~$8900 US
[g]CI-110 Imager, CID Bio-Science Inc, Camas, WA, USA (http://www.cid-inc.com/index.php/plant-canopy-analysis/ci-110), price ~$7600 US
[h]Institut für Tageslichttechnik, Stuttgart, Germany (http://www.horizontoscop.com/eng/index_eng.html)
[i]Solar Pathfinder, Linden, TN, USA (http://www.solarpathfinder.com/PF-TC.html), price ~$300 US
[j]FARO Focus^{3D} Laser scanner (http://www.faro.com/products/3d-surveying/laser-scanner-faro-focus-3d/overview), price ~$50,000 US

Bréda 2003). As is often the case with light transmittance measuring systems, the LAI-2000 provides reliable estimates of ground-based effective leaf area index (L_e), especially when two cross-calibrated units are synchronized to a data logger, with one devoted to above-canopy measurements and the other to below canopy measurements (Bréda 2003).

The Horizontoscope is an ingenious instrument of European origin that was designed to estimate the potential duration of sunlight (Schütz and Brang 1995; Brang 1998). It has the advantage of being small and simple enough to be used as a true decision-making device in the field. Measurements that are made with the Horizontoscope were found to be highly correlated ($r \sim 0.93$–0.95) and had very low bias when compared to corresponding hemispherical photographs (Schütz and Brang 1995; cf. also Swenson and Beilfuss 2001). The Horizontoscope and a nearly identical North American instrument called the Solar Pathfinder (Solar Pathfinder, Linden, TN, USA) have been used with some success in studies characterizing the light environment of seedlings (Brang 1998; Swenson and Beilfuss 2001; Brang et al. 2005) and in studies surveying stream shading (Teti and Pike 2005).

The Digital Plant Canopy Imager (CI-100, CI-110, CID Bio-Science, Camas, WA, USA) is a light intensity measuring device that is equipped with a self-leveling 150° fisheye lens, which is positioned at the end of a 0.8 m probe (Peper and McPherson 2003). The instrument can be used in either sunny or cloudy weather, and it has the advantage of processing color hemispherical images in real-time in the field (Bréda 2003). It comes with software that computes leaf area index, mean foliage inclination angle, foliage distribution, extinction coefficient of the canopy, and the sky view factor. The instrument has recently been used in comparative studies for the determination of the leaf area index of open-grown deciduous trees in urban forests (Peper and McPherson 1998, 2003). The results from these studies show a tendency for the CI-100 Digital Plant Canopy Imager to underestimate LAI of isolated trees.

Terrestrial laser scanners (TLS) have recently been used and tested for providing reliable gap fraction data (Seidel et al. 2011). Laser scanner sensors measure distances between the sensor emitting the laser pulse and the intercepted surface, thus returning the 3D coordinate of the position of that point on the surface (Pirotti et al. 2012). They are judged to be suitable for under canopy measurements and long term monitoring, measuring and data extraction of forest based data (Stanley 2013). Seidel and Ammer (2014) studied short rotation stands (*Populus* sp.) in Germany and concluded that TLS could be applied in the future to support monitoring growth developments in densely stocked stands. Seidel et al. (2012) analyzed forest canopies with ground-based laser scanning and compared TLS with hemispherical photography. They concluded that TLS in mature forests can overcome several of the methodological problems inherent to conventional canopy analysis with optical methods and thus may soon offer a promising tool for functional research in complex forest canopies.

As can be seen from this short review of other optical ground-based methods that are based on the fisheye approach, each one of these instruments has its own strengths and limitations. Some of them are better equipped for measuring L_e

(e.g., LAI-2000 PCA); others, such as the Digital Plant Canopy Imager, have a limited field of view (150°) compared to HP (generally set at 180°); some instruments have limited use that is related to the potential duration of sunlight (e.g., Horizontoscope). Users working in forest inventory or silviculture who are interested in a more versatile and relatively affordable instrument that can estimate both canopy cover and light transmittance should consider HP.

8.6 Conclusions and Recommendations

Examples of applications that have been illustrated in this review show that HP can be successfully used in support of forest inventory and silviculture. HP encompasses several ecological—and forestry-oriented measures that are already part of some inventory programs. The additional canopy variables that can be extracted from a single hemispherical image may also provide useful information for forest inventory programs, especially today when new information is desired but hindered by financial constraints. The HP technique is an efficient way to obtain rapid estimates of canopy openness or light transmittance in the field, and it has several advantages when compared to other methods. The increasing popularity of HP is due to several factors including recent advances in digital photography, affordability of the equipment, availability of excellent hemispherical image analysis programs, and development of highly effective automatic thresholding algorithms.

Despite these advantages, there are a number of issues of which the reader should be aware when considering whether to use HP or not. The environmental conditions for collecting HP data of adequate quality are generally limited to completely overcast days or to short periods of time at dusk or dawn. Furthermore, manual thresholding of digital images that is based on user experience and knowledge of field conditions should be replaced by automatic thresholding algorithms that are embedded into image analysis programs. These algorithms could help to reduce the bias and repetitiveness of tasks when compared to traditional light measurements using quantum sensors. The quality of affordable fisheye lenses should be checked carefully for the exact FOV and distortion. In addition, each lens should be calibrated using appropriate methods (e.g., Herbert 1987; Clark and Follin 1988). Reports of the relative imprecision of HP under low light conditions (e.g., <10%) should be taken into account when selecting a technique for measuring light transmittance under dense canopies. Depending upon the specific objective of the research project or inventory activity, HP may or may not be the optimal technique. Other types of instruments that are more or less similar to HP are available and, as a consequence, forest managers who are interested in canopy measurements for inventory or silvicultural purposes must select the best overall instrumentation or technique to accomplish a particular task. For instance, instruments such as the Horizontoscope or Solar Pathfinder (see Appendix) are portable, accurate, less expensive than HP equipment, and appear to be a viable alternative if one wants to obtain less canopy information than could be derived from HP

(Swenson and Beilfuss 2001). Clearly, there is room for further comparative studies between HP and other instruments that are used for estimating canopy cover and light transmittance.

Improvements in digital photography equipment have made it possible to carry much more compact cameras and fisheye lenses into the field. This is an important point since some of the most important characteristics of any forest inventory measuring device are its ruggedness and portability. In fact, the portable character of such an instrument, its affordability, and its ability to provide instantaneous readings appear to be crucial features when it comes to forest inventory programs. The CI-110 Digital Plant Canopy Imager provides instantaneous readings of fisheye images, but the apparatus is more expensive than entry-level digital HP and its FOV is limited to 150°. In contrast, a FOV of 180° can be obtained with standard HP equipment, but canopy images are not processed in the field and so direct readings are not yet possible. Even though HP is undoubtedly the most precise method that is currently available (Brown et al. 2000a), the user has to find ways of maintaining the equipment in good working order during prolonged field use. Until an affordable handheld electronic device capable of capturing and analyzing a 180° hemispherical image and displaying the results almost instantaneously in the field is developed, HP remains the most versatile, complete, and scalable technique for estimating canopy cover and light transmittance in support of forest inventory and silviculture.

Acknowledgements I thank Drs. Richard Fournier and Ron Hall for their helpful suggestions prior to the official submission and during the review and editing stages of the manuscript. Thanks are also due to Ms. Debra Christiansen-Stowe and Ms. Catherine Brown for editing the text. The comments made by two anonymous reviewers were helpful in improving the quality of the manuscript and are also acknowledged.

Appendix: Instruments not Based on the Fisheye Approach

In the main part of this chapter, HP and other optical ground-based methods using the fisheye approach were presented as potential techniques in support of forest inventory and silviculture for estimating canopy cover and light transmittance in forests. As has been shown, the use of HP for forest inventory and silvicultural applications is relatively new, but promising results can be obtained if the proper equipment is used and the technique is applied correctly. Before the advent of HP, however, several other techniques and instruments were available for estimating canopy cover and light transmittance in forests. The goal of this Appendix is to provide the reader with an overview of the main techniques and instruments that are not based on the fisheye approach and which may represent a valid alternative, if less detailed information is required about the structure and light regime of a forest canopy.

One of the most popular of these devices is the *convex spherical densiometer*, which was initially described by Lemmon (1956). This pocket-sized instrument measures canopy cover. It is designed for rugged field use and is comprised of a convex mirror divided into a cross-shaped grid of 24 squares. It must be held out in front of the body, about waist level, so that a clear view is given of the forest canopy without the observer shading the grid. Canopy cover is then estimated by counting the number of squares filled by vegetation. Baudry et al. (2014) found that densiometer readings were closely related to the relative light intensity assessed in overcast conditions, and suggested that the use of this inexpensive tool should be expanded. Conversely, some researchers have reported inconsistent results while using the convex spherical densiometer (e.g., Vales and Bunnell 1988; Cook et al. 1995). Despite these criticisms, convex spherical densiometers have remained a tool of choice because other cost-effective measurement devices are lacking (Forest Science Project 2000). However, according to Englund et al. (2000), spherical densiometers can be successfully used to characterize forest light environments as long as users have adequate experience and are trained to a consistent standard.

The *angular densiometer* is another instrument that is used to measure canopy cover. It consists of a hinged clamshell case containing a convex mirror, a bubble level, a magnetic compass, and a fixed eyesight. It is similar in design to a convex spherical densiometer but differs in that it provides a delineation of the sun path, a more accurate orientation relative to the celestial hemisphere, and a clearer image of the canopy (Teti 2001). It measures the percentage of time that shade is available during a 4-h period between 10:00 and 14:00, which is a period of time that have been judged to be critical for regulating stream temperature (Teti and Pike 2005).

The *vertical tube* is another device that is based on a similar principle for measuring canopy cover (Johansson 1985; Korhonen and Heikkinen 2009). It consists of a 20 cm-long and 1 cm-wide hand-held brass tube, which is mounted on a universal joint to make it hang vertically during operation. The vertical tube (or Cajanus tube) has not been widely used, but Johansson (1985) reported that the instrument is a reliable method for the measurement of crown projection, provided that the number of points sampled is adequate.

A *Moosehorn* is an instrument for measuring crown cover that was developed by Robinson (1947), and which was later redesigned by Garrison (1949). The instrument uses 25 dots in a 5 × 5 square array painted on a transparent screen and was designed to be very simple to use. Correct use of the instrument requires several steps: (i) holding the instrument in a vertical position, (ii) looking through a side aperture, and (iii) counting the number of dots that are covered by tree crowns. According to Fiala et al. (2006), the instrument is particularly well suited for rapid, efficient estimates of vertically projected canopy cover. However, Brown et al. (2000a) found the moosehorn to be cumbersome and fragile, and ended up redesigning it with several improvements.

The *canopy-scope* is an instrument that is used for the rapid assessment of canopy openness and it is relatively inexpensive, robust, accurate, and repeatable (Brown et al. 2000a). It is made of a transparent Perspex screen with a 20 cm cord attached to one corner. The cord allows the user to maintain the instrument at the same distance from the eye. As with the moosehorn, the screen of the canopy-scope is engraved with 25 dots in a 5 × 5 square array. The instrument is portable, robust, inexpensive to construct, and small enough to fit in a jacket pocket. Eight to ten measurements in a 0.25 ha plot seem to be sufficient for representing average canopy openness (Hale and Brown 2005). In comparison to other instruments, the canopy-scope was found to have lower within—and between-observer error than several other methods; it also had the highest correlation with canopy openness that was measured using hemispherical photographs (Brown et al. 2000a). For rapid assessments of canopy openness, the canopy-scope is the best instrument available when considering its low price, robustness, speed, and accuracy (Brown et al. 2000a).

The *crown illumination index* is a method for visually assessing the crown illumination of trees and was devised by Dawkins in 1958. Clark and Clark (1992) later modified the crown illumination index by subdividing class 2 (lateral light) into high, medium, and low lateral light. This widely used index allows the user to score the source and relative amount of light reaching a tree crown. It covers trees that do not receive any direct light either vertically or laterally (class 1) to trees that have completely exposed crowns (class 5; Clark and Clark 1992, p. 318). To overcome some of the limitations that are associated with using subjective indices such as the crown illumination index, a method based on crown illumination ellipses was developed. In this approach, a hole in the canopy is assessed by comparing it with a series of ellipses printed on a transparent Perspex screen (Brown et al. 2000a). As is the case with the canopy-scope, the instrument is held away from the observer with a 20 cm cord that is attached to the screen; the canopy cover can then be determined, based on the size of the ellipse that fits into the largest canopy opening that can be found in the canopy.

Single-point and *handheld light sensors* (e.g., AccuPAR, Decagon Devices, Pullman, WA, or Line Quantum Sensor, LI-COR Biosciences, Lincoln, NE) have long been used to estimate light transmittance through forest canopies (Pearcy 1989; Comeau 2000). Relative light transmittance can be calculated using two sensors. One is installed on a tower or in a large opening to record open sky light levels using a datalogger, and a second is used for light measurements in the understory. These measurements require clear-sky or completely overcast conditions, and variable conditions should be avoided.

Although the instruments that are described above for measuring canopy cover or canopy openness are convenient in the sense that they are simpler to use and technically less sophisticated than the equipment used in HP, they also have some drawbacks. For example, according to Brown et al. (2000a), resolution decreases for the crown illumination index, crown illumination ellipses, moosehorn, and canopy-scope when canopy openness exceeds 30%. Also, many of the aforementioned instruments have been shown to have noticeable within—and between-observer errors, although the canopy-scope is less prone to error and correlates well with canopy openness that is measured with HP (Brown et al. 2000a). Conversely, light measuring instruments such as single-point and handheld light sensors, while performing similarly to HP in various light environments

(Gendron et al. 1998), only measure light transmittance; as such, they are less versatile than HP for characterizing canopy structure in forested environments.

References

Astrup R, Larson BC (2006) Regional variability of species-specific crown openness for aspen and spruce in western boreal Canada. For Ecol Manag 228:241–250

Avery TE, Burkhart HE (2002) Forest measurements, 5th edn. McGraw-Hill, New York, 456 p

Battaglia MA, Mou P, Palik B, Mitchell RJ (2002) The effect of spatially variable overstory on the understory light environment of an open-canopied longleaf pine forest. Can J For Res 32:1984–1991

Baudry O, Charmetant C, Collet C, Ponette Q (2014) Estimating light climate in forest with the convex densiometer: operator effect, geometry and relation to diffuse light. Eur J For Res 133:101–110

Beaudet M, Messier C, Canham CD (2002) Predictions of understorey light conditions in northern hardwood forests following parameterization, sensitivity analysis, and tests of the SORTIE light model. For Ecol Manag 165:235–248

Bitterlich W (1948) Die Winkelzählprobe. Allg Forst und Holzwirtsch Zeitung 59:4–5

Blennow K (1995) Sky view factors from high-resolution fish-eye lens photographic negatives. J Atmos Ocean Tech 12:1357–1362

Blennow K (1998) Modelling minimum air temperatures in partially and clear felled forests. Agric For Meteorol 91:223–235

Brandeis TJ, Newton M, Cole E (2001) A comparison of overstory density measures for describing understory conifer growth. For Ecol Manag 152:149–157

Brang P (1998) Early seedling establishment of *Picea abies* in small forest gaps in the Swiss Alps. Can J For Res 28:626–639

Brang P, von Felten S, Wagner S (2005) Morning, noon, or afternoon: does timing of direct radiation influence the growth of *Picea abies* seedlings in mountain forests? Ann For Sci 62:697–705

Brazier JR, Brown GW (1973) Buffer strips for stream temperature control. Research Paper 15, Forest Research Laboratory, School of Forestry, Oregon State University, 9 pp

Bréda NJJ (2003) Ground-based measurements of leaf area index: a review of methods, instruments and current controversies. J Exp Bot 54:2403–2417

Brown N, Jennings S, Wheeler P, Nabe-Nielsen J (2000a) An improved method for the rapid assessment of forest understorey light environments. J Appl Ecol 37:1044–1053

Brown PL, Doley D, Keenan RJ (2000b) Estimating tree crown dimensions using digital analysis of vertical photographs. Agric For Meteorol 100:199–212

Canfield RH (1941) Application of the line interception method in sampling range vegetation. J Forest 39:388–394

Canham CD, Coates KD, Bartemucci P, Quaglia S (1999) Measurement and modeling of spatially explicit variation in light transmission through interior cedar-hemlock forests of British Columbia. Can J For Res 29:1775–1783

Chan SS, Walstad JD (1987) Correlations between overtopping vegetation and development of Douglas-fir saplings in the Oregon Coast Range. West J Appl For 2:117–119

Chan SS, McCreight RW, Walstad JD, Spies TA (1986) Evaluating forest vegetative cover with computerized analysis of fisheye photographs. For Sci 32:1085–1091

Chapman L (2007) Potential applications of near infra-red hemispherical imagery in forest environments. Agric For Meteorol 143:151–156

Clark DA, Clark DB (1992) Life history diversity of canopy and emergent trees in a neotropical rain forest. Ecol Monogr 62:315–344

Clark JA, Follin GM (1988) A simple "equal area" calibration for fisheye photography. Agric For Meteorol 44:19–25

Comeau PG (2000) Measuring light in the forest. BC ministry of forests, research branch, Victoria, BC. Extension Note 42, 7 p

Comeau PG, Gendron F, Letchford T (1998) A comparison of several methods for estimating light under a paper birch mixedwood stand. Can J For Res 28:1843–1850

Cook JG, Stutzman TW, Bowers CW, Brenner KA, Irwin LL (1995) Spherical densiometers produce biased estimates of forest canopy cover. Wildlife Soc Bull 23:711–717

Curtis RO, Marshall DD (2005) Permanent-plot procedures for silvicultural and yield research. USDA Forest Service, Pacific Northwest Research Station, Portland, OR. General Technical Report PNW-GTR-634, 86 pp

Dawkins HC (1958) The management of natural tropical high-forest with special reference to Uganda. Imperial Forestry Institute, University of Oxford, Oxford, UK. IFI Paper 34, 155 p

Decourt N (1956) Utilisation de la photographie pour mesurer les surfaces terrières. Rev For Fr 8:505–507

DeLong DL, Simard SW, Comeau PG, Dykstra PR, Mitchell SJ (2005) Survival and growth response of seedlings in root disease infected partial cuts in the Interior Cedar Hemlock zone of southeastern British Columbia. For Ecol Manag 206:365–379

Ducrey M, Bartoli M (1974) Utilisation des photographies hémisphériques pour le cubage des arbres sur pied. Ann Sci For 31:229–248

Englund SR, O'Brien JJ, Clark DB (2000) Evaluation of digital and film hemispherical photography and spherical densiometry for measuring forest light environments. Can J For Res 30:1999–2005

Evans GC, Coombe DE (1959) Hemispherical and woodland canopy photography and the light climate. J Ecol 47:103–113

Ferment A, Picard N, Gourlet-Fleury S, Baraloto C (2001) A comparison of five indirect methods for characterizing the light environment in a tropical forest. Ann For Sci 58:877–891

Fiala ACS, Garman SL, Gray AN (2006) Comparison of five canopy cover estimation techniques in the western Oregon Cascades. For Ecol Manag 232:188–197

Forest Science Project (2000) A fish-eye view of riparian canopy. FSP Technical Notes, vol III. Arcata, California

Fournier RA, Landry R, August NM, Fedosejevs G, Gauthier RP (1996) Modelling light obstruction in three conifer forests using hemispherical photography and fine tree architecture. Agric For Meteorol 82:47–72

Fournier RA, Mailly D, Walter J-MN, Jonckheere IGC (2017) Acquiring hemispherical photographs in forest environment: From planning to archiving photographs. In: Fournier RA, Hall RJ (eds) Hemispherical photography in forest science: theory, methods, applications. Springer, Berlin

Frazer GW, Fournier RA, Trofymow JA, Hall RJ (2001) A comparison of digital and film fisheye photography for analysis of forest canopy structure and gap light transmission. Agric For Meteorol 109:249–263

Garrison GA (1949) Uses and modifications for "moosehorn" crown closure estimation. J Forest 47:733–735

Gendron F, Messier C, Comeau P (1998) Comparison of various methods for estimating the mean growing season percent photosynthetic photon flux density in forests. Agric For Meteorol 92:55–70

Gougeon FA, Leckie DG (2003) Forest information extraction from high resolution images using an individual tree crown approach. Natural Resources Canada, Canadian Forest Service, Pacific Forestry Centre, Victoria, BC. Information Report BC-X-396. 27 p

Grosenbaugh LR (1952) Plotless timber estimates—new, fast, easy. J Forest 50:32–37

Hale SE (2001) Light regime beneath Sitka spruce plantations in northern Britain: preliminary results. For Ecol Manag 151:61–66

Hale SE (2003) The effect of thinning intensity on the below-canopy light environment in a Sitka spruce plantation. For Ecol Manag 179:341–349

Hale SE, Brown N (2005) Use of the canopy-scope for assessing canopy openness in plantation forests. Forestry 78:365–371

Herbert TJ (1987) Area projections of fisheye photographic lenses. Agric For Meteorol 39:215–223

Hu L, Zhu J (2008) Improving gap light index (GLI) to quickly calculate gap coordinates. Can J For Res 38:2337–2347

Inoue A, Yamamoto K, Mizoue N, Kawahara Y (2004) Effects of image quality, size and camera type on forest light environment estimates using digital hemispherical photography. Agric For Meteorol 126:89–97

Ishida M (2004) Automatic thresholding for digital hemispherical photography. Can J For Res 34:2208–2216. Software:RGBFisheye.exe available at: http://www1.gifu-u.ac.jp/~ishidam/RGBFisheye.htm

Jahnke LS, Lawrence DB (1965) Influence of photosynthetic crown structure on potential productivity of vegetation, based primarily on mathematical models. Ecology 46:319–326

Jennings SB, Brown ND, Sheil D (1999) Assessing forest canopies and understorey illumination: canopy closure, canopy cover and other measures. Forestry 72:59–74

Johansson T (1985) Estimating canopy density by the vertical tube method. For Ecol Manag 11:139–144

Jonckheere I, Fleck S, Nackaerts K, Muys B, Coppin P, Weiss M, Baret F (2004) Methods for leaf area index determination. Part I: theories, sensors and hemispherical photography. Agric For Meteorol 121:19–35

Jonckheere IGC, Macfarlane C, Walter J-MN (2017) Image analysis of hemispherical photographs, algorithms and calculation. In: Fournier RA, Hall RJ (eds) Hemispherical photography in forest science: theory, methods, applications. Springer, Berlin

Korhonen L, Heikkinen J (2009) Automated analysis of in situ canopy images for the estimation of forest canopy cover. For Sci 55:323–334

Kuusipalo J (1985) On the use of tree stand parameters in estimating light conditions below the canopy. Silva Fenn 19:185–196

Lacinová L, Zboja V (1985) The use of photographic technique for estimating the light regime within forest stands. Ecológia (CSSR) 4:377–386

Lakso AN (1976) Characterizing apple tree canopies by fisheye photography. HortScience 11:404–405

Lemmon PE (1956) A spherical densiometer for estimating forest overstory density. For Sci 2:314–320

Lemmon PE (1957) A new instrument for measuring forest overstory density. J For 55:667–669

Lhotka JM, Loewenstein EF (2008) Influence of canopy structure on the survival and growth of underplanted seedlings. New Forest 35:89–104

Lieffers VJ, Messier C, Stadt KJ, Gendron F, Comeau PG (1999) Predicting and managing light in the understory of boreal forests. Can J For Res 29:796–811

Machado J-L, Reich PB (1999) Evaluation of several measures of canopy openness as predictors of photosynthetic photon flux density in deeply shaded conifer-dominated forest understory. Can J For Res 29:1438–1444

Mailly D, Kimmins JP (1997) Growth of *Pseudotsuga menziesii* and *Tsuga heterophylla* seedlings along a light gradient: resource allocation and morphological acclimation. Can J Bot 75:1424–1435

Mailly D, Turbis S, Chazdon RL (2013) SOLARCALC 7.0: an enhanced version of a program for the analysis of hemispherical canopy photographs. Comput Electron Agric 97:15–20

Maltamo M, Hyyppä J, Malinen J (2006) A comparative study of the use of laser scanner data and field measurements in the prediction of crown height in boreal forests. Scand J For Res 21:231–238

Martins SV, Rodrigues RR (2002) Gap-phase regeneration in a semideciduous mesophytic forest, south-eastern Brazil. Plant Ecol 163:51–62

McIntosh ACS, Gray AN, Garman SL (2012) Estimating canopy cover from standard forest inventory measurements in western Oregon. Forest Science 58:154–167

Mitchell PL, Whitmore TC (1993) Use of hemispherical photographs in forest ecology. Oxford Forestry Institute, Department of Plant Sciences, University of Oxford. OFI Occasional Papers No. 44, 39 pp

Nobis M, Hunziker U (2005) Automatic thresholding for hemispherical canopy-photographs based on edge detection. Agric For Meteorol 128:243–250

North M, Chen J, Oakley B, Song B, Rudnicki M, Gray A, Innes J (2004) Forest stand structure and pattern of old-growth western hemlock/Douglas-fir and mixed-conifer forests. For Sci 50:299–311

Paletto A, Tosi V (2009) Forest canopy cover and canopy closure: comparison of assessment techniques. Eur J For Res 128:265–272

Pardé J, Bouchon J (1988) Dendrométrie, 2nd edn. Engref, Nancy

Pearcy RW (1989) Radiation and light measurements. In: Pearcy RW, Ehleringer J, Mooney HA, Rundel PW (eds) Plant physiological ecology: field methods and instrumentation. Chapman and Hall, New York, pp 95–116

Peper PJ, McPherson EG (1998) Comparison of five methods for estimating leaf area index of open-grown deciduous trees. J Arboric 24:98–111

Peper PJ, McPherson EG (2003) Evaluation of four methods for estimating leaf area of isolated trees. Urban For Urban Gree 2:19–29

Pirotti F, Grigolato S, Lingua E, Sitzia T, Tarolli P (2012) Laser scanner applications in forest and environmental sciences. Ital J Remote Sens 44:109–123

Puerta-Piñero C, Gómez JM, Valladares F (2007) Irradiance and oak seedling survival and growth in a heterogeneous environment. For Ecol Manag 242:462–469

Riaño D, Valladares F, Condés S, Chuvieco E (2004) Estimation of leaf area index and covered ground from airborne laser scanner (LIDAR) in two contrasting forests. Agric For Meteorol 124:269–275

Rich PM (1990) Characterizing plant canopies with hemispherical photographs. In: Goel NS, Norman JM (eds) Instrumentation for studying vegetation canopies for remote sensing in optical and thermal infrared regions. Remote Sens Rev 5:13–29

Robinson MW (1947) An instrument to measure forest crown cover. For Chron 23:222–225

Roxburgh JR, Kelly D (1995) Uses and limitations of hemspherical photography for estimating forest light environments. N Z J Ecol 19:213–217

Salminen R, Nilson T, Hari P, Kaipiainen L, Ross J (1983) A comparison of different methods for measuring the canopy light regime. J Appl Ecol 20:897–904

Schütz JP, Brang P (1995) L’horizontoscope: un étonnant outil pratique de sylviculture, notamment en haute montagne. Office National des Forêts, Bulletin Technique 28

Seidel D, Ammer C (2014) Efficient measurements of basal area in short rotation forests based on terrestrial laser scanning under special consideration of shadowing. iForest (early view). doi:10.3832/ifor1084-007

Seidel D, Fleck S, Leuschner C, Hammett T (2011) Review of ground-based methods to measure the distribution of biomass in forest canopies. Ann For Sci 68:225–244

Seidel D, Fleck S, Leuschner C (2012) Analyzing forest canopies with ground-based laser scanning: A comparison with hemispherical photography. Agric For Meteorol 154:1–8

Smith DM (1986) The practice of silviculture, 8th edn. Wiley, New York

Stadt KJ, Lieffers VJ (2000) MIXLIGHT: a flexible light transmission model for mixed-species forest stands. Agric For Meteorol 102:235–252

Stanley T (2013) Assessment of the FARO 3D focus laser scanner for forest inventory. BSc thesis, Faculty of Health, Engineering and Sciences, University of Southern Queensland, Toowooba, QLD, Australia, 86 p. http://eprints.usq.edu.au/id/eprint/24708

Swenson S, Beilfuss R (2001) Calculating light levels for savanna and woodland restorations. Ecol Restor 19:161–164

ter Steege H (1997) WINPHOT 5: a programme to analyze vegetation indices, light, and light quality from hemispherical photographs. Tropenbos-Guyana Programme and Utrecht University, Utrecht, 52 pp

Teti PA (2001) A new instrument for measuring shade provided by overhead vegetation. BC Ministry of Forests, Cariboo Forest Region, Extension Note 34, Williams Lake, BC, 8 p
Teti PA, Pike RG (2005) Selecting and testing an instrument for surveying stream shade. BC J Ecosyst Manag 6:1–17
Turner IM (1990) Tree seedling growth and survival in a Malaysian rain forest. Biotropica 22:146–154
Vales DJ, Bunnell FL (1988) Relationships between transmission of solar radiation and coniferous forest stand characteristics. Agric For Meteorol 43:201–223
van Gardingen PR, Jackson GE, Hernandez-Daumas S, Russell G, Sharp L (1999) Leaf area index estimates obtained for clumped canopies using hemispherical photography. Agric For Meteorol 94:243–257
van Leeuwen M, Hilker T, Coops NC, Frazer G, Wulder MA, Newnham GJ, Culvenor DS (2011) Assessment of standing wood and fiber quality using ground and airborne laser scanning: a review. For Ecol Manag 261:1467–1478
Welles JM, Norman JM (1991) Instrument for indirect measurement of canopy architecture. Agron J 83:818–825
Wilkinson DM (1995) Modelling tree crowns as geometric solids. Arboric J 19:387–393
Zhang Y, Chen JM, Miller JR (2005) Determining digital hemispherical photograph exposure for leaf area index estimation. Agric For Meteorol 133:166–181
Zhu J, Matsuzaki T, Lee F, Gonda Y (2003) Effect of gap size created by thinning on seedling emergency, survival and establishment in a coastal pine forest. For Ecol Manag 182:339–354

Chapter 9
Canopy Architectural Models in Support of Methods Using Hemispherical Photography

Jean-François Côté, Richard A. Fournier and Michel M. Verstraete

Abstract One of the most important limitations regarding the in situ measurement of forest canopies is the relative lack of knowledge of canopy structure. Yet the three-dimensional (3D) arrangement of canopy components is difficult to measure because of the significant spatial and temporal heterogeneity of canopy structure over a wide range of scales. In this context, models of plant and tree canopies represent our current simplified understanding of canopy structure, and provide a tool to analyze and interpret direct as well as indirect measurements. This chapter first reviews some of the architectural modeling approaches for simulating forest canopies and the extent to which they can account for the spatial variability present in hemispherical photographs (HPs). Then, a novel architectural model is described that is capable of explicitly representing conifer tree structure up to the level of individual branches, and which is called *VoxTreK* (*Vox*el *Tre*e with 3D *K*ites). Its ability to simulate HP is assessed through a comparative analysis with three other canopy architecture models of increasing complexity: from a simple turbid medium to detailed representation of individual shoots. Our comparative analysis of the four models consisted of simulations of forest canopy structure, reconstructions of several coniferous stands, and computer-generated simulations of actual HPs that were taken in forest stands. Results demonstrate that the *VoxTreK* model is capable of reproducing the zenith gap fraction distributions of conifer forest canopies with

---

J.-F. Côté (✉)
Natural Resources Canada, Canadian Forest Service, Canadian Wood Fibre Centre, 1055 Du P.E.P.S. Street, P.O. Box 10380, Québec, Québec G1V 4C7, Canada
e-mail: jean-francois.cote@canada.ca

R.A. Fournier
Centre d'applications et de Recherches en Télédétection (CARTEL), Département de Géomatique Appliquée, Université de Sherbrooke, Sherbrooke, QC J1K 2R1, Canada
e-mail: Richard.Fournier@USherbrooke.ca

M.M. Verstraete
Global Change and Sustainability Research Institute, The University of the Witwatersrand, Johannesburg, South Africa
e-mail: mmverstraete@gmail.com

R.A. Fournier and R.J. Hall (eds.), *Hemispherical Photography in Forest Science: Theory, Methods, Applications*, Managing Forest Ecosystems 28,
DOI 10.1007/978-94-024-1098-3_9

high correlations ($r \geq 0.92$) between the simulated and in situ HPs. They also suggest that *VoxTreK* can describe the canopy structure with greater detail than crown envelope, with accuracy comparable to a model including individual coniferous shoots for HP applications. Last, recent trends in architectural modeling are reviewed, thereby providing a summary of the state of the art in modeling canopy structure as well as outlining current applications and future developments.

Keywords Architectural model · Forest canopy structure · Gap distribution · Hemispherical photography · Ray tracing simulation

9.1 Introduction

9.1.1 Canopy Architectural Models and Hemispherical Photography

Canopy structure refers to the three-dimensional (3D) organization of forest vegetation components in space that is quantified by measurements of canopy element position, orientation, quantity, type, and connectivity (Ross 1981; Norman and Campbell 1989). It partially controls the exchanges of radiation, heat, moisture, carbon and other nutrient elements between the atmosphere and ecosystems, and exists over a wide range of spatial and temporal scales. Canopy structure also can control microclimates and ecological niches (Parker 1995). Physiological responses of the forest canopy depend upon canopy structure, which both promotes and regulates its own structural development as part of a bio-climatic feedback system, which in turn affects the growth, survival and reproduction of plants. Poor characterization of canopy structure can thus hinder our understanding of the processes and feedbacks controlling plant physiology and their interactions with climate.

The complexity of canopy structure greatly limits accurate retrieval of forest attributes, as well as identification of the spatial composition and structure of forest landscapes or patterns (Turner 1989). This is particularly the case of satellite remote sensing data interpretation, given that the typical spatial resolution of most of these instruments is generally too coarse (e.g., Landsat TM, 30 m; MISR, 250 m, nadir camera) to detect, let alone characterize, the structural properties of individual trees at all appropriate scales. While fieldwork constitutes a critical source of information, acquiring suitable datasets of ground measurements is often prohibitively expensive and time-consuming. Conversely, when adequate field-derived data are available, the accuracy with which remote sensing data are interpreted can be greatly improved. Nevertheless, issues of scale, vegetation phenology, data availability, and suitability of the spectral bands must be carefully weighed and addressed when analyzing space-borne data sources. Advances in technology have consistently improved the spatial resolution of satellite-based remote sensing systems; currently, operational systems deliver data with spatial resolutions ranging from about 1 m to several km. The lack of quantitative information about 3D canopy structure becomes a particularly critical limiting factor in the interpretation of remote sensing data when the horizontal spatial resolution of these sensors becomes comparable to the typical dimensions of the vegetation components and the gaps between them (e.g., Widlowski et al. 2008).

Architectural models that are applied in remote sensing have been developed to aid in the interpretation of imagery by establishing relationships between canopy structure and its reflectance, as observed by remote sensing instruments. These models represent our understanding of radiation transfer in vegetated environments and can be used to interpret passive (e.g., solar radiation; Pinty and Verstraete 1991) or active (e.g., synthetic aperture radar; Sun and Ranson 1998) measurements. They can be especially used to evaluate the roles of canopy structural elements in various radiative processes. With respect to forest remote sensing, models are widely used to study the light regime within the canopy (e.g., Fournier et al. 1996; van Leeuwen et al. 2013), to measure biophysical parameters (e.g., Goel 1988; Myneni et al. 1995; Disney et al. 2006), to simulate radiative transfer (e.g., Govaerts and Verstraete 1998), and to study ecological and climate processes (e.g., Sun and Ranson 1998). Three-dimensional canopy models are thus expected to improve reflectance calculations for vegetated areas, especially when fine spatial resolution is required, but their complexity (measured by the very large number of internal parameters) increases the need for detailed empirical characterizations of canopy structure.

Hemispherical photography (HP) is a field-based remote sensing method that characterizes the spatial distribution and geometric properties of canopy elements (e.g., Nilson and Ross 1979; Chen and Black 1991; Andrieu et al. 1994; Chen and Cihlar 1995; Chen et al. 1997; Planchais and Pontailler 1999; White et al. 2000; Law et al. 2001; Leblanc et al. 2002; Fournier et al. 2003; Jonckheere et al. 2004). HP analysis relies upon inversion methods to describe a complex 3D situation that is observed through a two-dimensional (2D) window (Leblanc and Fournier 2017). Specifically, the gap fraction that is observed in 2D is interpreted in terms of a spatial distribution of foliage, which yields an estimate of the Leaf area index (LAI), as well as information on the clumping of canopy elements (Walter et al. 2003; Leblanc et al. 2005; Govind et al. 2013; Frazer et al. 2017). HP thus provides a useful basis for the development of 3D canopy architecture models.

9.1.2 Modeling Forest Canopy Structure

The architectural models that are currently available provide canopy representations over a range of scales and resolutions for various applications (e.g., forest remote sensing, ecology). They can be divided into four model classes: (i) turbid medium; (ii) geometric crowns; (iii) hybrid; and (iv) morphological. Each one requires accurate environmental datasets at the appropriate scale and resolution to specify plant structures. In the case of remote sensing, these models document the structure of canopies for the purposes of simulating the transfer of radiation. All reflectance models assume, implicitly or explicitly, the structural properties of the medium. Thus, early architectural models addressed this need but aimed at describing only the broad features of canopies at coarse scales. They were based upon 'turbid medium' theory (Ross 1981), where all foliage elements are assumed to be randomly dispersed in a horizontally infinite homogeneous layer. These were (and still are) adequate for analyzing remote sensing data that have been collected at resolutions coarser than 100 m, where the horizontal radiation contribution is negligible relative to the measured reflectance signature of a target pixel (Widlowski et al. 2008). However, technological advances in measurement hardware and computer performance, together with progress in developing more sophisticated retrieval algorithms, have progressively led to more refined models that are capable of representing the structure of individual trees, down to the shoot-level. These more detailed descriptions of canopy (and individual tree) structure not only provide an explicit representation of the structure for (3D) radiative transfer models, but also improve our understanding of plant development and ecological processes at a finer spatial scale.

Turbid medium models (e.g., Ross and Nilson 1966; Verhoef and Bunnik 1981; Verstraete 1987, 1988) represent the broad features of radiation transfer in plant canopies. The latter are idealized as an infinite cloud of many dimensionless scattering particles that are homogeneously distributed within a semi-infinite medium, possibly layered, with a top surface that is subjected to solar irradiance (Fig. 9.1a). These models rely upon a limited number of key structural variables (e.g., particle density and scattering properties) to compute how leaves interact with radiation. This type of model was later expanded to account, first, for the effect of a limited number of scattering elements of finite dimension (e.g., Verstraete et al. 1990; Pinty et al. 1990), which is known as the "hot spot" and, second, for the impact of canopies of finite depth (e.g., Gobron et al. 1997), in which case the main variables are LAI and the leaf normal distribution function that describes the orientation of the leaves' normal vectors. A different approach is required to account for significant gaps or clumping of plant elements in the canopy; canopy 3D structure must then be either parameterized clumping (e.g., Pisek et al. 2011) or described more explicitly (e.g., Kimes and Kirchner 1982; Gastellu-Etchegorry et al. 1996).

Geometric models constitute the next level of complexity. In this case, the forest canopy is considered an assemblage of individual plants (trees), each characterized by a distinct volume, which is typically a simple geometrical shape such as a sphere, a cone, or a cylinder. These volumes, in turn, may contain a turbid medium and, therefore, be characterized by properties such as transmittance, reflectance, and absorbance (Fig. 9.1b). For example, Horn (1971) proposed a geometric model that was later developed by Koop (1989), and which defined crown shape with three parameters: absolute dimension, width/height ratio, and convexity. Other models use mathematical functions or surfaces of revolution that are generated by 2D curves to represent individual crowns, including conical (Li and Strahler 1985), cylindrical (Pacala et al. 1993), or ellipsoidal shapes (Nilson and Peterson 1991). Different shapes can be used simultaneously within a scene to model different

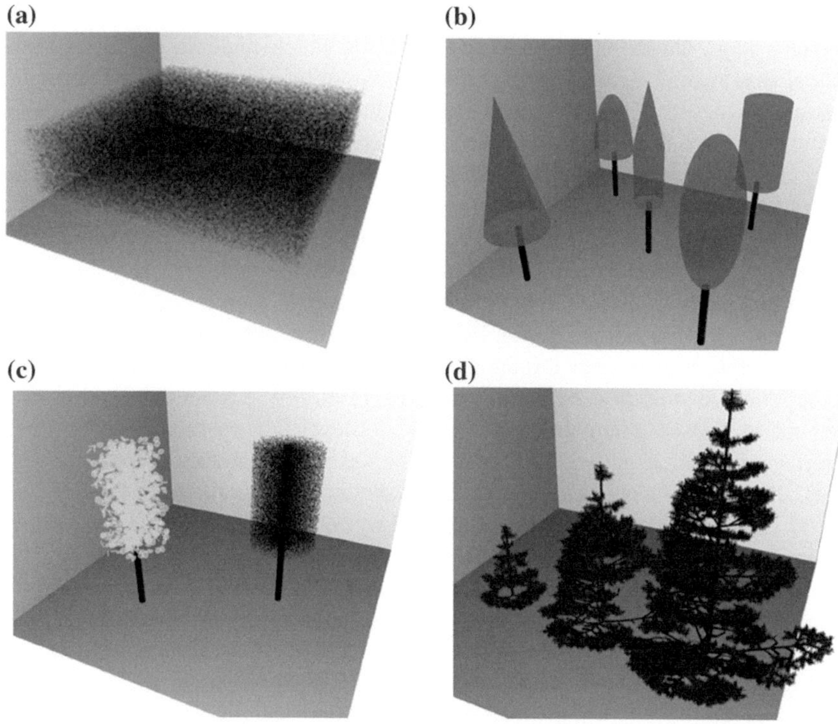

Fig. 9.1 Model representations of forest canopies: **a** infinite horizontal layer that is treated as a gas of scattering point particles (or a turbid medium). **b** Geometric shapes that are used to represent individual tree crowns. **c** Hybrid model combining the approaches of (**a**) and (**b**). **d** Morphological models that describe structure down to the smallest canopy element

species (St–Onge et al. 1991) or are joined together to form more complex and realistic crown shapes (Chen and Leblanc 1997; Stadt and Lieffers 2000). Convex polyhedrons have been used to adjust the natural profile of tree crowns (Cescatti 1997). Geometric models are suitable for sparse canopies where multiple scattering and mutual shadowing within the crowns can be ignored. However, they still assume a uniform spatial distribution of foliage material within the crowns. This assumption can be relaxed by using more complex models, which are referred to as *hybrid* models in the context of this chapter.

Hybrid models improve upon geometric models by allowing for more realistic spatial distributions of canopy elements and foliage within individual crowns (Fig. 9.1c), which are treated as either (i) a gas of scattering point particles (turbid medium) or (ii) geometric objects (e.g., disks). Forest canopies have been modeled, for instance, by using cones, cylinders, ellipsoids, and paraboloids to represent the embedding volumes of tree crowns (e.g., Wang and Jarvis 1990). In the first sub-class, the foliage distribution within individual crowns is random or follows a specific leaf area spatial profile (Larsen and Kershaw 1996; Brunner 1998). In the second sub-class, the use of geometric shapes may represent individual foliage elements (e.g., leaves, shoots, needles, etc.) for which multiple interactions and mutual shadowing between canopy elements within crowns are taken into account. In some cases, sets of simple shapes, such as disks, can adequately approximate the spatial distribution of foliage at the crown-level (Borel et al. 1991). *Hybrid* models are more versatile than turbid medium or geometric models, but they cannot describe the complex distribution of material at the branch- or shoot-level. This limitation is addressed with the last class of models, which explicitly simulate all canopy elements, down to individual leaves and needles.

The precise description of the morphological aspects of the vegetation canopy in these detailed, realistic models can be categorized into two broad sub-classes. The first represents tree topology through an attribute dataset and requires measurements of the foliage distribution. Sinoquet and Rivet (1997) have proposed 3D digitizing methods to measure plant structure at the shoot-level. The method simultaneously describes the spatial pattern of branches, the spatial coordinates of plant components, the basal diameter, and the number of shoots. At the same time, Landry et al. (1997) developed the *Vectorization* method to characterize fine tree structure of mature forests. This method permits the statistical reconstruction of a tree in three dimensions by selectively sampling its structural components. A constraint on the applicability of this model sub-class is clearly the availability of in situ datasets that are required to reconstruct every tree.

The second sub-class of morphological models describes plant structure on the basis of methods or algorithms that are derived from such diverse fields as botany, computer graphics, and language theory. Many approaches have been attempted to link knowledge of growth processes and plant genetics to the structural development of plants (Měch and Prusinkiewicz 1996; Perttunen et al. 1996; de Reffye et al. 1997). Such models aim to describe the structure of plants as a response of vegetation canopies to environmental factors (e.g., climate, water and nutrient availability) and ecosystem management (Kurth and Sloboda 1997; de Reffye et al. 1998). For example, the LIGNUM model (Perttunen et al. 1996) considers a tree as a 3D collection of simple units representing tree organs, which grow in direct response to specific metabolic rules. Another example, AMAP (*botAnique et bioinforMatique de l'Architecture des Plantes*, de Reffye 1988; Godin et al. 1999), simulates plant growth by using probabilities that are associated with mortality, pauses in development, ramification, and shoot reiteration. The AMAP model exploits species-specific knowledge regarding plant growth and key environmental variables for the accurate structural reconstruction of vegetation canopies. The effective parameterization of morphological models requires a considerable amount of information, so it may be available only for a limited number of species, although it might be possible at some point to derive generic rules. Growth grammars[1] can

[1]Parallel rewriting systems that consist of a set of modules, each representing plant components, a collection of production rules which expand each module at every step of the simulation into some larger string of modules, an initial state from which the simulation begins, and a mechanism for translating the modules that are generated into geometric structures.

serve as a formal framework for representing and encoding plant structure, and simulating their growth with production rules that are applied to shoot segments (Prusinkiewicz and Lindenmayer 1990; Kurth 1994, 2000). Lindenmayer systems, or L-Systems, are a particular case of growth grammars that encode plant topological structure in a string of symbols. Subsequent developments have improved the formalism of parameterization (Prusinkiewicz and Lindenmayer 1990) and the effects of environmental factors (Prusinkiewicz et al. 1994). More recently, Open L-System formalism has increased the complexity and realism of modeled structures, together with the interactions between the plant and its environment (Měch and Prusinkiewicz 1996). Growth grammars with a relatively limited knowledge of functional processes can be used to simulate realistic plant structure (Fig. 9.1d), but reproducing realistic plant canopies in natural environments, such as forests, remains difficult due to the numerous factors influencing the final structure of plants. As a simplification, growth and yield models (e.g., Tree and Stand Simulator [TASS], Mitchell 1975) simulate crowns of individual trees to expand and contract asymmetrically in a 3D growing space in response to internal growth processes and the physical restrictions that are imposed by competitors. However, these models do not explicitly focus upon descriptions of the individual tree structure.

Selecting the most suitable architectural model for inclusion in a reflectance model or for generating a synthetic image ultimately depends upon the requirements of the application in terms of its accuracy and level of detail. Nevertheless, all models that aim to simulate the structure of individual trees or forest canopies face the following challenges: (i) significant spatial and temporal variability over a wide range of scales; (ii) the need for a considerable number of field measurements to represent the geometry and the physiological properties of plants; and (iii) practical constraints concerning the acquisition of measurements for mature trees in natural forests (Fournier et al. 1997). HP offers a simple and convenient means of acquiring data on some aspects of plant canopy structure. Although limited in complex canopies, HP can assess canopy structure over the entire hemisphere and demands few hardware resources, when compared to other techniques such as terrestrial LiDAR (Seidel et al. 2012). Fournier et al. (1997) suggest that a 3D architectural model, which provides details up to the branch-level, should be capable of reproducing hemispherical photographs that have been acquired in coniferous forest. It is not known whether this level of detail is required for this purpose, and what advantages this type of model may offer compared to models with a coarser or finer level of detail. Intuitively, it is often assumed that the finest level of detail should lead to the best results, but this assumption needs to be evaluated against the costs of achieving this goal. In practice, we must establish which level of detail provides the best compromise between the modeling effort and the performance that is required for the application at hand. The following comparative analysis provides some initial answers to this question.

9.1.3 Objectives

This chapter explores to what extent a novel canopy architectural model, *VoxTreK*, which is capable of describing canopy structure up to the branch-level, can replicate an HP and how it might be used to retrieve the structural parameters of a stand, once it is constrained by HP observations. Specifically, this study assessed the benefits and limitations of *VoxTreK* canopy models relative to three other extant models with different levels of detail. These three models have been configured specifically to simulate coniferous forest canopies with an explicit characterization of (i) forest cover (mono-layer model), (ii) tree crown model, and (iii) segment-shoot-level model (*Vectorization*), respectively. The first two models have used a coarser level of detail compared to *VoxTreK* and correspond to turbid medium and hybrid models, respectively. The most detailed model uses the *Vectorization* architectural model, which has been described by Landry et al. (1997). The following two specific objectives are addressed:

1. To develop and evaluate an architectural model at the branch level, which we have called *VoxTreK* (*Vox*el *Tre*e with 3D *K*ites).
2. To simulate HPs for several test sites using the four architectural models for a comparative analysis, with an emphasis on *VoxTreK* model performance.

The relative performance of *VoxTreK* was evaluated through a comparison of in situ and simulated HPs for four test sites, since more direct validation was problematic given the availability of data, time and practical/logistical constraints. This also allowed us to determine how the level of detail of a given architectural model influenced its performance in terms of simulating realistic HPs of coniferous forest canopies.

9.2 Materials and Methods

9.2.1 Site Description

Four sites were selected to represent a variety of conifer canopy types and species for the evaluation of the proposed architectural model: closed and clumped (ECOLEAP-FM), open and clumped (BOREAS NSA-OBS; SSA-OJP), and closed and random (Le Bray). These sites have been extensively characterized by other researchers (Table 9.1).

The deciduous forest site of the ECOLEAP project in Forêt Montmorency (**ECOLEAP-FM**), which was dominated by balsam fir (*Abies balsamea* [L.] Miller), had been selected previously for a study of structural measurements (Ung et al. 2001) and for testing the applicability of ecophysiological models (Bernier et al. 1999). The project is located within the Université Laval experimental forest,

Table 9.1 Main characteristics of the test sites

Validation site characteristics	ECOLEAP-FM	BOREAS NSA-OBS	BOREAS SSA-OJP	Le Bray
Cover type	Coniferous	Coniferous	Coniferous	Coniferous
Canopy type	Closed and clumped	Open and clumped	Open and clumped	Closed and random
Location	Québec, Canada	Manitoba, Canada	Saskatchewan, Canada	Bordeaux, France
Dominant species	91% balsam fir; 9% paper birch	Black spruce interspersed with tamarack	Jack pine	Maritime pine
Central longitude	71°06′00″W	98°28′53″W	104°41′31″W	00°46′00″ W
Central latitude	47°19′00″N	55°52′48″N	53°54′58″N	44°42′00″ N
Mean elevation (m)	800	629	579	60
Number of trees	49	642	551	2711
Number of HPs	7	7	10	10
Total area (m^2)	437	2400	5400	42,920
Basal area (m^2/ha)	44.5	30.0	33.0	34.2
Stand density (stems/ha)	1121	2537	1019	621
Mean DBH (cm)	20.0	10.5	11.9	26.0
Mean tree height (m)	15	11	13	18
Slope (°)	15–25 NW	<3	<3	0

which is 100 km north of Québec City, within the Boreal Shield ecozone (Ecological Stratification Working Group 1995). Average slope angle is 15°.

Two coniferous sites from the BOREAS experiment (Sellers et al. 1995), which is in the Boreal Shield ecozone of central Canada, were also selected: the Northern Study Area of Old Black Spruce (**BOREAS NSA-OBS**) and the Southern Study Area of Old Jack Pine (**BOREAS SSA-OJP**). BOREAS NSA-OBS is located about 50 km from Thompson, Manitoba, and is composed almost exclusively of black spruce (*Picea mariana* [Miller] BSP), interspersed with tamarack or eastern larch (*Larix laricina* [Du Roi] K. Koch). The terrain is essentially flat (altitude variations <15 m). BOREAS SSA-OJP is located about 70 km from Prince Albert, Saskatchewan, in stands that were composed exclusively of jack pine (*Pinus banksiana* Lambert), with essentially flat terrain (slope <3°).

The fourth site, **Le Bray** (Porté et al. 2000), is located 20 km southwest of Bordeaux, France, and is a Mediterranean plantation of maritime pine (*Pinus pinaster* Aiton). The climate is Mediterranean-montane, and the site is characterized by flat relief.

In situ measurements were taken during previous, independent field campaigns at each test site and included, for each selected tree on each site: species identification, tree location within a cartesian coordinate system (*x, y*), trunk diameter at breast height (DBH, 1.3 m), and hierarchical height class (i.e., dominant, co-dominant, intermediate, or suppressed). Several allometric relationships were established from stratified sampling, including (1) tree height as a function of DBH (Fournier et al. 1997; Bernier et al. 1999) and (2) the leaf area of branches as a function of basal diameter (Porté and Lousteau 1997; Ung and Melanson 1998 personal communication; Laroche and Dagneault 2000). Destructive sampling of a small number of trees at each test site was required to explicitly describe tree structural parameters for use in the proposed architectural model: five trees at ECOLEAP-FM, four at BOREAS NSA-OBS, three at BOREAS SSA-OJP, and four at Le Bray. For each sample tree, the position and dimensions of all branches were measured. Trees were selected from all relevant hierarchical ecological classes. A series of HPs was acquired along transects during the field campaigns: seven HPs at ECOLEAP-FM, seven at BOREAS NSA-OBS, ten at BOREAS SSA-OJP, and ten at Le Bray. All photographs were taken with a traditional Nikon F-2 camera (Nikon Corporation, Tokyo, Japan), which was equipped with a Nikkor 8 mm fisheye lens. The equipment provided an image with near-perfect equiangular projection, did not require lens correction (Herbert 1987; Rich 1990), and used 400 ASA black-and-white film. The photos were digitized with the Kodak Photo CD system (Eastman Kodak, Rochester, NY).

9.2.2 Architectural Models for the Forest Canopy

This section provides an in-depth description of the four architectural models with levels of detail ranging from the forest cover to tree crowns, branches, and segment-shoots. The models were evaluated and compared through HP simulations with a specialized ray-tracing algorithm (*ULooPS*, Fournier et al. 1996), which only supports forest scene descriptions in voxel format. Therefore, all reconstructed trees and forest canopies were discretized in voxel space (or *voxelized*) to be compatible with the HP simulator. This representation consisted of dividing the virtual (forest or tree) scene into identical 3D cubic cells, or *voxels*, where each voxel contained a value for the material density. *ULooPS* takes advantage of this representation to decide whether a pixel becomes a gap or not using a stochastic process, which is based upon calculating a probability of occlusion (see Sect. 9.2.3). Each test site was simulated by using in situ measurements of tree attributes, together with their individual positions within the stand. Voxel representation offered a practical means of describing the heterogeneous biophysical properties of the medium (vegetation) without considering the intricate arrangement of fine-scale canopy components.

9.2.2.1 Mono-Layer Model

Coarse modeling at the forest cover-level, hereafter called *Layer*, was a turbid medium approach where all foliar surfaces at the test site were distributed within a homogeneous horizontal layer of vegetation. This layer was represented as a rectangular prism extending over the entire forest canopy. The average density of the foliar surfaces from the catalog of all trees that were modeled for the site was used as the density value in the *Layer* model. The thickness of the *Layer* model was calculated as the average of all live crown heights for the simulated forest site. The *voxelization* process was straightforward and involved division of the rectangular prism into voxels of predefined dimensions.

9.2.2.2 Tree Crown Model

Two architectural models were used to simulate the canopy—*(Layers)* and crown-levels of detail for the ECOLEAP-FM site. The modeling approach that was employed at the tree crown-level, hereafter called *Ellipsoids*, was a hybrid approach using a geometric shape, within which the canopy components were distributed and considered to be a turbid medium. The foliar surface was uniformly distributed within an ellipsoidal crown containing random gap inclusions. Tree crown dimensions and the quantity of foliar surface that was present in the crown were identical to those used in the branch-level model. *Voxelization* checked which voxels, approximated by their center points, intersected with the ellipsoidal crown.

9.2.2.3 Segments-Shoots-Level Model *(Vectorization)*

The *Vectorization* method (Landry et al. 1997) is the most detailed architectural model of the four that were considered and was developed to provide local estimates of wood and foliage biomass for an individual tree. It generates a statistical 3D reconstruction of a tree at the level of wood segments and shoots from a selective sampling of its structural components. A sampling design was proposed for obtaining spatial information for the branches and wood segments. It involved data acquisition at four levels: trunk inventory, branching structure, foliage distribution, and the dimensions and arrangements of the leaves. The result was a 3D reconstruction of the branches and tree, where: (i) the spatial variation of wood segments was preserved, and (ii) no assumptions were made regarding branch spatial arrangement. The 3D reconstruction was not perfect since similarity principles were used to rebuild whole branches from sampled portions. However, this reconstruction procedure was evaluated in Fournier et al. (1996), which resulted in a 3D representation of the fine structure of forest trees with a known spatial distribution of their components. A *Vectorization* dataset was available for only one of our four test sites, viz., SSA-OJP. For this site, one jack pine tree, which was located just beyond the plot map, was selected for simulation with the *Vectorization* model (see Fournier et al. 1996 for a complete description).

9.2.2.4 Branch-Level Model (*VoxTreK*)

VoxTreK was developed to simulate the structure of a forest canopy by using geometric and biometric tree attributes that were measured in situ at the branch-level (Côté 2006). *VoxTreK* is more detailed than hybrid models but less laborious to construct than morphological models that explicitly represent the shoot-level.

Tree reconstruction using *VoxTreK* proceeded in 3 steps: in situ measurements, branch generation, and *voxelization*. A set of trees from each dominance class (dominant, co-dominant, intermediate, and suppressed) was selected for in situ measurements. DBH was measured and magnetic north was marked to locate the branch positions on the trunk before felling the tree. Each branch was described by its basal diameter, elevation angle (θ), azimuth angle (φ), length of the major axis $|A_{major}|$, width of the minor axis $|A_{minor}|$, and distance between the point of attachment to the trunk and maximal width (d_{mw}) (Fig. 9.2). Total foliar surface for each branch geometric unit (S) was calculated with the allometric relationship: $S = aD^b$, where the basal diameter of the branch (D) was measured in situ and the

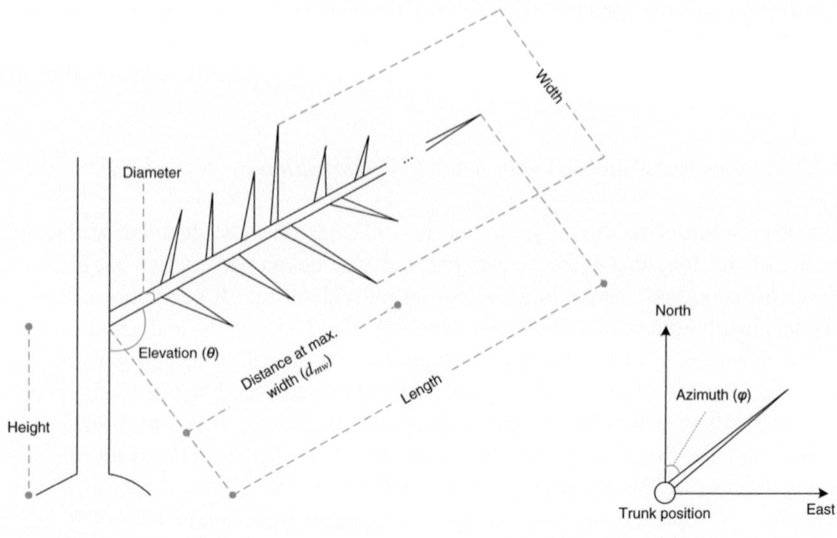

Fig. 9.2 Measured branch parameters that were used for modeling individual trees. [*Left*] Side view of a schematic representation of a tree. [*Right*] Top view of a schematic representation of a tree

coefficients a and b were estimated by regression. This allometric relationship parallels what is frequently used to describe the correlation between tree volume and DBH (Avery and Burkhart 1983, p. 103). We assumed that it also applied to branches and that the total foliar surface of a branch was linked to its estimated volume. This functional dependence is also supported by pipe model theory (Mäkelä 1986) and is based upon a vigor index, using water and nutrient transport considerations to explain shoot growth within tree crowns (Nikinmaa et al. 2003). A 3D kite shape (Fig. 9.3) was used to represent each branch because of its similarity to the shape of coniferous branches. Overall, we adopted an asymmetrical distribution of foliage that simulates what is normally found on a tree. Leaves are distributed according to mechanical constraints while also seeking the location that best captures incoming solar radiation. This approach for simulating spatial density of the foliage distribution is subjective but follows common sense. It differs from a purely random simulation of foliage elements within a kite, an approach which was discarded after preliminary tests led to poor results. Branch thickness varied from zero to *thk* on segment AO, while there was a constant value *thk* for segment OB (Fig. 9.3). Foliage was distributed on the branches following (i) a truncated normal distribution along the minor axis of the 3D kite, and (ii) a truncated Weibull distribution (Weibull 1951) along the major axis of the 3D kite (Fig. 9.4). The normal and Weibull distributions covered the domains $[-|A_{minor}|/2, |A_{minor}|/2]$ and $[0, |A_{major}|]$, respectively. This configuration distributed most of the foliage towards

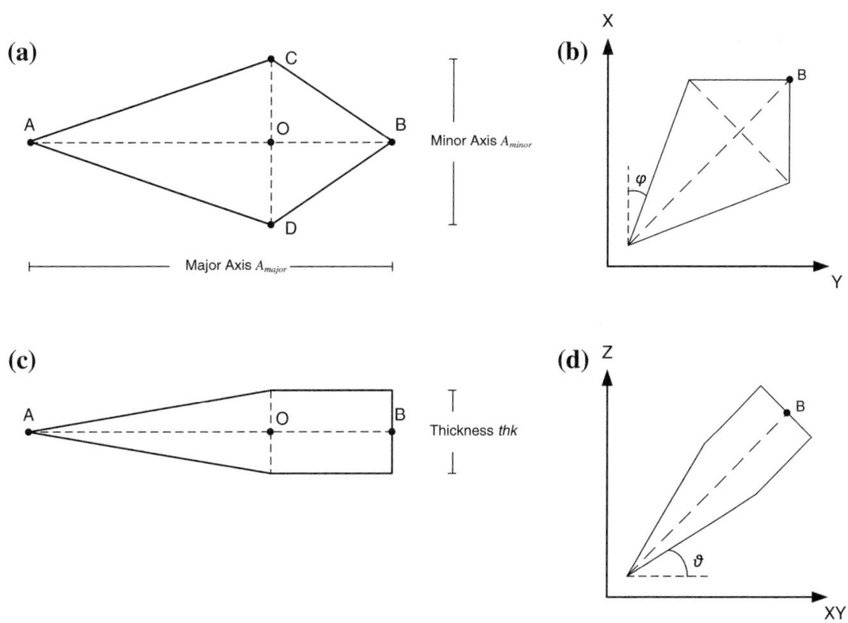

Fig. 9.3 The 3D kite representation of a branch that is used in the *VoxTreK* model: **a** and **b** *top view*, **c** and **d** *side view*

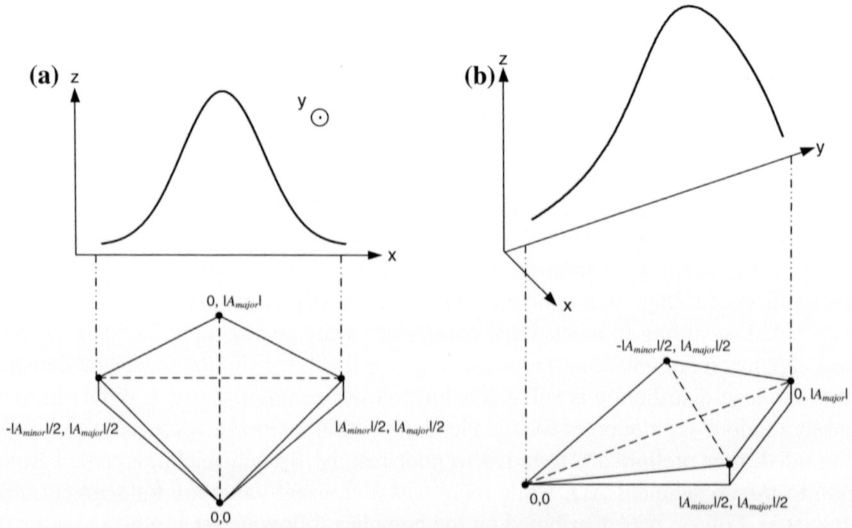

Fig. 9.4 The foliar surface is distributed in a 3D kite following **a** a normal distribution superposed over the minor axis and **b** a Weibull distribution superposed over the major axis

the branch tip. The local foliage area corresponding to the parameter values for both distributions depended upon the branch dimensions and the overall shape of the distributions. In practice, both distributions were truncated to eliminate 30% of values at the extremes of the functions, which avoided assigning excessively small amounts of foliage to the branch extremities. This renormalization realistically distributed the foliage along the simulated branches. The foliage surface distribution across branch thickness *e* of the 3D kite was assumed to be constant for the sake of simplicity.

A catalog of modeled trees was produced by using *VoxTreK* with trees in all dominance classes that were represented, thereby covering the range of tree dimensions at each site: five modeled trees for ECOLEAP-FM, four for BOREAS NSA-OBS, three for BOREAS SSA-OJP, and four for Le Bray. Another catalog was created for the 3D matrices of voxels for all trees that were simulated, where each voxel contained a foliar surface density value. The *voxelization* process transferred the foliar surface that was simulated for each branch in the *VoxTreK* 3D kites to the 3D matrix of voxels for each tree in the catalog. *Voxelization* calculated the foliar surface density for each voxel (Fig. 9.5). To do so, its algorithm identified which voxels intersected with all of the modeled branches of a tree. An intersection test was performed with the *(x, y, z)*-positions of the center and eight vertices of each voxel. The foliage probability value P_f was computed for all branch-voxel intersections according to the foliage density at this location in the 3D kite, as was defined by the truncated normal and Weibull distributions:

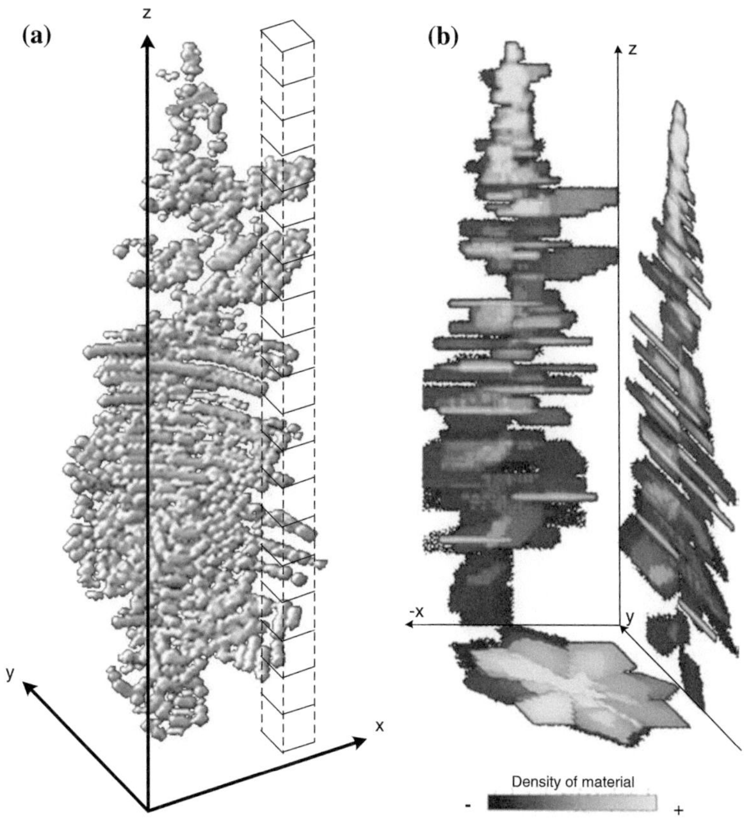

Fig. 9.5 Example of the voxel representation of a tree. **a** The tree is modeled as a 3D matrix of voxels, where each voxel contains the foliar surface density. Voxel size is not to scale. **b** The projection of the density of material (foliar surface) over the *xy-plane*, *xz-plane*, and *yz-planes*

$$P_f = W(p_x) * G(p_y) \tag{9.1}$$

where *W(t)* and *G(t)* are the respective Weibull and normal probability distributions that were evaluated at *t* and p_x, p_y is the *(x, y)*-position of the voxel's center or vertex that intersected with the branch. The foliage density value was altered, as follows, by introducing gaps to increase the realism of the modeled branches. The gap probability P_g was applied to each voxel that was influenced by foliage, where P_g was assigned a random value between 0 and the maximum possible value of P_f for the 3D kite. P_f was calculated according to the global maximum of the normal and Weibull distributions over their respective domains. If $P_g > P_f$ was satisfied for all nine intersection points of a voxel, its associated mean probability of foliage $\left(\overline{P_f}\right)$ was set to zero. Then, the total foliar surface that was allocated to the branch was redistributed to the other intercepted voxels for which the foliage density had been allocated. When all voxels intersecting a branch were identified, *voxelization* distributed the total quantity of foliar surface for the branch to the remaining voxels, each weighted by their respective mean probability value $\overline{P}_f$. This ensured that all leaf area was distributed among the voxels, according to the branch distribution. The same process was repeated for all branches until the entire tree model was included in the matrix *M*. The application of *VoxTreK* resulted in the simulation of all test sites, thereby providing the density of foliar material for all voxels.

9.2.3 Simulation of HPs with ULooPS

A ray-tracing algorithm was required to simulate HPs from simulated forest stands, using a 3D matrix of voxels as the scene representation. The simulation process modeled the voxel content as a turbid medium; voxel dimensions could be adjusted according to the spatial scale of interest, thereby limiting memory usage. All trees were placed in the simulated stand according to their original stand locations and recorded attributes. The ray tracer that was developed for this purpose, *ULooPS* (*Upward Looking Photography Simulator*, adapted from Fournier et al. 1996), was used (i) to reproduce the camera position that was used for HP acquisition, (ii) to simulate the technical specifications of the fisheye lens and camera system, (iii) to use the exact positions of the trees in each test site, (iv) to identify the DBH of each tree and, accordingly, (v) to place the 3D matrix of voxels that provided foliage densities at every location within the scene. *ULooPS* launched rays from the focal point of the fisheye lens. The geometric configuration of the camera-fisheye system was replicated, using height above the ground in the system (focal point), azimuth orientation, and lens viewshed. The central pixel of all simulated HPs corresponded to the zenith angle 0°, and the field of view (FOV) covered the entire hemisphere. The (*x*, *y*)-position of the camera was specified using the same reference system that was used to locate the trees. Tree heights were obtained from available species-specific allometric relationships, which related height to DBH. These relationships were calculated from regressions using in situ measurements from the test sites (Table 9.2).

Tree crowns were represented by 3D matrices of voxels that had been generated by the four architectural models. Reconstruction of the simulated forest site within *ULooPS* was performed by placing a modeled tree at all tree locations in the scene that had been identified from field measurements. For each (*x*, *y*)-position, a tree model was selected from the catalog according to its species and DBH. A random azimuthal rotation was applied to every tree in the virtual forest site to decrease the chance of having repetitive structural patterns in the simulated forest canopy.

Table 9.2 Allometric relationships that were used to calculate branch leaf area (vs. branch diameter D_b) and tree height (vs. *DBH*) for each test site

Validation site	Branch leaf area (m^2)	Tree height (m)
ECOLEAP-FM	$\exp(2.245 \ln(D_b) - 6.294)$ [*RMSE* = 0.34]	$8.9066 \ln(DBH) - 11.087$ [$R^2 = 0.79$]
BOREAS NSA-OBS	$\exp(1.592 \ln(D_b) - 0.5854)$ [*RMSE* = 0.16]	$0.00002(DBH)^3 - 0.041(DBH)^2 + 1.5855(DBH) - 1.0696$ [$R^2 = 0.91$]
BOREAS SSA-OJP	$\exp(1.152 \ln(D_b) - 2.0189)$ [*RMSE* = 0.24]	$0.0018(DBH)^3 - 0.0853(DBH)^2 + 0.6687(DBH) + 0.8551$ [$R^2 = 0.52$]
Le Bray	$N_b(1) = (D_b^2)^{1.205}(0.254\ h + 0.026)^{1.205}$ [*RMSE* = 0.36] $N_b(2) = (D_b^2)^{1.296}(0.159\ h + 0.051)^{1.296}$ [*RMSE* = 0.27] $N_b(3) = 0.133 - 0.153\ N_b(1) + 0.230\ N_b(2)$ [*RMSE* = 0.07]	–

$N_b(i)$ is the leaf area of a branch of age *i*, *RMSE* is the root mean square error and R^2 is the coefficient of determination

Except for the *Layer* model, a trunk was automatically added to the scene at the (*x*, *y*)-position; it was represented by a conical shape that was parameterized using tree DBH and total height. The area outside of the documented test site was populated with a clone of the test site, assuming that the test sites were representative of a larger area. Therefore, the extended virtual forest was used to simulate HPs with the same site statistics (i.e., DBH, species and inter-tree distances, among other variables), which allowed us to reproduce the strong light attenuation that is observed at high zenith angles (near the horizon). The probability of light transmission through the canopy (P_T) was computed for each ray intercepting one or multiple tree crowns, following a Poisson distribution:

$$P_T = \prod_{voxel=1}^{n} \exp\left[-G \cdot S_{voxel} \cdot D_{voxel}\right] \tag{9.2}$$

where *n* is the number of voxels that were intercepted by the ray, $G = (2/\pi)*\sin(\varphi)$ is the projection coefficient for vertical leaves/needles (Ross 1981; Oker-Blom and Kellomäki 1982), $S_{voxel} = z/\cos(\varphi)$ is the path length (m), where *z* is the vertical distance and φ is the zenith angle, and *D* is the density (m^{-1}) of foliar surface material. A uniform random variable *U* between 0 and 1 was generated for each pixel to determine if there was a gap in the canopy. This took into consideration the probability of obstruction: $1 - P_T < U$. A binary image resulted, with values of 1 ($1 - P_T < U$) and 0 (otherwise) for the gaps and interceptions, respectively.

9.2.4 Evaluation of Architectural Models and HP Simulations

The potential benefits and limitations of the four architectural models were assessed by comparing the in situ and simulated HPs for the four test sites. For the evaluation of *VoxTreK*, we used the tree model catalog and the corresponding 3D matrices of voxels for each test site. Models for the four levels of detail were used to simulate HPs with the same scene attributes, i.e., site configuration, model characteristics, and voxel representations. Comparisons of the architectural models allowed us to evaluate their abilities to simulate realistic HPs and, thus, document the specific benefits of increasingly complex canopy representations. The *Layer*, tree crowns, and *VoxTreK* models were used for the four test sites. In addition, the BOREAS SSA-OJP site was also reproduced with the *Vectorization* method. HPs were generated with *ULooPS* for all stations within the test sites where photographs were taken. In situ and simulated HPs from all four modeling approaches were compared by analyzing gap fraction distributions, thereby documenting their performance, strengths, and weaknesses.

The primary physical properties of canopy structure that can be retrieved from HP measurements are the number, size, and shape of 2-D projected gaps in the canopy; these characteristics can indirectly provide information about canopy structure. The statistical distribution of gaps is often used to estimate the LAI of the canopy. Gap fractions (Leblanc and Fournier 2017) were calculated with *Gap Light Analyzer* (GLA) software[2] for zenith angular intervals, or rings, of 5°. A single threshold value was used to segment each HP that was described in Sect. 9.2 and its value was selected based upon the shape of the image histogram. Normally, a histogram of pixel gray levels from an HP would show two major peaks corresponding to opaque objects and a light background. A "gray" zone between these two peaks characterizes the pixels, or *mixels* that contain both the dark objects and light background (usually the sky). The threshold value was selected as the center of this gray zone (Fig. 9.6). Relatively few pixels in all HPs were identified as *mixels* and the resulting gap fraction per ring did not exhibit sensitivity with respect to this threshold value, just as long as it was located near the center of the gray zone.

9.2.5 Parameterization of *VoxTreK*

Parameterization of the *VoxTreK* forest architecture model was based on measurements whenever these were available, and otherwise, was based upon reasonable working assumptions. Parameters requiring attention included (i) the quantity

[2]http://www.rem.sfu.ca/forestry/publications/downloads/gaplightanalyzer.htm.

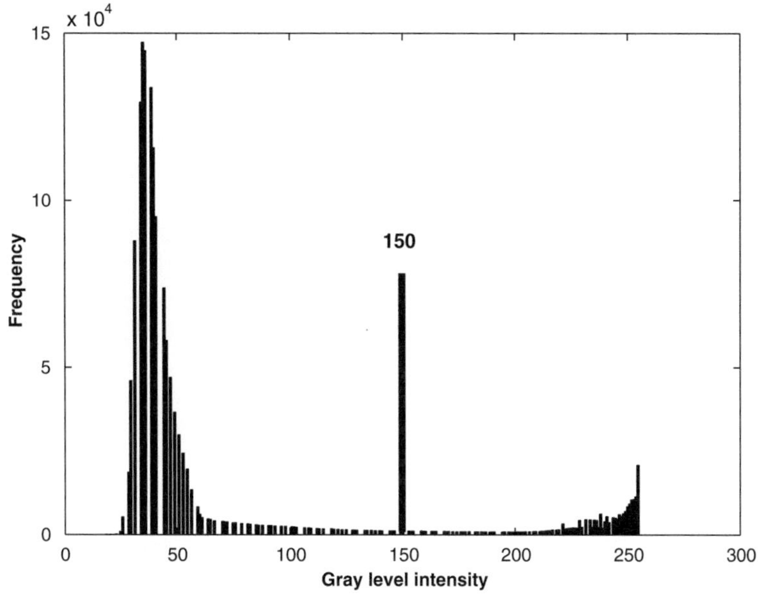

Fig. 9.6 Histogram of light intensity for all of the pixels in one particular HP, and the threshold value of 150 that was used to segment this HP. The *dark peak* (low values) corresponds to opaque objects while the *bright peak* (high values) is generated by the sky background

of foliar surface per branch and tree height, (ii) modeled branch thickness, (iii) voxel size, and (iv) the dimensions of the scene. First, the foliar surface and tree height for the reconstructed tree models were calculated following the allometric relationships that are presented in Table 9.2. An allometric relationship for tree height for the Le Bray site was unavailable, which led us to generate tree heights randomly by considering height values that were measured in situ ranged between 13 and 15 m. Second, the thickness *e* of the 3D kites was set according to simple rules of thumb linking branch thickness to the typical size of foliage tufts or foliage clusters. After comparing different branch thicknesses at each test site, values for *e* of 0.25, 5, 10, and 12 cm were used for the ECOLEAP-FM, BOREAS SSA-OJP, BOREAS NSA-OBS, and Le Bray sites, respectively. Third, the size of a voxel was fixed at 5 × 5 × 5 cm for every test site to maintain a common scale at a fine level of detail in the description of the canopy for the evaluation of the model. It corresponds approximately to the pixel size of a HP when considering a distance of 15 m from the lens focal point and a matrix of 1024 × 1024 pixels. Fourth, the simulation of the test sites required the extension of the spatial limits for the plots where the measurements were taken. The scene that was viewed in each HP encompassed a large area which was centered on the plot. Therefore, we extended the measurement plot to be able to simulate light attenuation by the canopy as seen in HPs at all zenith angles, particularly those close to the horizon. Scene extension was performed by cloning simulated forest plots over a very large spatial extent for three of the test sites: ECOLEAP-FM, BOREAS NSA-OBS, and Le Bray. Site statistics were used to increase the spatial extent of BOREAS SSA-OJP so that this site was consistent with the configuration used in Fournier et al. (1996). In the case of *Vectorization*, the spatial extent that was used in Fournier et al. (1996, 1997) was limited to 160 m × 160 m, which resulted in an undesired exposed area near the horizon (large number of gaps) for zenith angles greater than 75°. We decided to mask these areas in black to remove this artifact from the HPs that were simulated with *Vectorization*. Lastly, in situ position and orientation of the camera for HP acquisition was measured during fieldwork at ECOLEAP-FM, BOREAS NSA-OBS and SSA-OBS. In the case of Le Bray, the exact camera positions were unknown and, consequently, were estimated by triangulation using the spatial configuration of the trees surrounding the camera. With all known model and scene parameters, *VoxTreK* was used to reconstruct each test site and *ULooPS* was used to simulate HPs.

9.3 Results and Interpretation

9.3.1 Simulated Forest Stands and HPs

Several practical issues had to be resolved prior to production of the scenes and the simulation of HPs. For example, the translation from the tree geometric representation (3D kite branch models) to the scene that was composed of voxels (3D matrix of foliage density) resulted in some loss of leaf area (Table 9.3) because of algorithmic considerations. Likewise, the branch-voxel intersection method (Sect. 9.2.2) often neglected branches with dimensions smaller than half of the voxel size, since no voxel vertices or centers intersected the branch geometry. Thus, the 3D matrix of voxels contained a similar, but smaller, quantity of tree material compared to the original tree model that was produced by *VoxTreK*. This effect resulted in an average reduction of about 7% for plant material, with a maximum (worst case scenario) of 20% loss for the test sites. Simulation of the ECOLEAP-FM site yielded the greatest differences because a large number of very small branches were discarded. Compared to average size branches, those that were discarded did not contribute substantially to the intercepted surface that was observed in the simulated HPs. Correcting this anomaly by redistributing the lost leaf area throughout the remaining branches was not an option since the updated value for leaf area per branch would not agree with the allometric relationships that were used in this study. A solution would require precise calculation of the intersected volumes with the voxels, according to their branch geometry, and including gaps. These recalculations would have required significant changes in the computer code, which was beyond the scope of our study. Another modeling issue was the limited number of trees that were available in the tree model catalog for

Table 9.3 Characteristics of the tree models that were used to reconstruct each test site

Test Site	DBH (cm)	Hierarchical height class	Crown volume (m^3)	Leaf area (m^2)	Loss of leaf area (in m^2 and %)
ECOLEAP-FM	11.8	Suppressed	25.3	12.4	1.6 (11%)
	23.7	Intermediate	207.7	111.4	27.8 (20%)
	16.3	Co-dominant	129.9	121.7	18.0 (13%)
	26.2	Co-dominant	231.5	218.3	14.0 (6%)
	29.3	Dominant	210.9	219.4	43.6 (17%)
BOREAS NSA-OBS	7.3	Suppressed	18.8	8.4	0.1 (1%)
	12.9	Co-dominant	38.8	21.5	0.4 (2%)
	11.8	Co-dominant	19.2	25.0	0.2 (1%)
	17.5	Dominant	31.9	25.2	0.3 (1%)
BOREAS SSA-OJP	11.8	Suppressed	43.1	12.3	0.2 (1%)
	12.8	Co-dominant	32.6	13.0	0.6 (5%)
	10.2	Dominant	73.8	13.5	0.2 (2%)
	10.4*	Suppressed	–	18.89	–
	12.5*	Co-dominant	–	38.15	–
	17.9*	Dominant	–	76.61	–
Le Bray	20.5	Suppressed	42.9	46.5	4.9 (10%)
	24.5	Intermediate	82.1	69.4	1.0 (1%)
	33.5		288.0	155.6	6.0 (4%)
	38.3	Dominant	346.7	197.6	6.3 (3%)

The loss of leaf area due to small branches that were neglected in the voxelization process is indicated for each tree model
*Constructed with the *Vectorization* method

representing the entire scene. The main characteristics of all tree models for the test sites are presented in Table 9.3. Nevertheless, structural variety in the tree models from different hierarchical classes, combined with random azimuthal rotation that was applied to each tree to avoid spurious frequencies, was assumed to be adequate for attaining visual realism close to what was observed in the in situ HPs.

Finally, image size was selected to provide detail that was sufficient for estimating structural properties and suitable model representations of canopy structure. HPs were produced using *ULooPS*, with image dimensions of 1024 × 1024 pixels (Figs. 9.7 and 9.8), as described in Fournier et al. (1996). Visual comparisons among the different architectural models are illustrated in Fig. 9.9. Direct evaluation of the architectural model would have required many forest structural attribute measurements, which was impracticable due to several factors, including time, cost and alterations in the original forest stands. Instead, our modeling approach was evaluated by comparing the gap fraction distributions of the in situ and simulated HPs.

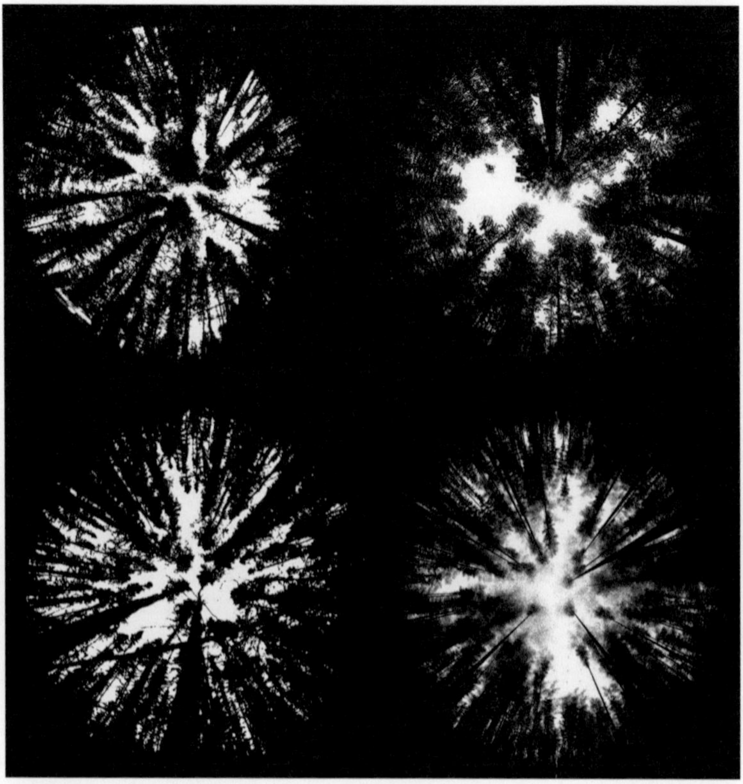

Fig. 9.7 Examples of in situ HPs (*left*) and HPs that were simulated using the *VoxTreK* model (*right*). The HPs were segmented using a single value threshold for the ECOLEAP-FM (*top*) and BOREAS NSA-OBS (*bottom*) sites. Fuzziness that is observed in the simulated photographs is due to the random assignment of gaps during simulations with *ULooPS*

9.3.2 Gap Fraction Analysis

The angular distribution of the gap fractions between zenith angles of 0° and 75° was used to evaluate the capability of the *VoxTreK* model to reproduce canopy structure. Figures 9.10, 9.11, 9.12 and 9.13 show comparisons of the gap fractions that were calculated for zenith angles over the entire hemisphere for in situ and simulated HPs. The estimated standard error *se* is used for error bars in these figures. The gap fraction values that were calculated at every ring between 0° and 75°, and averaged over the HPs in the transect (Table 9.4), $\overline{G_f}$, were estimated for each validation site for both in situ and simulated HP data sets. Mean absolute differences between the in situ values, $\overline{G_f}^{in\ situ}$, and values that were simulated with the *VoxTreK* model, $\overline{G_f}^{sim}$, were 11, 18, 10, and 16% for the ECOLEAP-FM, BOREAS NSA-OBS, BOREAS SSA-OJP, and Le Bray validation sites, respectively. $\overline{G_f}^{sim}$ values that were estimated

Fig. 9.8 Examples of in situ HPs (*left*) and HPs that were simulated using the *VoxTreK* model (*right*). The HPs were segmented using a single value threshold for the Le Bray (*top*) and BOREAS SSA-OJP (*bottom*) sites. Fuzziness that is observed in the simulated photographs is due to the random assignment of gaps during simulations with *ULooPS*

from the *VoxTreK* model show correlations of 0.95, 0.96, 0.96, and 0.92 ($P < 0.0003$) with $\overline{G_f}^{in\ situ}$ values for the ECOLEAP-FM, BOREAS NSA-OBS, BOREAS SSA-OJP, and Le Bray validation sites, respectively (Table 9.4). These results demonstrate that the *VoxTreK* model was capable of reproducing the typical gap fraction distributions $\left(\overline{G_f}\right)$ of complex coniferous canopies that were observed with HPs. The simulated gap fraction values for the rings with zenith angles greater than 70° were outside of the error bars, but the greatest variability was observed in the interval 0°–20° for every validation site. A fisheye lens produces strong visual distortion when creating a hemispherical image (polar projection). As a consequence, objects that are located near the optical axis of the fisheye lens are amplified due to that distortion. The gap fractions that are calculated for the HPs in this particular zone were subject to high variability due to: (i) the natural variation among tree species in the actual forest, since only one species was used for the simulated

Fig. 9.9 Examples of HPs that were simulated showing the gradient of complexity in the four architectural models. *Top* ECOLEAP-FM with the *Layer* (*left*), *Ellipsoid* (*center*) and *VoxTreK* (*right*) model. *Bottom* SSA-OJP with the *Ellipsoid* (*left*), *VoxTreK* (*center*) and *Vectorization* (*right*) model. They correspond to the same in situ HPs that are shown in Figs. 9.7 and 9.8

site; (ii) dead branches and trees were not measured, and, thus were not considered in the simulations; (iii) segmentation error for the HPs; (iv) positioning errors for the simulated trees, due to the limited accuracy in measuring their locations in the field; (v) errors in the geometric configuration of the fisheye lens; and (vi) errors that were associated with the allometric relationships which were used to build the tree models. In contrast, a substantial reduction in the number of gaps was observed for measurements near the horizon where light attenuation by canopy elements was very high. As a result, standard errors tended to decrease with zenith angle, since less variability in gap fractions was observed towards the horizon. Thus, zenith angles greater than 70° offered less relevant information for retrieving canopy structural parameters from HPs or for evaluating model capabilities in reproducing the canopy structure.

Figure 9.10 shows how $\overline{G_f}^{sim}$ varies the zenith angle for the (i) *Layer*, (ii) *Ellipsoids*, and (iii) *VoxTreK* models. Mean absolute differences between $\overline{G_f}^{in\ situ}$ and $\overline{G_f}^{sim}$ values were of 23, 16, and 11% for the *Layer*, *Ellipsoids*, and *VoxTreK* models, respectively. The results that are depicted in Fig. 9.10 indicated: (1) a weak correspondence between $\overline{G_f}^{in\ situ}$ and $\overline{G_f}^{sim}$ for the *Layer* model, which did not reflect the structural variability of the ECOLEAP-FM canopy; (2) a rapid decrease for $\overline{G_f}^{sim}$

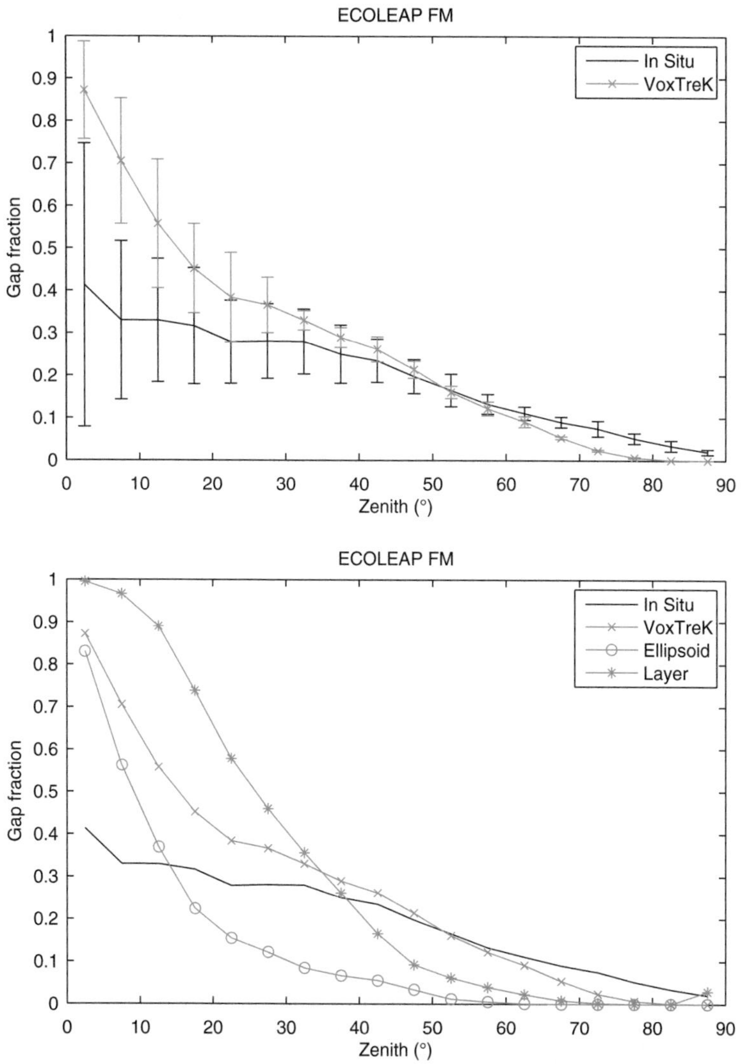

Fig. 9.10 Comparisons of gap fraction values for simulated and in situ HPs for the ECOLEAP-FM validation site. [*Top*] *VoxTreK* architectural model (×) and in situ (*solid line*) comparison. The error bars corresponding to the standard error are shown for the in situ and *VoxTreK* HPs. [*Bottom*] Three different architectural models are compared with in situ (*solid line*): *VoxTreK* (×), *Ellipsoids* (O), and *Layer* (*). Error bars were omitted to improve visual clarity

of the *Ellipsoids* model compared to $\overline{G_f}^{in\ situ}$ for angles >30°, due to the absence of details (and clumping) that were finer than those of tree crown; and (3) a good capability of the *VoxTreK* model to reproduce the $\overline{G_f}^{in\ situ}$ values and simulate the details of HPs, especially at zenith angles greater than 20°. Figure 9.11 shows that

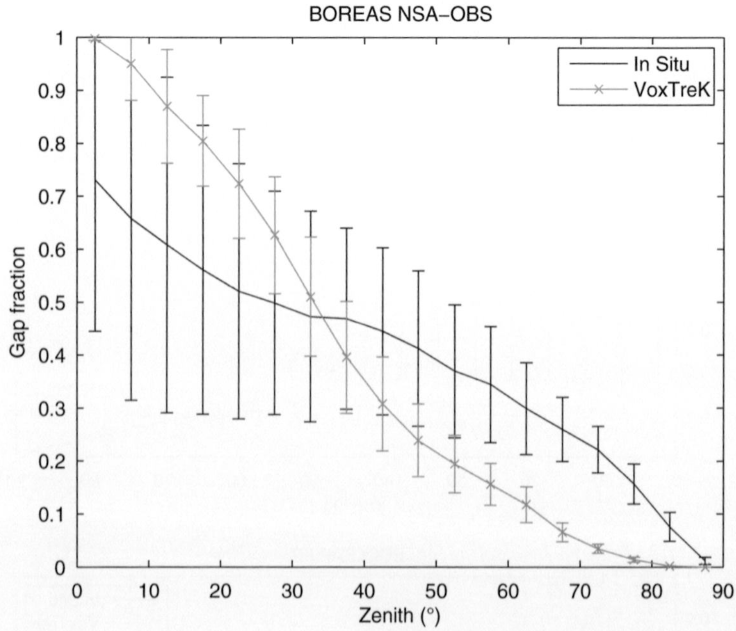

Fig. 9.11 Comparisons of gap fraction values for simulated and in situ HPs for the BOREAS NSA-OBS validation site. HPs from the *VoxTreK* architectural model are displayed with × –markers. The error bars corresponding to the standard error are shown for the in situ and *VoxTreK* HPs

$\overline{G_f}^{sim}$ values were significantly different from in situ values around a zenith angle of 50°, which means that the spatial variability and high foliage clumping of black spruce at BOREAS NSA-OBS was difficult to reproduce with the *VoxTreK* model. Figure 9.12 (top) indicates good agreement between $\overline{G_f}^{sim}$ with the *VoxTreK* model and $\overline{G_f}^{in\ situ}$ for zenith angles larger than 20°. Figure 9.12 (bottom) shows the relationship between $\overline{G_f}^{in\ situ}$ and $\overline{G_f}^{sim}$ values for the BOREAS SSA-OJP forest sites that were reconstructed with the *VoxTreK* model and the *Vectorization* method. The mean absolute differences between $\overline{G_f}^{in\ situ}$ and $\overline{G_f}^{sim}$ values were 10% for the *VoxTreK* model and 5% for the *Vectorization* method, respectively. The Le Bray site was modeled with a lower material density, on average, than was observed in the field, which yielded higher $\overline{G_f}^{sim}$ values (Fig. 9.13). Maritime pines generally exhibited less clumping and lower foliage density compared to jack pine, balsam fir and black spruce. The 3D kite model for branches may not have been the most suitable representation for that particular species. In addition, wood material was not taken into account in the model and could have contributed to increasing the gap fraction for branches with low foliage density.

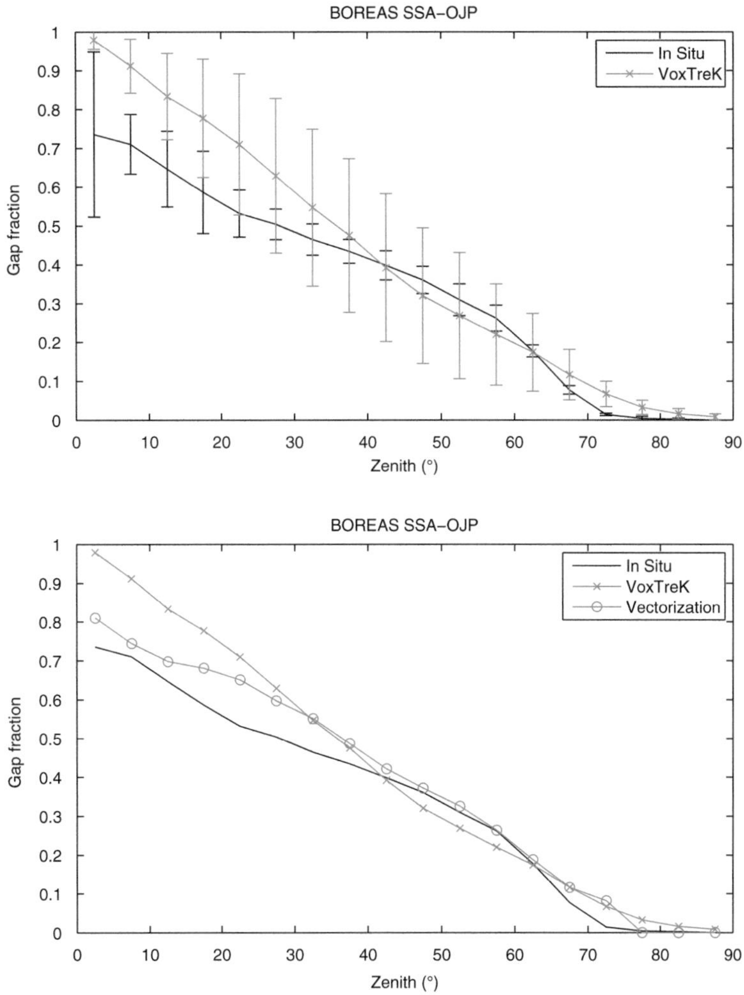

Fig. 9.12 Comparisons of gap fraction values for simulated and in situ HPs for the SSA-OJP validation site. [*Top*] *VoxTreK* architectural model (×) and in situ (*solid line*) comparison. The error bars corresponding to the standard error are shown for the in situ and *VoxTreK* HPs. [*Bottom*] Two different architectural models are compared with in situ (*solid line*): *VoxTreK* (×) and *Vectorization* (○). Error bars were omitted to improve visual clarity

The *VoxTreK* model and the two architectural models with the lowest levels of detail were compared with respect to their performance in reproducing forest canopies of the ECOLEAP-FM site. The similarity of results that were obtained with the *VoxTreK* and *Vectorization* models suggests that a model that was detailed enough to represent individual branches may be sufficient to reproduce gap fraction distributions within forest canopies with the same accuracy as a model that included individual coniferous shoots.

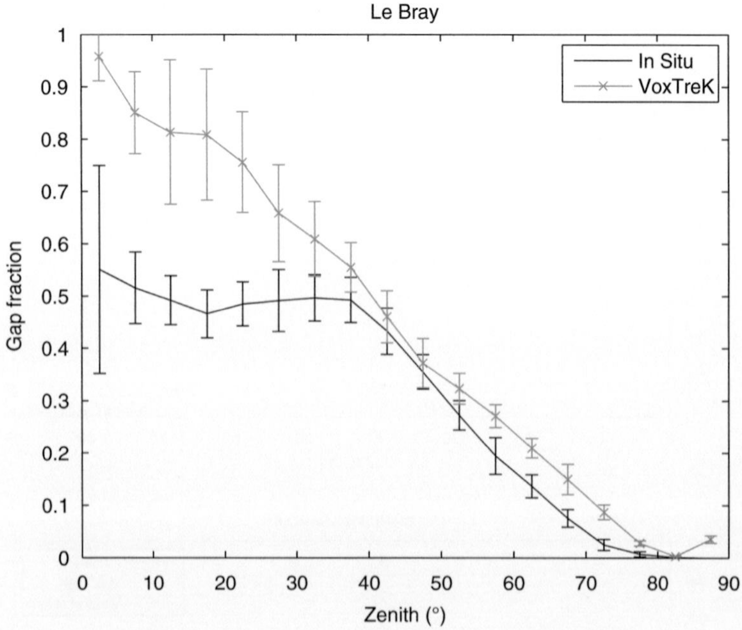

Fig. 9.13 Comparisons of gap fraction values for simulated and in situ HPs for the Le Bray validation site. HPs from the *VoxTreK* architectural model are displayed as ×s. Error bars corresponding to the standard error are shown for in situ and *VoxTreK* HPs

Table 9.4 Correlations between the in situ and simulated gap fraction values that were calculated for the zenith angle interval [0°–75°] for all validation sites and the architectural models

Validation Site	Architectural model	Correlation†
ECOLEAP-FM	*VoxTreK*	0.95
	Ellipsoids	0.81
	Layer	0.92
BOREAS NSA-OBS	*VoxTreK*	0.96
BOREAS SSA-OJP	*VoxTreK*	0.96
	Vectorization	0.99
Le Bray	*VoxTreK*	0.92

†Statistical significance $P < 0.003$

9.4 Discussion and Conclusion

This study compared the performance of four models that are capable of representing the structure of coniferous forest canopies with widely different degrees of detail, in terms of their abilities to simulate hemispheric photographs that had been acquired in the same type of environment. The *VoxTreK* model, representing canopy structure up to the branch-level, offers an intermediate solution between simple geometric models that only deal with tree crown envelopes and highly detailed models that account for twigs, leaves, and other small-scale elements. Not surprisingly, this model offers an improved ability to simulate HPs in such forests compared to simpler models, but what is interesting is that the much greater efforts that were required to implement the most detailed approach (*Vectorization*) did not yield appreciable improvements. The results that were obtained with the *VoxTreK* model underscored the advantage of describing canopy structure with a level of detail greater than just the crown envelope for the purposes of comparison with HPs. At the same time, *VoxTreK* requires substantially fewer measurements (or data) to reconstruct the vegetation canopy compared to the *Vectorization* approach, while conserving a comparable accuracy in the context of HP applications.

Characterizing the structure of a forest canopy implies collecting information such as foliage clumping at all relevant spatial scales (Frazer et al. 2017), the extent of the wood surface area, and the number, size, shape, and orientation distribution of leaves or needles. Some of these variables may be retrieved at broad spatial scales from satellite remote sensing platforms, but direct field observations are required for the most part. Field data are typically acquired using a variety of methods, including HPs, since no single technique can provide all of the necessary information at once. In the context of HP applications, a 3D forest structure model provides an opportunity to integrate observations of canopy structure and, thus, calculate other vegetation attributes. Moreover, HPs would offer considerable advantages for developing novel measurement techniques that can be tested against explicit and unbiased datasets describing vegetation structure. For instance, developing practical algorithms to calculate clumping (Frazer et al. 2017) could be significantly improved by using an architectural model that provides known values of LAI, wood area ratio, and leaf normal distribution.

The application domain of architectural models depends upon the conditions under which they are used. For instance, using a coarse model to aid in the interpretation of HPs where the intricate structural elements are relevant for evaluating vegetation attributes would lead to incorrect results. In contrast, going from an architectural model at the branch-level of detail to another model at the shoot-(or leaf) level of detail would be tedious and might provide relatively little improvement in accuracy for extracting attributes from HPs. Trying to reproduce the 3D spatial arrangement of canopy elements at the scale of shoots generally involves the use of complex algorithms and efficient memory management strategies. These architectural models, therefore, have been underexploited (e.g., Côté et al. 2013). *VoxTreK* has the advantage of reproducing the spatial distribution of gaps at a level of complexity that is manageable using a personal computer, while providing results that are comparable to shoot-level models. It is suitable for the voxel representation of 3D space, which in turn can facilitate explicit measurements of tree structure like the gap fraction, LAI, the wood-to-foliage ratio, and foliage clumping. Therefore, *VoxTreK* could provide valuable information for better understanding the link between the gap distribution that is observed with HPs and ecological processes that are influenced by the light regime in natural coniferous forests.

Based on the observations and results that were obtained in this study, two improvements could be suggested for the next implementation of *VoxTreK*. First, we could address the issue of small branch omissions during the *voxelization* process by having an explicit representation of the branch geometry with gaps using constructive solid geometry (Hoffmann 1989). Then, it would be possible to know precisely for each branch which voxel(s) it intersects, together with the volume fractions (Sen and Srikanth 2008). Second, the inclusion of a variety of 3D shapes (e.g., 3D kites, spheroids, asymmetric hulls) for modeling the individual branches would potentially increase the capacity of *VoxTreK* to simulate different conifer species. The conclusions reached in this chapter were derived from simulation studies that were conducted for coniferous forests, but similar results are to be expected for deciduous forests. In this latter case, a different statistical distribution of foliage around the branches may be needed to reflect the evolutionary strategies that have been developed by these trees to capture light inside the crown.

Acknowledgements We are grateful to the ECOLEAP group of the Canadian Forest Service (Natural Resources Canada), the BOREAS project, and the Institut National de la Recherche Agronomique (INRA) at Pierroton for providing access to critical datasets that were used in this study. The work of Jean-François Côté was made possible by a Natural Sciences and Engineering Research Council of Canada (NSERC) research grant awarded to Richard A. Fournier.

References

Andrieu B, Sohbi Y, Ivanov N (1994) A direct method to measure bidirectional gap fraction in vegetation canopies. Remote Sens Environ 50:61–66

Avery TE, Burkhart HE (1983) Forest measurements, 3rd edn. McGraw-Hill, New York 331 pp

Bernier PY, Fournier RA, Ung C-H, Robitaille G, Larocque GR, Lavigne MB, Boutin R, Raulier F, Paré D, Beaubien J, Delisle C (1999) Linking ecophysiology and forest productivity: an overview of the ECOLEAP project. For Chron 75:417–421

Borel CC, Gerstl SAW, Powers BJ (1991) The radiosity method in optical remote sensing of structured 3-D surfaces. Remote Sens Environ 36:13–44

Brunner A (1998) A light model for spatially explicit forest stand models. For Ecol Manag 107:19–46

Cescatti A (1997) Modelling the radiative transfer in discontinuous canopies of asymmetric crowns. I. Model structure and algorithms. Ecol Model 101:263–274

Chen JM, Black TA (1991) Measuring leaf area index of plant canopies with branch architecture. Agric For Meteorol 57:1–12

Chen JM, Cihlar J (1995) Quantifying the effect of canopy architecture on optical measurements of leaf area index using two gap size analysis methods. IEEE T Geosci Remote 33:777–787

Chen JM, Leblanc SG (1997) A four-scale bidirectional reflection model based on canopy architecture. IEEE T Geosci Remote 35:1316–1337

Chen JM, Rich PM, Gower ST, Norman JM, Plummer S (1997) Leaf area index of boreal forests: theory, techniques, and measurements. J Geophys Res 102(D24):29429–29443

Côté J-F (2006) Un modèle d'architecture 3D par voxels pour simuler les paramètres structuraux des couverts forestiers de conifères. MSc thesis, Université de Sherbrooke, Sherbrooke, Quebec. 85 pp

Côté J-F, Fournier RA, Luther JM (2013) Validation of L-Architect model for balsam firs and black spruce with structural measurements. Can J Remote Sens 39:S41–S59

de Reffye P (1988) Plant models faithful to botanical structure and development. Comp Graph 22:151–158
de Reffye P, Blaise F, Houllier F (1998) Modelling plant growth and architecture: some recent advances and applications to agronomy and forestry. Acta Hortic 456:105–116
de Reffye P, Fourcaud T, Blaise F. Barthélémy D, Houllier F (1997) A functional model of tree growth and architecture. Silva Fenn 31:297–311
Disney M, Lewis P, Saich P (2006) 3D modelling of forest canopy structure for remote sensing simulations in the optical and microwave domains. Remote Sens Environ 100:114–132
Ecological Stratification Working Group (1995) A national ecological framework for Canada. Agriculture and Agri-Food Canada, Research Branch, Centre for Land and Biological Resources Research and Environment Canada, State of Environment Directorate, Ecozone Analysis Branch, Ottawa/Hull. 125 pp and national map at 1:7 500,000 scale
Fournier RA, Landry R, August NM, Fedosejevs G, Gauthier RP (1996) Modelling light obstruction in three conifer forests using hemispherical photography and fine tree architecture. Agric For Meteorol 82:47–72
Fournier RA, Mailly D, Walter J-MN, Soudani K (2003) Indirect measurement of forest structure from *in situ* optical sensors. In: Wulder MA, Franklin SE (eds) Methods and applications for remote sensing of forests: concepts and case studies. Kluwer Academic Publishers, Dordrecht, The Netherlands, pp 77–113
Fournier RA, Rich PM, Landry R (1997) Hierarchical characterization of canopy architecture for boreal forest. J Geophys Res 102(D24):29445–29454
Frazer GW, Fournier RA, Leblanc SG, Walter J-MN (2017) View angle-dependent clumping indices for indirect LAI estimation. In: Fournier RA, Hall RJ (eds) Hemispherical photography in forest science: theory, methods, applications. Springer, Berlin Heidelberg New York
Gastellu-Etchegorry J-P, Demarez V, Pinel V, Zagoliski F (1996) Modeling radiative transfer in heterogeneous 3-D vegetation canopies. Remote Sens Environ 58:131–156
Gobron N, Pinty B, Verstraete MM, Govaerts Y (1997) A semidiscrete model for the scattering of light by vegetation. J Geophys Res 102(D8):9431–9446
Godin C, Costes E, Sinoquet H (1999) A method for describing plant architecture which integrates topology and geometry. Ann Bot-London 84:343–357
Goel NS (1988) Models of vegetation canopy reflectance and their use in estimation of biophysical parameters from reflectance data. Remote Sens Rev 4:1–212
Govaerts YM, Verstraete MM, Raytran (1998) A Monte Carlo ray-tracing model to compute light scattering in three-dimensional heterogeneous media. T Geosci Remote 36:493–505
Govind A, Guyon D, Roujean J-L, Yauschew-Raguenes N, Kumari J, Pisek J, Wigneron J-P (2013) Effects of canopy architectural parameterizations on the modeling of radiative transfer mechanism. Ecol Model 251:114–126
Herbert TJ (1987) Area projections of fisheye photographic lenses. Agric For Meteorol 39: 215–223
Hoffmann CM (1989) Geometric and solid modeling: an introduction. Morgan Kaufmann, San Francisco, California 338 pp
Horn HS (1971) The adaptive geometry of trees. Princeton University Press, Princeton, New Jersey 141 pp
Jonckheere I, Fleck S, Nackaerts K, Muys B, Coppin P, Weiss M, Baret F (2004) Methods for leaf area index determination. Part I: theories, sensors and hemispherical photography. Agric For Meteorol 121:19–35
Kimes DS, Kirchner JA (1982) Radiative transfer model for heterogeneous 3-D scenes. Appl Opt 21:4119–4129
Koop H (1989) Forest dynamics. A comprehensive monitoring system. Springer, Berlin Heidelberg New York, SILVI-STAR
Kurth W (1994) Morphological models of plant growth: possibilities and ecological relevance. Ecol Model 75(76):299–308
Kurth W (2000) Towards universality of growth grammars: models of Bell, Pagès, and Takenaka revisited. Ann For Sci 57:543–554

Kurth W, Sloboda B (1997) Growth grammars simulating trees—An extension of L-systems incorporating local variables and sensitivity. Silva Fenn 31:285–295

Landry R, Fournier RA, Lang R, Ahern FJ (1997) Tree vectorization: a methodology to characterize fine tree architecture in support of remote sensing models. Can J Remote Sens 23:91–107

Laroche G, Dagnault S (2000) Rapport d'inventaire Forestier des Sites ECOLEAP. Resources naturelles Canada, Service canadien des forêts, Centre de foresterie des Laurentides, Quebec

Larsen DR, Kershaw JA Jr (1996) Influence of canopy structure assumptions on predictions from Beer's law. A comparison of deterministic and stochastic simulations. Agric For Meteorol 81:61–77

Law BE, Cescatti A, Baldocchi DD (2001) Leaf area distribution and radiative transfer in open-canopy forests: implications for mass and energy exchange. Tree Physiol 21:777–787

Leblanc SG, Chen JM, Fernandes R, Deering DW, Conley A (2005) Methodology comparison for canopy structure parameters extraction from digital hemispherical photography in boreal forests. Agric For Meteorol 129:187–207

Leblanc SG, Fernandes R, Chen JM (2002) Recent advancements in optical field leaf area index, foliage heterogeneity, and foliage angular distribution measurements. In: IGARSS 2002, Proceedings of the IEEE International Geoscience and Remote Sensing Symposium, 24–28 June, Toronto, Ontario. CD-ROM. IEEE, Piscataway, New Jersey. 5:2902–2904

Leblanc SG, Fournier RA (2017) Measurement of forest structure with hemispherical photography. In: Fournier RA, Hall RJ (eds) Hemispherical photography in forest science: theory, methods, applications. Springer, Berlin Heidelberg New York

Li X, Strahler AH (1985) Geometrical-optical modeling of a conifer forest canopy. IEEE T Geosci Remote 23:705–721

Mäkelä A (1986) Implications of the pipe model theory on dry matter partitioning and height growth of trees. J Theor Biol 123:103–120

Měch R, Prusinkiewicz P (1996) Visual models of plants interacting with their environment. Proceedings of SIGGRAPH 96 (New Orleans, Louisiana, August 4–9, 1996). In *Computer Graphics* Proceedings, Annual Conference Series, ACM SIGGRAPH, 397–410

Mitchell KJ (1975) Dynamics and simulated yield of Douglas-fir. For Sci Monogr 17:1–39

Myneni RB, Maggion S, Iaquinta J, Privette JL, Gobron N, Pinty B, Kimes DS, Verstraete MM, Williams DL (1995) Optical remote sensing of vegetation: modeling, caveats and algorithms. Remote Sens Environ 51:169–188

Nilson T, Peterson U (1991) A forest canopy reflectance model and a test case. Remote Sens Environ 37:131–142

Nilson T, Ross V (1979) Characterization of the transparency of a forest canopy by fish-eye photographs [Vooremaa Forest Ecology Station]. Estonian contributions to the International Biological Programme progress report: 2) 117–130

Nikinmaa E, Messier C, Sievänen R, Perttunen J, Lehtonen M (2003) Shoot growth and crown development: effect of crown position in three-dimensional simulations. Tree Physiol 23: 129–136

Norman JM, Campbell GS (1989) Canopy structure. In: Pearcy RW, Ehleringer J, Mooney HA, Rundel PW (eds) Physiological plant ecology: field methods and instrumentation. Chapman & Hall, London, pp 301–325

Oker-Blom P, Kellomäki S (1982) Effect of angular distribution of foliage on light absorption and photosynthesis in the plant canopy: theoretical computations. Agr Meteorol 26:105–116

Pacala SW, Canham CD, Silander JA Jr (1993) Forest models defined by field measurements: I. The design of a northeastern forest simulator. Can J For Res 23:1980–1988

Parker GG (1995) Structure and microclimate of forest canopies. In: Lowman MD, Nadkarni NM (eds) forest canopies. Academic Press, San Diego, CA, pp 73–106

Perttunen J, Sievänen R, Nikinmaa E, Salminen H, Saarenmaa H, Väkevä J (1996) LIGNUM: A tree model based on simple structural units. Ann Bot-London 77:87–98

Pinty B, Verstraete MM (1991) Extracting information on surface properties from bidirectional reflectance measurements. J Geophys Res 96(D2):2865–2874

Pinty B, Verstraete MM, Dickinson RE (1990) A physical model of the bidirectional reflectance of vegetation canopies: Part 2. Inversion and validation. J Geophys Res 95(D8):11767–11775
Pisek J, Lang M, Nilson T, Korhonen L, Karu H (2011) Comparison of methods for measuring gap size distribution and canopy nonrandomness at Järvselja RAMI (RAdiation transfer Model Intercomparison) test sites. Agric For Meteorol 151:365–377
Planchais I, Pontailler J-Y (1999) Validity of leaf areas and angles estimated in a beech forest from analysis of gap frequencies, using hemispherical photographs and a plant canopy analyzer. Ann For Sci 56:1–10
Porté A, Bosc A, Champion I, Loustau D (2000) Estimating the foliage area of maritime pine (*Pinus pinaster* Ait.) branches and crowns with application to modelling the foliage area distribution in the crown. Ann For Sci 57:73–86
Porté A, Loustau D (1997) Needle surface area and density distributions in a 25-year-old maritime pine canopy (*Pinus pinaster* Ait.). Institut National de la recherche Agronomique (INRA). Internal Report. France. 14 pp
Prusinkiewicz P, James M, Měch R (1994) Synthetic topiary. Proceedings of SIGGRAPH 94 (Orlando, Florida, 24–29 July 1994), In Computer Graphics Proceedings, Annual Conference Series, 1994, ACM SIGGRAPH. pp 351–358
Prusinkiewicz P, Lindenmayer A (1990) The algorithmic beauty of plants. Springer, Berlin Heidelberg New York
Rich PM (1990) Characterizing plant canopies with hemispherical photographs. In: Goel NS, Norman JM (eds) Instrumentation for studying vegetation canopies for remote sensing in optical and thermal infrared regions. Remote Sens Rev 5:13–29
Ross J (1981) The radiation regime and architecture of plant stands. Junk publishers, The Hague, Dr W
Ross JK, Nilson TA (1966) A mathematical model of the radiation regime of vegetation. In: Pyldmaa VK (ed). Actinometry and atmospheric optics: Proceedings of the 6th Interdepartmental Symposium on Actinometry and Atmospheric Optics, June 1966, Tartu. [Mezhvedomstvennoe soveshchanie po aktinometrii i optike atmosfery (Tartu) (1966), in Russian] Translated by Israel Program for Scientific Translations, Cat. No. 5835, 253–274, Jerusalem, 1971
Seidel D, Fleck S, Leuschner C (2012) Analyzing forest canopies with ground-based laser scanning: a comparison with hemispherical photography. Agric For Meteorol 154–155:1–8
Sellers P, Hall F, Margolis H, Kelly B, Baldocchi D, den Hartog G, Cihlar J, Ryan MG, Goodison B, Crill P, Ranson JK, Lettenmaier D, Wickland DE (1995) The boreal ecosystem-atmosphere study (BOREAS): an overview and early results from the 1994 field year. Bull Am Meteorol Soc 76:1549–1577
Sen D, Srikanth TK (2008) Efficient computation of volume fractions for multi-material cell complexes in a grid by slicing. Comput Geosci 34:754–782
Sinoquet H, Rivet P (1997) Measurement and visualization of the architecture of an adult tree based on a three-dimensional digitising device. Trees 11:265–270
St-Onge BA, Cavayas F, Teillet PM (1991) Étude de la signature spatiale des couverts forestiers par modélisation géométrique-optique. In: Proceedings of the 5th International Colloqium on Physical Measurements and Signatures in Remote Sensing, Courchevel, France, 14–18 January, ESA SP-319. pp 671–674
Stadt KJ, Lieffers VJ (2000) MIXLIGHT: a flexible light transmission model for mixed-species forest stands. Agric For Meteorol 102:235–252
Sun G, Ranson KJ (1998) Radar modelling of forest spatial patterns. Int J Remote Sens 19: 1769–1791
Turner MG (1989) Landscape ecology: the effect of pattern on process. Annu Rev Ecol Syst 20:171–197
Ung C-H, Melanson P (1998) Canadian Forest Service, Laurentian Forestry Centre, Québec, Canada. Personal communication
Ung C-H, Bernier PY, Raulier F, Fournier RA, Lambert M-C, Régnière J (2001) Biophysical site indices for shade tolerant and intolerant boreal species. For Sci 47:83–95

van Leeuwen M, Coops NC, Hilker T, Wulder MA, Newnham GJ, Culvenor DS (2013) Automated reconstruction of tree and canopy structure for modeling the internal canopy radiation regime. Remote Sens Environ 136:286–300

Verhoef W, Bunnik NJJ (1981) Influence of crop geometry on multispectral reflectance determined by the use of canopy reflectance models. In: Proceedings of international colloquium on signatures of remotely sensed objects. Avignon, France, 8–11 September 1981. pp 273–290

Verstraete MM (1987) Radiation transfer in plant canopies: transmission of direct solar radiation and the role of leaf orientation. J Geophys Res 92(D9):10985–10995

Verstraete MM (1988) Radiation transfer in plant canopies: scattering of solar radiation and canopy reflectance. J Geophys Res 93(D9):9483–9494

Verstraete MM, Pinty B, Dickinson RE (1990) A physical model of the bidirectional reflectance of vegetation canopies: Part 1. Theory. J Geophys Res 95(D8):11755–11765

Walter J-MN, Fournier RA, Soudani K, Meyer E (2003) Integrating clumping effects in forest canopy structure: an assessment through hemispherical photographs. Can J Remote Sens 29:388–410

Wang YP, Jarvis PG (1990) Description and validation of an array model—MAESTRO. Agric For Meteorol 51:257–280

Weibull W (1951) A statistical distribution function of wide applicability. J Appl Mech-T ASME 18:293–297

White MA, Asner GP, Nemani RR, Privette JL, Running SW (2000) Measuring fractional cover and leaf area index in arid ecosystems: digital camera, radiation transmittance, and laser altimetry methods. Remote Sens Environ 74:45–57

Widlowski J-L, Lavergne T, Pinty B, Gobron N, Verstraete MM (2008) Towards a high spatial resolution limit for pixel-based interpretations of optical remote sensing data. Adv Space Res 41:1724–1732

Chapter 10
Hemispherical Photography in Forest Science: Conclusions, Applications, Limitations, and Implementation Perspectives

Richard A. Fournier and Ronald J. Hall

Abstract The purpose of this chapter is to summarize the key messages from each chapter of the book and the rationale for using hemispherical photography (HP) in the discipline of forest ecology. The chapters describing the theory and methods presented the state of the science and the opportunities for developing applications to forestry and natural resources, as well as helping to identify limitations that require further investigation. This chapter also provides a literature review that expands the topic of applications beyond those already covered in Chaps. 8 and 9. Forest ecology, like many other fields, has a rich pool of published studies that illustrate the applications of HP technology. By understanding the relevant theory and methods for acquiring and analyzing hemispherical photographs, the book provides a fundamental basis for new users to understand the published material, from which to plan and undertake their own studies.

Keywords Limitations of hemispherical photography · Applications of hemispherical photography · Perspectives of hemispherical photography · Architectural models · Leaf area index

Abbreviations

HP	Hemispherical photography
LAI	Leaf area index
LiDAR	Light detection and ranging
RGB	Red-green-blue
2D	Two-dimensional
3D	Three-dimensional
TLS	Terrestrial laser scanner

R.A. Fournier (✉)
Département de Géomatique Appliquée, Centre d'Applications et de Recherches en Télédétection (CARTEL), Université de Sherbrooke, Sherbrooke, QC J1K 2R1, Canada
e-mail: Richard.Fournier@USherbrooke.ca

R.J. Hall
Canadian Forest Service, Northern Forestry Centre, Natural Resources Canada, 5320—122 Street, Edmonton, AB T6H 3S5, Canada
e-mail: Ron.Hall@Canada.ca

R.A. Fournier and R.J. Hall (eds.), *Hemispherical Photography in Forest Science: Theory, Methods, Applications*, Managing Forest Ecosystems 28,
DOI 10.1007/978-94-024-1098-3_10

10.1 Progress in the Use of HP in Forestry

The use of HP was inspired by the initiative of several innovators in the late 1950s and 1960s who appreciated the potential in this technique, including Evans and Coombe (1959), Anderson (1964a, b), Brown and Worley (1965), and Magdwick and Brumfield (1969), to name a few. However, HP began to be used more extensively in forestry in the 1970s. Early use was based on the working premise that a hemispherical photograph taken from the ground looking upward would provide a means to quantify the light regime and to characterize attributes of forest structure where the photo was taken. The history of HP in forestry provided by Hall et al. (2017a) provides a vivid account of the rapid evolution of the use of HP in forestry. Technological advances, such as the advent of the digital camera, have greatly improved the capabilities and popularity of this technique, resulting in increased demand for knowledge about the theory, equipment, and methods of analysis. While much of the past work used analog film and single lens reflex cameras, hemispherical photographs are now largely taken by digital cameras that allow for immediate availability of the photographs in electronic format. With the use of digital cameras and the rapid improvement in image-analysis methods, conditions for the use of HP in forestry are broadening.

As an indicator of the use of HP, we have determined the number of published articles in which the title or abstract explicitly contained the words "*hemispherical photograp**" in the TREECD and SCOPUS databases for studies related to forest science. The search found only eight articles before 1978, but this number increased gradually in the following five-year periods to three (in 1979–1984), seven (in 1985–1989), 14 (in 1990–1994), 36 (in 1995–1999), 70 (in 2000–2004), 112 (in 2005–2009), and 150 for the 2010–2014 period. Further verification using known articles selected at random confirmed that these numbers only represent about 40 to 50% of the scientific articles published on the subject. The rising trend in the number of articles in which HP was used as a tool is evidence of the significant level of interest in the extended forest scientific community since the mid-1990s. The increasing number of publications on the application of HP in forestry demonstrates that adoption of this technique is on the rise.

While studies of HP applications are extremely diverse and easily accessible through the published material in scientific journals, this is not the case for the theory and the methods for HP, which are not necessarily covered in length in peer-reviewed journals. This book therefore focused primarily on the theory and methods, but also included two chapters on the application of HP in forestry. The book was designed as a base document that presents the useful theory, explains existing methods, covers limitations, and establishes perspectives for those interested in using HP in forestry.

10.2 Theory

Scientific reports regarding the use of HP in forestry are markedly different, depending on whether they focus on developing a new method or on acquiring data to improve information for a particular ecological problem. Despite these differences, there are two major and distinctive sets of applications of HP that dominate scientific reports: the study of the light regime within the canopy and the quantification of stand structural attributes. Methods and applications using HP are best appreciated if there is an understanding of the background theory on which they are based. That is why this book was divided into three sections (theory, methods, and applications) that created a natural continuum from which a deeper understanding can be drawn. Consequently, two theory chapters were needed: the first describing the theory underlying the light regime, energy balance, and measurement of radiation within the canopy (Schleppi and Paquette 2017), and the second focusing on the estimation of forest canopy structural attributes from HP (Leblanc and Fournier 2017). Both applications benefit from the main advantage that HP provides—a spatially explicit view of the forest canopy from a particular point near the ground, looking up. These two categories of HP applications, however, involve very distinct sets of theoretical backgrounds.

Estimating the light regime within a forest canopy with HP involves the size and orientation of the visible gaps (Schleppi and Paquette 2017). However, while HP has strong capabilities in the angular assessment of the light regime, it does not have the precision, accuracy, and repeatability of the more expensive specialized instruments designed to measure solar or sky radiation. For example, specific spectral range and calibrated intensity values are difficult to extract from HP. It therefore requires the use of several models that (1) simulate solar and sky radiation in the visible and near-infrared spectra in either clear or overcast sky conditions, while also taking into account terrain slope; (2) quantify the position of the sun year-round (in the case of the clear-sky radiation models); (3) estimate the proportion of clear versus overcast days for a time period; (4) account for radiation propagation inside the canopy; and (5) describe the energy budget from the visible through to the thermal range of the spectrum. HP is often used in conjunction with these models and other instruments to estimate the light regime in the forest canopy. Fortunately, many of these models are available in specialized software for processing hemispherical photographs, but a basic knowledge of how these models work and their assumptions is required to interpret the results from photos. While the theory underlying these models is relatively mature, there are continuous improvements based on research, particularly the development of canopy radiation models that explicitly incorporate knowledge of the structural composition of the canopy.

The motivation for assessing forest structure depends heavily on the application. For example, a forest inventory application emphasizes estimation of stand and tree attributes required for forest management. Assessing wood timber volume is an application that is important for the forest industry. Measurements of stand tree density and average stand height can be used to estimate volume or biomass, if the stand composition is known (Fournier et al. 2003). Such estimates can be improved if stem diameter and the height of the trees composing a stand are also known. In ecological studies, the complexity of forest canopy structure can be used to assess habitat or microclimate, or to provide inputs to models of ecosystem functions. A great deal of work using HP focuses on the estimation of the leaf surface, most often expressed as the leaf area index (LAI). The LAI of a forest canopy is an important driver for estimates of light interception and for models that describe foliage and tree functions such as canopy interception, transpiration, and net primary productivity. Describing LAI with HP is currently based on a theoretical framework in which LAI is defined from the angular distribution of gap fractions, as seen from a hemispherical view of the canopy. Development of methods to estimate canopy structural attributes using HP requires such a theoretical framework based on the physical representation of the canopy and the distribution of its components (Norman and Campbell 1989; Leblanc and Fournier 2017). Aside from the gap fraction, the main elements of this framework account for woody components (stem and branches), occlusion of canopy components, leaf angle distribution (Leblanc and Fournier 2017), and the need to adjust for foliage clumping (Frazer et al. 2017). Early representations of the physical structure of the forest canopy considered only the foliage (Goel 1988). Gradually, models of canopy structure included tree stems and branches because these components govern how foliage is distributed within a tree and forest stand, providing a more holistic representation (Côté et al. 2011; Leblanc and Fournier 2014). Accounting for each of these structural attributes to improve the estimation of LAI is a challenge, and the availability of reliable methods to do so is limited. For example, foliage clumping methods are intended to transform the estimate of effective LAI to true LAI by accounting for the non-randomness of the gap patterns in hemispherical photos. While the effect of the woody components has been included in the general theoretical framework, quantifying its contribution is considered a limitation with current methods. Specialized instruments such as visible–near-infrared cameras (Kucharik et al. 1997; Chapman 2007) or terrestrial LiDAR (Béland et al. 2011) may be useful to resolve this ambiguity. Similarly, the ability to quantify leaf angle distribution is still not fully understood because it is an attribute that is very difficult to measure in the field. To resolve this problem, an increasing number of studies estimate leaf angle distribution from architectural models of trees that simulate stands (Leblanc and Fournier 2014). Our ability to assess forest structure depends strongly on both our capability to expand the current theoretical framework and the capabilities of current instrumentation. Both are likely to evolve quickly, as a result of combining new equipment such as terrestrial LiDAR, which can effectively assess three-dimensional (3D) structures (Hackenberg et al. 2015), with new methods that can contribute to more precise estimates (Béland et al. 2011).

10.3 Methods

Using HP to quantify the solar regime in a stand or to estimate structural attributes of a stand requires several methodological steps, including obtaining suitable equipment, designing the experiment to achieve statistical significance, acquiring hemispherical photos, processing the images, and applying algorithms to estimate the attributes of interest. Any weak aspect in that chain of steps is likely to reduce the meaningfulness of the results from HP. Therefore, all steps must be attended to carefully.

The first critical steps for acquiring suitable hemispherical photos involve obtaining the equipment, planning the experiment, and adopting an efficient strategy to acquire and archive the photos (Fournier et al. 2017). The two main pieces of equipment are the hemispherical lens and the camera. Rapid technological improvements have resulted in an ever-decreasing compromise between equipment quality and cost. Today, almost all cameras sold are digital, with image resolution approaching that of film images. The easy availability of useful auxiliary equipment for HP acquisition, such as self-levelling mounts, monopods, and light meters, allows configuring a system for specific experiment or application. Once the equipment has been selected, the next challenge is planning a data acquisition strategy that can be defended statistically. For instance, the number of HPs required to characterize one site can vary from a few ($\sim$6–12) to many, depending on stand heterogeneity (Leblanc and Fournier 2017). The acquisition strategy depends heavily on the spatial configuration of the forest stand. Thus, only general strategies can be suggested, and these must be adapted to the situation at hand. Once a rigorous acquisition strategy is selected, data acquisition has several challenges, including adopting an efficient data acquisition sequence to ensure all photos are well documented and suited for processing (e.g., document camera position and height, and identification of magnetic north on the image) and selecting a day with suitable meteorological conditions for differentiation of sky-objects, to name but a few. Efficient data acquisition requires planning and testing in the forest environment being studied.

The second critical steps include applying image processing to raw HP images to derive meaningful results (Jonckheere et al. 2017). Raw images are first processed to classify pixels to one of two values: white (1), assigned to sky, or black (0), related to an object. Classifying pixels into sky/canopy state was originally done through a simple threshold applied to the grey value of the pixel. Implementation of the methods in different programs shows that the algorithm used in thresholding the images for classification will influence the results (Hall et al. 2017b). Gradually, advanced image-processing techniques allow adaptive procedures such as angular dependence of the grey value threshold or the integration of "gray" pixels combining sky and object within the viewshed. Development of specialized image-analysis procedures to separate sky and canopy objects in the hemispherical photographs is currently an active field of research, because it is a key aspect to improve quantitative estimates (Macfarlane 2011; Macfarlane et al. 2014;

Jonckheere et al. 2017). The second stage of processing, beyond pixel classification, involves applying an algorithm to the white, black, or gray pixels of the image to estimate light regime or a structural attribute of the stand. HP is a two-dimensional (2D) representation of a complex 3D distribution of forest components, and this leads to several distortions. However, the effectiveness of these methods has been difficult to evaluate in the context of forest measurement (Demarez et al. 2008; Ryu et al. 2010) and is best quantified with the use of simulated environments. The work of Leblanc and Fournier (2017) and Frazer et al. (2017) guided our selection of a method to assess foliage clumping in estimating LAI. Other potential methods to address distortions in hemispherical photos include the ability to separate wood from foliage.

10.4 Applications

A large number of specialized studies have used HP in a wide array of fields related to forestry, including forest ecology (e.g., light regime—Canham et al. 1990; soil moisture—Gray et al. 2002; plant distribution—Connel et al. 2006; plant diversity—Thomas et al. 1999; seedling regeneration—Nicotra et al. 1999; and amphibian distribution—Halverson et al. 2003), forest inventory (e.g., tree growth—Opuni-Frimpong et al. 2008), stand dynamics (Lin et al. 2003), hydrology (e.g., snow distribution—López-Moreno and Latron 2008), effects of flooding (Battaglia and Sharitz 2006), stream temperature (Isaak et al. 2010), biophysical parameter estimation (e.g., LAI—Riaño et al. 2004; spring phenology—White et al. 2014), and canopy openness (Machado and Reich 1999), to name a few. The information extracted from HP provides a useful complement to standard inventory methods (Mailly 2017), in addition to supplying biophysical measures that would not otherwise be available. For example, variables such as tree volume or biomass (Clark and Murphy 2011; Lindner and Sattler 2012), canopy gap distribution (Danson et al. 2007), stem diameter or sapwood area (Vertessy et al. 1995), tree regeneration (Martinez Pastur et al. 2011), and LAI (Weiss et al. 2004) can be derived from HP. There are also specialized topics for which these methods are well suited, such as the assessment of potential forest fuels in order to analyze fire behavior (Pierce et al. 2012).

However, two topics dominate the scientific publications using HP in forest ecology: the assessment and modeling of the light regime in a stand, and the estimation of forest structural attributes (Hall et al. 2017a). HP has been used extensively to aid in developing models of the light regime within the forest canopy (e.g., Lieffers et al. 1999; Bertin et al. 2011; Schleppi and Paquette 2017). For example, HP has been used to characterize a specific environment, such as riparian zones (Dignan and Bren 2003; Bulliner and Hubbart 2012) and tropical rain forests (Ediriweera et al. 2008), as well as to assess the rate of snow melt beneath various forest canopy types (Essery et al. 2008; Musselman et al. 2012). Moreover, HP has been a primary data source to study the light regime, providing either an account of light conditions beneath the forest canopy (Cunningham et al. 2011; Hu et al. 2010), or an assessment of the effect of local variables, such as the viewing geometry and topography, on gap fraction or light regime (Liu et al. 2004). HP has also been widely explored for the estimation of LAI in a variety of forested environments and, thus, the theoretical framework supporting LAI estimation is widely available (e.g., Wang et al. 1992; Jonckheere et al. 2004; Leblanc and Fournier 2017). It has also been well documented in application studies, in which plot-level LAI estimation using HP is linked with data from other methods, such as the MODIS optical satellite sensor (Marsden et al. 2010), SAR satellite (Canisius and Fernandes 2012) or airborne LiDAR (Morsdorf et al. 2006; Zhao and Popescu 2009), for spatially explicit analyses of large areas. The importance of an HP method to provide plot-level estimates of biophysical variables has been a recurring topic, and many studies have compared different plot-level methods, including HP, to estimate LAI (e.g., Bréda 2003; Weiss et al. 2004; Leblanc et al. 2005). Finally, another important trend involves the use of HP to support architectural models for the assessment of the light regime or stand level attributes (Côté et al. 2017). Overall, the best way to ascertain the applications of HP is to conduct an extensive search of the available scientific literature. The diversity of application fields for HP is evidence of its versatility in a wide range of situations. The diversity of scientific publications in forest ecology indicates one active field in which the use of HP is well developed.

To assess potential new applications for HP in forestry, it is useful to examine the evolution of HP and its future direction. HP started nearly a century ago and evolved gradually, with adoption across a wide range of disciplines. In its early stage, HP was used to look at underwater sites (Wood 1906) and to estimate the percentage of clouds in the sky (Hill 1924; Slater 1932). Hemispherical lenses were used extensively in meteorological stations to assess sky cloudiness: a paper behind the lens was used to link the time of day (or sun trajectory) and sunny periods with burnt portions of the paper. Detection of solar energy and clouds with sensors under fisheye lenses has been a continuing tradition in meteorology (e.g., Martins et al. 2003) and has been extended to assess the occurrence of shooting stars (Luoma 2002). Hemispherical lenses have also been used in industrial applications to guide machinery and in robotics to guide robots (e.g., Kannala and Brandt 2006). Similarly, wide-angle and hemispherical lenses are frequently used for security and surveillance, with an emphasis on improving image-processing methods (Shiroma et al. 2004). Urban environments are particularly suitable for hemispherical view systems, and specialized methods using HP have been developed to assess urban canyons, energy loss, and view-factors (Eliasson 1996; Grimmond et al. 2001; Chapman et al. 2007). With the advent of new mapping strategies for urban environments, HP methods can also be applied in studies of augmented reality (e.g., Kuchelmeister et al. 2009). New sensor support systems, such as unmanned airborne vehicles, are also stimulating development of HP methods (e.g., Hrabar et al. 2005). It is difficult to predict the future applications of HP within and beyond forestry, but the scientific literature shows that its applications have diversified and are continuing to grow.

10.5 Limitations on the Use of HP in Forest Science

The use of HP in forest science has increased as knowledge of the theory underlying HP has improved and as technological advances in both equipment and methods have been made. The evolution of digital photography and automated image processing in recent years are two important factors that have led to improved methods, which in turn have increased the use of HP for a wider range of applications. For example, the assessment of below-canopy light regime or canopy structure using HP has benefited from the gradual improvement of both the spatial resolution of the photographs and of the methods for image processing. However, several important limitations remain:

- The need for suitable meteorological conditions limits the acquisition of good-quality photographs.
- The limited spectral discrimination of HP may be a constraint for certain applications.
- The local view of each photograph means that a sampling strategy must be selected and adapted to the question addressed and to the forest environment.
- HP is a 2D medium used to infer information from a 3D environment.

All of these limitations can constrain the use of HP in forest science.

Waiting for suitable meteorological conditions to acquire photos has always been the most serious limitation. It is strongly suggested that photos be acquired under one of two sky illumination conditions: (1) under cloudless sky conditions at dusk or dawn, or (2) under uniform cloud conditions during the daytime. In practice, most sky illumination conditions diverge from these two ideal cases. For example, overcast skies rarely provide illumination conditions that are ideal (e.g., isotropic) or similar to those modelled (e.g., with a predictable angular dependence). Also, clear-sky conditions, including direct solar illumination, cause uneven sky lighting and over-exposed sunlit foliage. As a result of field and cost logistics, photographs are often acquired under suboptimal conditions. However, there are several ways to resolve this problem. Widening the range of environmental conditions in which photos can be acquired is probably the most practical means, and several alternatives have been proposed. For example, it may be useful to acquire photos at night, using a flash to identify object and sky pixels. Another alternative involves image-processing procedures such as the use of color information or local thresholding to reduce the effects of variable sky illumination or high-contrast leaf reflectance on the analysis of photos. Overall, extending the acquisition conditions for photos is a critical aspect of development, and advances may increase the feasibility of HP for forest applications.

Photographs provide red-green-blue (RGB) information but not calibrated spectrometric values. HP has poor spectral discrimination compared with specialized instruments used in light spectrometry. The use of spectral information from HP is therefore limited. There are many situations in which relative or absolute (calibrated) spectral intensities values are important, such as the measurement of the light regime and calculation of the energy balance. There are also cases in which photo acquisition in a specific spectral range is preferable. For instance, LICOR® instruments (LAI-2000/LAI-2200) collects images in the low blue and ultraviolet portions of the spectrum to take advantage of vegetation that appears relatively opaque (low reflection and transmission) in that portion of the electromagnetic spectrum (Kucharik et al. 1997). Chapman (2007) has also suggested several advantages when there is simultaneous use of red and near-infrared information. These innovations could be applied to HP if an effective band separation were possible. Fortunately, the limitation on spectral information can sometimes be offset by using an empirical model in which some spectral distribution of the light regime is assumed. Alternatively, cameras can be adapted to collect calibrated spectral measurements in RGB or even transformed to become sensitive to near-infrared radiation. Technical advances are therefore predicted to enable image acquisition in specific spectral bands, which would reduce the current limitations on spectral measures in images acquired with a camera.

Each hemispherical photo provides only a localized view of a physical site, resulting in the need for a sampling strategy to define the number and position of the photos. The sampling strategy needs to be adapted to the question addressed and also to the environment under study, and is therefore fundamental to field preparation for most studies in ecology or forest science. Leblanc and Fournier (2017) demonstrated that the number of photos acquired depends not only on the variable being studied (e.g., gap fraction or LAI), but also on site heterogeneity. HP can provide a good local estimate of several variables, but the sampling strategy needs to be adapted according to stand complexity. There are too many potential situations to adopt a rigid strategy, and the choice of sampling strategy must be guided by the operator's experience along with general guidelines. Although the limitation related to sampling will always remain, adopting a strategy suited to stand characteristics will remain an important preparation step.

HPs provide a simplified, 2D representation of forest canopies that, in reality, are complex and distributed in 3D. In this simplified representation, any object produces an occluded area from the viewpoint of the origin (the position of the focal point in the camera). If canopy elements were randomly distributed, then it would be simple to model the occlusion effect. However, canopy structural organization is very complex, including levels of structure varying from coarser to finer, such as tree, canopy, whorl, branch, tuft, and leaf. Canopy elements are generally assumed to be clumped; hence, methods to estimate the clumping index of a stand have been developed (Frazer et al. 2017). Unfortunately, the clumping index is difficult to quantify, in addition to being intrinsically variable for most natural forest stands. One way to deal with the clumping caused by the 2D representation of forests is to model their 3D architecture. There are two alternatives to obtain explicit 3D representations of a forest: the first is to use the available information in combination with a forest architecture model to establish a spatially explicit 3D representation (Côté et al. 2017), and the second is to use an instrument with the hemispherical view, similar to HP, but with 3D sampling information, including distance. Such an instrument is the terrestrial laser scanner (TLS), which scans millions of points using a hemispherical geometry, with distance and intensity value at every point (Danson et al. 2007; Dassot et al. 2011; Hancock et al. 2014; Seidel et al. 2012). Some TLSs have a single return, but others allow multiple returns or even a full waveform signal. This ability of TLS to provide precise 3D information using a hemispherical view makes them compatible with many HP methods, several of which may be applied to TLS datasets (Hopkinson et al. 2004; Seidel et al. 2012; Zhao et al. 2012).

10.6 Perspectives on the Use of HP in Forest Science

The use of HP in forest science is anticipated to grow, owing to its relatively low cost and proven effectiveness in quantifying the light regime and structural attributes of forests. The assessment of forest light regimes through the analysis of hemispherical photos is an important input to light models, and HP provides useful estimates of forest structural attributes at the plot level. HP can contribute to complementary and comprehensive multi-scale monitoring strategies using a variety of sensors, methods, and models.

The current limitations on using HP are being addressed through technological advances in sensors, algorithmic solutions, and methods. For example, the methods being developed for using HP in forest science are highly compatible with those for using TLS data, as hemispherical scanning is being adopted for both. However, the use of TLS breaks the 2D-to-3D barrier that is a limitation on the use of HP. Therefore, many methods can be expanded from HP to TLS. In return, this ability to break the 2D-to-3D barrier will also benefit from the improved algorithms adapted for HP. Continual improvements in image-processing capabilities will help to resolve the limitations on using HP in forestry. Algorithms to improve the processing of raw hemispherical photos into binary (sky–object) images are one example of advances in image processing.

Developing methods to acquire reliable field data with known error can be challenging. This is particularly the case for complex forest structural variables such as LAI (Weiss et al. 2004), 3D distribution of leaf and woody surface (Béland et al. 2011), and tree topology (Landry et al. 1997; Côté et al. 2013), and these variables can be costly to estimate in situ. This limitation explains why many HP methods are assessed through inter-comparison of methods (e.g., Garrigues et al. 2008; Gendron et al. 1998; Leblanc et al. 2005). However, methods inter-comparison provides error estimation based on a reference method rather than an absolute estimation of measurement error. Often, the absolute error of the reference method is unknown. This limitation on measurement of complex forest structural variables also explains why there is increasing interest in the use of forest architectural models to alleviate the problem of acquiring sufficient, reliable field data at the level required for model development. For example, the architectural models proposed by Côté et al. (2011, 2017) and Leblanc and Fournier (2014) provide a platform to develop and test new methods to infer light regime or structural variables from HP. The availability of a modeled scene made from architectural models provides a flexible benchmarking environment for method development because light conditions or forest structural conditions are known. Simulated hemispherical photos made in such environments can be used to retrieve variables of the well documented scene. Therefore, the practical benchmarking of new methods in a simulated environment is made possible with architectural models that favor the use of HP because it allows for documented evaluation of method error and limitations.

All HP methods provide estimates of the light regime and structural attributes at the point location where images are collected. The data collected serve two major purposes: (1) supporting development of models and methods to gain a deeper understanding of forested environments, and (2) scaling plot-level estimates to spatially explicit estimates over large areas. In this context, comparisons between monitoring methods are often useful to assess the methods' relative strengths and limitations (e.g., Mussche et al. 2001). Plot-level estimates of forest structural attributes using HP data can be used in scaling up to large areas. It is anticipated that such an approach will become more prevalent, as a larger number of methods are using plot-level estimates from HP methods in conjunction with other remote sensing data from satellite optical (Morisette et al. 2006; Ryu et al. 2012), radar (Lang et al. 2008), or LiDAR data (Solberg et al. 2006; Zhao and Popescu 2009). Similarly, local variable estimation using HP will continue to support model development. For instance, modeling tree growth requires the ability to effectively measure the light regime and several structural variables that are relevant to tree competition for local resources.

In summary, although the options being developed to monitor the forest environment are becoming more diverse, HP remains a viable tool, with its main advantages being a favorable cost–quality ratio, readily available equipment suitable for field work, spatially explicit plot-level 2D views of the canopy, and plot-level estimates at an accuracy comparable with the other alternatives. It is anticipated that these advantages will only increase with time.

Acknowledgements The realization of this book took many years, from the original idea to its publication. Such a project relies on the good will of those passionate enough to volunteer their time and effort toward scientific issues they value. We observed that it was difficult to ask scientists to participate actively in a book when their scientific contribution was based mainly on articles published in peer-reviewed journals. We are therefore grateful to all of those who decided to prioritize their commitment to a book and who kept their word to the end. We wish to mention the important contribution of two English editors, Catherine A. Brown from the Centre d'Applications et de Recherches en Télédétection (CARTEL) at the Université de Sherbrooke, Canada, and William Parsons from the Centre d'étude de la forêt in Québec, Canada. Both supported the authors to improve their text and to ensure the book was uniform among most all chapters. We are grateful for funding support from CARTEL. We are also indebted to all the reviewers, whose names are listed below. Their contribution made a great difference to the quality of the book. Finally, we wish to extend our appreciation to all authors, who demonstrated their passion for this field and its scientific advancement.

Reviewers in alphabetical order: David Coates, Phil Comeau, Richard Fournier, Gordon Frazer, Alemu Gonsamo, Ron Hall, Chris Hopkinson, Sylvain Leblanc, Michael Leuchner, Daniel Mailly, Craig Macfarlane, Paul Osmond, Alain Paquette, Paul Rich, Kamel Soudani, and Jean-Michel Walter.

References

Anderson MC (1964a) Studies of the woodland light climate. I. The photographic computation of light conditions. J Ecol 52:27–41

Anderson MC (1964b) Studies of the woodland light climate. II. Seasonal variation in the light climate. J Ecol 52:643–663

Battaglia LL, Sharitz RR (2006) Responses of floodplain species to spatially condensed gradients: a test of the flood-shade tolerance tradeoff hypothesis. Community Ecology 147(1):108–118

Béland M, Widlowski JL, Fournier RA, Côté JF, Verstraete MM (2011) Estimating leaf area distribution in savanna trees from terrestrial LiDAR measurements. Agric For Meteorol 151:1252–1266

Bertin S, Palmroth S, Kim HS, Perks MP, Mencuccini M, Oren R (2011) Modelling understorey light for seedling regeneration in continuous cover forestry canopies. Forestry. doi:10.1093/forestry/cpr026

Bréda N (2003) Ground-based measurements of leaf area index: a review of methods, instruments and current controversies. J Exp Bot 54:2403–2417

Brown HE, Worley DP (1965) Some applications of the canopy camera in forestry. J Forest 63:674–680

Bulliner E, Hubbart JA (2012) An improved hemispherical photography model for stream surface shortwave radiation estimations in a central U.S. hardwood forest. Hydrol Process 27(26):3885–3895

Canham CD, Denslow JS, Platt WJ, Runkle JR, Spies TA, White PS (1990) Light regimes beneath close canopies and tree-fall gaps in temperate and tropical forests. Can J For Res 20(5):620–631

Canisius F, Fernandes R (2012) ALOS PALSAR L-band polarimetric SAR data and in situ measurements for leaf area index assessment. Remote Sens Lett 3(3):221–229

Chapman L (2007) Potential applications of near infra-red hemispherical imagery in forest environments. Agric For Meteorol 143(1–2):151–156

Chapman L, Thornes JE, Muller J-P, McMuldroch S (2007) Potential applications of thermal fisheye imagery in urban environments. IEEE Trans Geosci Remote Sens Lett 4(1):56–59

Clark J, Murphy G (2011) Estimating forest biomass components with hemispherical photography for Douglas-fir stands in northwest Oregon. Can J For Res 41(5):1060–1074

Connel L, Redman R, Craig S, Rodriguez R (2006) Distribution and abundance of fungi in the soils of Taylor Valley. Antarctica. Soil Biol Biochem 38(10):3083–3094

Côté JF, Fournier RA, Égli R (2011) An architectural model of trees to estimate forest structural attributes using terrestrial LiDAR. Environ Model Softw 26:761–777

Côté JF, Fournier RA, Luther JE (2013) Validation of L-Architect model for balsam fir and black spruce tree with structural measurements. Can J Remote Sens 39(S1):S41–S59

Côté JF, Fournier RA, Verstraete MM (2017) Canopy architectural models in support of methods using hemispherical photography. In: Fournier RA, Hall RJ (eds) Hemispherical photography in forest science: theory, methods, applications. Springer, New York

Cunningham KK, Peairs SE, Ezell AW, Belli KL, Hodges JD (2011) Understory light conditions associated with partial overstory removal and midstory/understory control applications in a bottomland hardwood forest. Forests 2(4):984–992

Danson FM, Hetherington D, Morsdorf F, Koetz B, Gower A (2007) Forest canopy gap fraction from terrestrial laser scanning. IEEE Trans Geosci Remote Sens Lett 4(1):157–160

Dassot M, Constant T, Fournier M (2011) The use of terrestrial LiDAR technology in forest science: application fields, benefits and challenges. Annuals For Sci 68(5):959–974

Demarez V, Duthoit S, Baret F, Weiss M, Dedieu G (2008) Estimation of leaf area and clumping indexes of crops with hemispherical photographs. Agric For Meteorol 148(4):644–655

Dignan P, Bren L (2003) A study of the effect of logging on the understorey light environment in riparian buffer strips in a south-east Australian forest. For Ecol Manage 172(2–3):161–172

Ediriweera S, Snighakumara BMP, Ashton MS (2008) Variation in canopy structure, light and soil nutrition across elevation of a Sri Lankan tropical rain forest. For Ecol Manage 256(6): 1339–1349

Eliasson I (1996) Urban nocturnal temperatures, street geometry and land use. Atmos Environ 30 (3):379–392

Essery R, Pomeroy J, Ellis C, Link T (2008) Modelling longwave radiation to snow beneath forest canopies using hemispherical photography or linear regression. Hydrol Process 22(15): 2788–2800

Evans GC, Coombe DE (1959) Hemispherical and woodland canopy photography and the light climate. J Ecol 47:103–113

Fournier RA, Mailly D, Walter J-MN, Soudani K (2003) Indirect measurement of forest structure from in situ optical sensors. In: Wulder MA, Franklin SE (eds) Methods and applications for remote sensing of forests: concepts and case studies. Kluwer Academic Publishers, Dordrecht, The Netherlands, pp 77–113

Fournier RA, Mailly D, Walter J-MN, Jonckheere IGC (2017) Acquiring hemispherical photographs in forest environments: from planning to archiving photographs. In: Fournier RA, Hall RJ (eds) Hemispherical photography in forest science: theory, methods, applications. Springer, New York

Frazer GW, Fournier RA, Leblanc SG, Walter J-MN (2017) View angle-dependent clumping indices for indirect LAI estimation. In: Fournier RA, Hall RJ (eds) Hemispherical photography in forest science: theory, methods, applications. Springer, New York

Garrigues S, Shabanov NV, Swanson K, Morisette JT, Baret F, Myneni RB (2008) Intercomparison and sensitivity analysis of Leaf Area Index retrievals from LAI-2000, AccyPAR, and digital hemispherical photography over croplands. Agric For Meteorol 148(8–9): 1193–1209

Gendron F, Messier C, Comeau PG (1998) Comparison of various methods for estimating the mean growing season percent photosynthetic photon flux density in forests. Agric For Meteorol 92(1):55–70

Goel NS (1988) Models of vegetation canopy reflectance and their use in estimation of biophysical parameters from reflectance data. Remote Sens Rev 4:1–212

Gray AN, Spies TA, Easter MJ (2002) Microclimatic and soil moisture responses to gap formation in coastal Douglas-fir forests. Can J For Res 32:332–343

Grimmond CSB, Potter SK, Zutter HN, Souch C (2001) Rapid methods to estimate sky-view factors applied to urban areas. Int J Climatol 21(7):903–913

Hackenberg J, Spiecker H, Calders K, Disney M, Raumonen P (2015) Simple Tree—an efficient open source tool to build tree models from TLS clouds. Forests 6(11):4245–4294

Hall RJ, Fournier RA, Rich PM (2017a) Introduction. In: Fournier RA, Hall RJ (eds) Hemispherical photography in forest science: theory, methods, applications. Springer, New York

Hall RJ, Côté JF, Mailly D, Fournier RA (2017b) Comparison of software tools for analysis of hemispherical photographs. In: Fournier RA, Hall RJ (eds) Hemispherical photography in forest science: theory, methods, applications. Springer, New York

Halverson MA, Skelly DK, Kiesecker JM, Freidenburg LK (2003) Forest mediated light regime linked to amphibian distribution and performance. Popul Ecol 134(3):360–364

Hancock S, Essery R, Reid T, Carle J, Baxter R, Rutter N, Huntley B (2014) Characterising forest gap fraction with terrestrial lidar and photography: an examination of relative limitations. Agric For Meteorol 189–190(1):105–114

Hill R (1924) A lens for whole sky photographs. Quart J Roy Meteorol Soc 50:227–235

Hopkinson C, Chasmer L, Young-Pow C, Treitz P (2004) Assessing forest metrics with a ground based scanning lidar. Can J For Res 34(3):573–583

Hrabar S, Sukhatme G, Corke P, Usher K, Roberts J (2005) Combined optic-flow and stereo-based navigation of urban canyons for a UAV. In: International conference on intelligent robots and systems. 2–6 Aug. pp 3309–3316. doi:10.1109/IROS.2005.1544998

Hu L, Yan B, Wu X, Li J (2010) Calculation method for sunshine duration in canopy gaps and its application in analyzing gap light regimes. For Ecol Manage 259(3):350–359

Isaak DJ, Luce CH, Rieman BE, Nagel DE, Peterson EE, Horan DL, Parkes S, Chandler GL (2010) Effects of climate change and wildfire on stream temperatures and salmonid thermal habitat in a mountain river network. Ecol Appl 20(5):1350–1371

Jonckheere IGC, Fleck S, Nackaerts K, Muys B, Coppin P, Weiss M, Baret F (2004) Methods for leaf area index determination. Part I: theories, sensors and hemispherical photography. Agric For Meteorol 121:19–35

Jonckheere IGC, Macfarlane C, Walter J-MW (2017) Image analysis of hemispherical photographs, algorithms and calculation. In: Fournier RA, Hall RJ (eds) Hemispherical photography in forest science: theory, methods, applications. Springer, New York

Kannala J, Brandt SS (2006) A generic camera model and calibration method for conventional, wide-angle, and fish-eye lenses. IEEE Trans Pattern Anal Mach Intell 28(8):1335–1340

Kucharik CJ, Norman JM, Murdock LM, Gower ST (1997) Characterizing canopy non-randomness with a multiband vegetation imager (MVI). J Geophys Res 102(D24): 29455–29473

Kuchelmeister V, Shaw J, McGinity M, Del Favero D, Hardjono A (2009) Immersive mixed media augmented reality applications and technology. In: Proceedings of the 10th pacific rim conference on multimedia, Bangkok, Thailand, Dec 15–18, Vol 5879:1112–1118

Landry R, Fournier RA, Ahern FJ, Lang RH (1997) Tree vectorization: a methodology to characterize fine tree architecture in support of remote sensing models. Can J Remote Sens 23(2):91–107

Lang MW, Kasischke ES, Prince SD, Pittman KW (2008) Assessment of C-band synthetic aperture radar data for mapping and monitoring coastal plain forested wetlands in the Mid-Atlantic Region, U.S.A. Remote Sens Environ 112(11):4120–4130

Leblanc SG, Fournier RA (2014) Hemispherical photography simulations with an architectural model to assess retrieval of leaf area index. Agric For Meteorol 194:64–76

Leblanc SG, Fournier RA (2017) Measurement of forest structure with hemispherical photography. In: Fournier RA, Hall RJ (eds) Hemispherical photography in forest science: theory, methods, applications. Springer, New York

Leblanc SG, Chen JM, Fernandes R, Deering DW, Conley A (2005) Methodology comparison for canopy structure parameters extraction from digital hemispherical photography in boreal forests. Agric For Meteorol 129:187–207

Lieffers VJ, Messier C, Stadt KJ, Gendron F, Comeau PG (1999) Predicting and managing light in the understory of boreal forests. Can J For Res 29:796–811

Lin T-C, Hamburg SP, Hsia Y-J, Lin T-T, King H-B, Wang L-J, Lin K-C (2003) Influence of typhoon disturbances on the understory light regime and stand dynamics of a subtropical rain forest in northern Taiwan. J For Res 8(3):139–145

Lindner A, Sattler D (2012) Biomass estimations in forests of different disturbance history in the Atlantic Forest of Rio de Janeiro, Brazil. New For 43(3):287–301

Liu J, Melloh RA, Woodcock CE, Davis RE, Ochs ES (2004) The effect of viewing geometry and topography on viewable gap fractions through forest canopies. Hydrol Process 18(18): 3595–3607

López-Moreno JI, Latron J (2008) Influence of canopy density on snow distribution in a temperate mountain range. Hydrol Process 22(1):117–126

Luoma D (2002) Shooting Shooting Stars. J Roy Astron Soc Can 96:96–99

Macfarlane C (2011) Classification method of mixed pixels does not affect canopy metrics from digital images of forest overstorey. Agric For Meteorol 151:740–833

Macfarlane C, Ryu Y, Ogden GN, Sonnentag O (2014) Digital canopy photography: exposed and in the raw. Agric For Meteorol 197:244–253

Machado J-L, Reich PB (1999) Evaluation of several measures of canopy openness as predictor of photosynthetic photon flux density in deeply shaded conifer-dominated forest understory. Can J For Res 29(9):1438–1444

Magdwick HAI, Brumfield GL (1969) The use of hemispherical photographs to assess light climate in the forest. J Ecol 57:537–542

Mailly D (2017) Hemispherical photography in support of forest inventory and silviculture. In: Fournier RA, Hall RJ (eds) Hemispherical photography in forest science: theory, methods, applications. Springer, New York

Marsden C, Le Maire G, Stape J-L, Seen DL, Roupsard O, Cabral O, Epron D, Nascimento Lime AM, Nouvellon Y (2010) Relating MODIS vegetation index time-series with structure, light absorption and stem production of fast-growing *Eucalyptus* plantations. For Ecol Manage 259(9):1741–1753

Martinez Pastur GJ, Peri PL, Cellini JM, Lencinas MV, Berrera M, Ivancich H (2011) Canopy structure analysis for estimating forest regeneration dynamics and growth in *Nothofagus pumilio* forests. Annuals of For Sci 68(3):587–594

Martins FR, Souza MP, Pereira EB (2003) Comparative study of satellite and ground techniques for cloud cover determination. Adv Space Res 32(11):2275–2280

Morisette JT, Baret F, Privette JL, Myneni RB, Nickeson J, Garrigues S, Shabanov N, Weiss M, Fernandes R, Leblanc S, Kalacska M, Sánchez-Azofeifa GA, Chubey M, Rivard B, Stenberg P, Rautiainen M, Voipio P, Manninen T, Pilant AN, Lewis TE, Iiames JS, Colombo R, Meroni M, Busetto L, Cohen W, Turner DP, Warner ED, Petersen GW, Seufert G, Cook R (2006) Validation of global moderate-resolution LAI products: a framework proposed within the CEOS land product validation subgroup. IEEE Trans Geosci Remote Sens 44(7):1804–1817

Morsdorf F, Kötz B, Meier E, Itten KI, Allgöwer B (2006) Estimation of LAI and fractional cover from small footprint airborne laser scanning data based on gap fraction. Remote Sens Environ 104(1):50–61

Mussche S, Samson R, Nachtergale L, Schrijver AD, Lemeur R, Lust N (2001) A comparison of optical and indirect methods for monitoring the seasonal dynamics of leaf area index in deciduous forests. Silva Fennica 35(4):373–384

Musselman KN, Molotch NP, Margulis SA, Kirchner PB, Bales RC (2012) Influence of canopy structure and direct beam solar irradiance on snowmelt rates in a mixed conifer forest. Agric For Meteorol 161:46–56

Nicotra AB, Chazdon RL, Iriarte SVB (1999) Spatial heterogeneity of light and woody seedling regeneration in wet tropical forests. Ecology 80(6):1908–1926

Norman JM, Campbell GS (1989) Canopy structure. In: Pearcy RW, Ehleringer J, Mooney HA, Rundel PW (eds) Physiological plant ecology: field methods and instrumentation. Chapman Hall, London, pp 301–325

Opuni-Frimpong E, Karnosky DF, Storer AJ, Cobbinah JR (2008) Silvicultural systems for plantation mahogany in Africa: influences of canopy shade on tree growth and pest damage. For Ecol Manage 255(2):328–333

Pierce AD, Farris CA, Taylor AH (2012) Use of random forests for modeling and mapping forest canopy fuels for fire behavior analysis in Lassen Volcanic National Park. California, USA. For Ecol Manage 279:77–89

Riaño D, Valladares F, Condés S, Chuvieco E (2004) Estimation of leaf area index and covered ground from airborne laser scanner (Lidar) in two contrasting forests. Agric For Meteorol 124(3–4):269–275

Ryu Y, Nilson T, Kobayashi H, Sonentag O, Law E, Baldocchi D (2010) On the correct estimation of leaf area index: does it reveal information on clumping effects? Agric For Meteorol 150(3): 463–472

Ryu Y, Verfaillie J, Macfarlane C, Kobayashi H, Sonentag O, Vargas R, Ma S, Baldocchi D (2012) Continuous observation of tree leaf area index at ecosystem scale using upward-pointing digital cameras. Remote Sens Environ 126:116–125

Schleppi P, Paquette A (2017) Solar radiation in forests: theory for hemispherical photography. In: Fournier RA, Hall RJ (eds) Hemispherical photography in forest science: theory, methods, applications. Springer, New York

Seidel D, Fleck S, Leuschner C (2012) Analyzing forest canopies with ground-based laser scanning: a comparison with hemispherical photography. Agric For Meteorol 154–155:1–8

Shiroma N, Sato N, Yu-huan C, Matsuno F (2004) Study on effective camera images for mobile robot teleoperation. 13th International IEEE workshop on robot and human interactive communication. 20–22 Sept, 107–112 doi:10.1109/ROMAN.2004.1374738

Slater JM (1932) Photography with the whole-sky lens. American photographer pp 580–583

Solberg S, Naesset E, Hanssen KH, Christiansen E (2006) Mapping defoliation during a severe insect attack on Scots pine using airborne laser scanning. Remote Sens Environ 102(3–4): 364–376

Thomas SC, Halpern CB, Falk DA, Liguori DA, Austin KA (1999) Plant diversity in managed forest: understory responses to thinning and fertilization. Ecol Appl 9(3):864–879

Vertessy RA, Benyon RG, O'Sullivan SK, Gribben BR (1995) Relationship between stem diameter, sapwood area, leaf area and transpiration in a young mountain ash forest. Tree Physiol 15(9):559–567

Wang YS, Miller DR, Welles JM, Heisler G (1992) Spatial variability of canopy foliage in an oak forest estimated with fisheye sensors. For Sci 38(4):854–865

Weiss M, Baret F, Smith GJ, Jonckheere I, Coppin P (2004) Review of methods for in situ leaf area index (LAI) determination: Part II. Estimation LAI, errors and sampling. Agric For Meteorol 121:37–53

White K, Pontius J, Schaberg P (2014) Remote sensing of spring phenology in northeastern forests: a comparison of methods, field metrics and source of uncertainty. Remote Sens Environ 148:97–107

Wood RW (1906) Fish-eye views, and vision underwater. Philosophical magazine series 6 12(68): 159–162

Zhao K, Popescu S (2009) Lidar-based mapping of leaf area index and its use for validating GLOBCARBON satellite LAI product in a temperate forest of the southern USA. Remote Sens Environ 113(8):1628–1645

Zhao F, Strahler AH, Schaaf CL, Yao T, Yang X, Wang Z, Schull MA, Román MO, Woodcock CE, Olofsson P, Ni-Meister W, Jupp DLB, Lovell JL, Culvenor DS, Newnham GJ (2012) Measuring gap fraction, element clumping index and LAI in sierra forest stands using a full-waveform ground-based lidar. Remote Sens Environ 125:73–79

Index

R.A. Fournier and R.J. Hall (eds.), *Hemispherical Photography in Forest Science: Theory, Methods, Applications*, Managing Forest Ecosystems 28,
DOI 10.1007/978-94-024-1098-3

MIX
Papier aus verantwortungsvollen Quellen
Paper from responsible sources
FSC® C105338

FSC
www.fsc.org

If you have any concerns about our products,
you can contact us on
ProductSafety@springernature.com

In case Publisher is established outside the EU,
the EU authorized representative is:
Springer Nature Customer Service Center GmbH
Europaplatz 3, 69115 Heidelberg, Germany

Printed by Libri Plureos GmbH
in Hamburg, Germany